Advancing UAV-Based Remote Sensing: Innovations, Techniques and Applications

Advancing UAV-Based Remote Sensing: Innovations, Techniques and Applications

Guest Editors

Rui Li

Yinxia Cao

Haoyang Yang

Dongyu Li

MDPI

Basel • Beijing • Wuhan • Barcelona • Belgrade • Novi Sad • Cluj • Manchester

Guest Editors
Rui Li
School of Engineering
University of Warwick
Coventry
UK

Yinxia Cao
Institute of Geographic
Sciences and Natural
Resources Research
Chinese Academy of Sciences
Beijing
China

Haoyang Yang
Department of Aeronautical
and Aviation Engineering
The Hong Kong Polytechnic
University
Hong Kong SAR
China

Dongyu Li
School of Cyber Science
and Technology
Beihang University
Beijing
China

Editorial Office
MDPI AG
Grosspeteranlage 5
4052 Basel, Switzerland

This is a reprint of the Special Issue, published open access by the journal *Remote Sensing* (ISSN 2072-4292), freely accessible at: https://www.mdpi.com/journal/remotesensing/special_issues/MW3422LNSO.

For citation purposes, cite each article independently as indicated on the article page online and as indicated below:

Lastname, A.A.; Lastname, B.B. Article Title. *Journal Name* **Year**, *Volume Number*, Page Range.

ISBN 978-3-7258-7673-0 (Hbk)
ISBN 978-3-7258-7674-7 (PDF)
https://doi.org/10.3390/books978-3-7258-7674-7

Contents

About the Editors

Rui Li

Rui Li is a research fellow at the University of Warwick. His research focuses on intelligent perception, physics informed machine learning, and autonomous systems, with the aim of developing a unified spatial intelligence and digital twin framework for complex engineering systems. His work spans UAV-based vision language intelligence and multi-source remote sensing, enabling advanced visual recognition and analysis for low-altitude platforms and smart infrastructure monitoring. He also integrates advanced artificial intelligence algorithms with physics-based modeling to address high-dimensional state estimation and prediction problems in large scale systems, bridging the gap between data driven approaches and physical system understanding. He has published more than 20 academic papers in leading journals across remote sensing, artificial intelligence, and energy systems, with over 5000 citations and an h index of 19. His publications include highly cited and hot papers, and his work has received international recognition, including a best paper award from a leading journal in photogrammetry and remote sensing. He has served as a Guest Editor for several international journals and has contributed extensively as a reviewer for varied high-impact journals in the fields of remote sensing, engineering, and artificial intelligence.

Yinxia Cao

Yinxia Cao is an assistant researcher at the State Key Laboratory of Geographic Information Science and Technology, Institute of Geographic Sciences and Natural Resources Research, Chinese Academy of Sciences. Her research focuses on high-resolution remote sensing image interpretation, remote sensing foundation models, and three-dimensional urban reconstruction. Her work aims to advance intelligent analysis of satellite imagery through the integration of artificial intelligence and geospatial data, supporting accurate and efficient understanding of complex urban environments. She is actively engaged in developing novel methodologies for large-scale remote sensing data processing and contributes to research on next-generation geospatial technologies for urban analysis and mapping.

Haoyang Yang

Haoyang Yang is a research assistant professor in the Department of Aeronautical and Aviation Engineering and the Research Centre for Low-Altitude Economy, The Hong Kong Polytechnic University. He received a BEng degree in Measurement and Control Techniques and Instruments from Harbin Institute of Technology, Harbin, China, in 2017 and a PhD degree in Navigation Guidance and Control from Beihang University, Beijing, China, in 2023, undertaking an exchange to National University of Singapore during his PhD studies. Before becoming a research assistant professor, he was a postdoctoral fellow in AAE and a research fellow at the University of Warwick, UK. His research interests include safety-critical control, intelligent control, and their applications in UAV and spacecraft systems.

Dongyu Li

Dongyu Li is a professor and doctoral supervisor at Beihang University and a participant in a national level young talent program. His research focuses on intelligent planning and control of unmanned systems, large model-based multi-source perception and reasoning, and network security for aerospace systems. He has published more than 50 papers in leading international journals in the fields of control science, computational intelligence, and aerospace engineering. His work has received significant academic recognition, with multiple papers selected as highly cited publications and the publication of two international academic monographs. His research outcomes have been applied to several major national aerospace systems. He has led numerous research projects funded by national agencies, including the National Natural Science Foundation and major programs from the Ministry of Science and Technology. He has received several prestigious awards, including a major young researcher award from an international control society, first-class prizes from national academic societies in automation and artificial intelligence, a national award for outstanding doctoral dissertation, and a top gold award with special recognition at an international invention exhibition.

Preface

This Reprint presents a collection of peer reviewed contributions that reflect recent progress in UAV-based remote sensing, with a focus on emerging technologies, methodological innovation, and practical applications. The rapid development of unmanned aerial vehicles has significantly expanded the capabilities of remote sensing by enabling flexible deployment, high spatial and temporal resolution, and cost-effective data acquisition across varied environments.

This Reprint will provide a comprehensive overview of current research trends and challenges in this field. The included studies address key aspects such as advanced sensing systems, photogrammetric reconstruction, data fusion techniques, and intelligent algorithms for information extraction. Particular attention is given to improving data quality, enhancing processing efficiency, and ensuring robustness under complex operational conditions.

The motivation for this Reprint arises from the increasing demand for accurate, timely, and scalable geospatial information in applications such as environmental monitoring, precision agriculture, infrastructure assessment, and hazard detection. By bringing together contributions from different research communities, this Reprint highlights the interdisciplinary nature of UAV-based remote sensing and its growing role in addressing real-world problems.

This Reprint is intended for researchers, engineers, and practitioners working in remote sensing, geoinformatics, environmental science, and related fields. It is also relevant to those interested in the integration of UAV platforms with advanced data analysis methods. The editors hope that this collection will serve as a useful reference and inspire further developments in this rapidly evolving area.

Rui Li, Yinxia Cao, Haoyang Yang, and Dongyu Li
Guest Editors

remote sensing

MDPI

Article

High Resolution UAV-Based Monitoring of Ambient Methane: Field Deployment and Intercomparison with Reference Standards

Daja Elum [1,*], Nakul N. Karle [1], Ricardo K. Sakai [1], Xinrong Ren [2], Phillip Stratton [2,3], Nicholas R. Nalli [4], Monique Walker [4], Adrian Flores [1], Johan R. Villanueva [5] and Joseph Wilkins [1]

1 Department of Earth, Environment, and Equity, Howard University, Washington, DC 20059, USA; nakul.karle@howard.edu (N.N.K.); aflores@howard.edu (A.F.)

2 NOAA Air Resources Laboratory, College Park, MD 20740, USA; xinrong.ren@noaa.gov (X.R.)

3 Department of Atmospheric and Oceanic Science, University of Maryland, College Park, MD 20742, USA

4 National Geospatial-Intelligence Agency, Springfield, VA 22150, USA; nicholas.r.nalli@nga.mil (N.R.N.)

5 Department of Physics, University of Puerto Rico at Mayagüez, Mayagüez, PR 00680, USA; johan.villanueva@upr.edu

* Correspondence: daja.elum@bison.howard.edu

Highlights

What are the main findings?

- The MIRA Strato Leak Detection System (LDS) methane and ethane midwave-infrared analyzer demonstrates high laboratory and field precision against NOAA Standard Reference Materials and during intercomparison with a Picarro gas analyzer.
- The MIRA Strato LDS resolves diurnal variability during ground deployment and microscale variability of ambient methane through UAV-based profiling.

What is the implication of the main finding?

- Ambient methane structure transitions from near-surface accumulation to convective mixing, driven primarily by local meteorological conditions, with temperature identified as the strongest predictor of variability.
- UAV-based midwave-infrared analyzers can characterize boundary-layer methane dynamics for ambient monitoring applications, independent of direct emission source quantification.

Abstract

This study investigates the spatiotemporal variability of ambient methane (CH_4) using a drone-deployable Aeris Technologies MIRA Strato LDS midwave-infrared analyzer. Laboratory calibration with NOAA-certified gas standards and Standard Reference Material (SRM) for CH_4 demonstrated high measurement precision across a range of concentrations (R^2 = 0.9986, slope = 0.9678). Field validation conducted during a two-week intercomparison with a Picarro G2301 in September 2023 confirmed the MIRA Strato's reliability under ambient conditions (R^2 = 0.9845; slope = 0.9438), indicating strong agreement with the reference analyzer. Diurnal patterns revealed peak CH_4 concentrations (~2.2 ppm) between 04:00–08:00 LT and minima (~2.1 ppm) between 13:00–17:00 LT, consistent with nocturnal boundary-layer stability and daytime convective mixing. Across 14 midday UAV flights from October 2023 to September 2024, CH_4–altitude slopes ranged from -3.05×10^{-4} to $+1.41 \times 10^{-4}$ ppm/m, reflecting variable stratification and uplift regimes. The highest flight concentration (2.23 ppm) was observed on 19 October under stable conditions, while the lowest (2.03 ppm) was observed on 14 August under elevated vertical mixing. These

https://doi.org/10.3390/rs18040549

extremes reflect seasonal background accumulation and convective transport effects. Temperature was the most consistent predictor, with regression coefficients ranging from −0.021 to +0.008 ppm/°C, while ethane (C_2H_6) coefficients were significant but confounded due to measurements below detection limits. The analyzer maintained strong signal stability throughout (mean CV ≈ 0.0066; max = 0.0114), and remote sensing validation with TROPOMI supported observed seasonal accumulation trends. These results demonstrate the MIRA Strato's capability to resolve near-surface CH_4 dynamics and characterize convective transport in complex atmospheric environments.

Keywords: thermal infrared (TIR); midwave IR (MWIR); methane; UAV measurements; environmental security; air quality; thermogenic emissions; biogenic emissions

1. Introduction

Methane (CH_4) is a chemically active atmospheric gas with critical implications for air quality, environmental security, and environmental monitoring. With an average atmospheric lifetime of 12 years [1], methane indirectly contributes to ground-level ozone (O_3) formation through tropospheric oxidation processes, impacting human health, ecosystems, and agriculture [2,3]. A balance of natural and anthropogenic sources and sinks influences its atmospheric concentration. Since the pre-industrial period, global methane has increased by approximately 160% [4,5], with a +4.8% increase over the period of 2003 to 2020 [6] and overall global emissions of 575 Tg CH_4 yr^{-1} from 2010–2019 [7]. Baseline emissions will increase by 20–50 Mt annually by 2030 [7,8].

Methane can be classified by its formation process into biogenic, thermogenic, and pyrogenic categories, and by emission sector (e.g., agriculture, waste, and fossil-fuel systems). Biogenic methane is produced by methanogenic archaea via anaerobic decomposition of organic matter [9,10], typically formed at low temperatures below 50 °C, and is composed almost entirely of methane [11]. It is prevalent in natural environments, such as wetlands and ruminant digestive systems, as well as in anthropogenically influenced settings, such as rice paddies and landfills [12]. Thermogenic methane is formed through geological processes via thermal cracking of organic matter in sedimentary basins, under high-pressure and temperatures (typically 157–221 °C) at depths > 1000 m [11,12]. It often includes a mix of hydrocarbons and can be distinguished by isotopic signatures or compositional indicators such as the ethane (C_2H_6)/CH_4 ratio [13]. Pyrogenic methane is produced during the incomplete combustion of biomass and fossil fuels, for example, in wildfires and some biofuel or energy combustion sources [14]. In terms of emission sources, anthropogenic methane encompasses emissions from human activities involving biogenic, thermogenic, and pyrogenic processes, including fossil fuel extraction, agricultural practices, and waste management.

Methane removal strategies vary significantly depending on source type; however, most tropospheric methane is removed through oxidation of hydroxyl radicals (OH) and chlorine to a lesser degree [15]. This naturally occurring process accounts for 95% of global methane destruction, with an estimated rate of approximately 500 Tg yr^{-1} [7,16]. Rising emissions disrupt the natural removal process by depleting OH, reducing its effectiveness, and extending its atmospheric lifetime [16–19]. Biogenic sources can be removed by soil methanotrophs, i.e., bacteria in aerated soil that oxidize methane into carbon dioxide or methanol. However, it is only effective at high concentrations (>500 ppm) [20,21] and accounts for ~5% of global destruction [7,16]. Thermogenic methane is managed through industrial processes such as thermal-catalytic oxidation and gas scrubbing, which are

https://doi.org/10.3390/rs18040549

also most effective at high concentrations (>5000 ppm) [21]. Anthropogenic sources are addressed through leak detection and repair, targeted removal technologies, and improved land management practices.

Institutional and research-based initiatives have supported the development of tools to better characterize methane emissions. Advances in airborne, ground-based, and satellite-based remote sensing, along with improved emissions inventories, are progressively enhancing our abilities to identify methane sources and temporal trends. Traditional ground-based and stationary methane monitoring networks are often limited in spatial coverage and can miss diffuse or unexpected emission sources [22]. Among mobile sensing technologies, UAV-based approaches have grown in prominence. Open-path systems such as tunable diode laser absorption spectroscopy (TDLAS) offer path-integrated CH_4 concentrations and enable dense spatial sampling for site-scale assessments [23]. In parallel, unmanned aerial vehicles (UAVs) equipped with in situ gas analyzers can detect and characterize methane concentrations in the lower atmosphere with high temporal resolution. To this end, in this work, we examine the field performance of a UAV-mounted midwave infrared (MWIR) methane analyzer in terms of its ability to capture ambient boundary-layer temporal variability and reproduce diurnal patterns observed in co-located reference instruments.

Drone-Deployed Measurement Techniques

Small-scale UAVs offer flexibility in detection and quantification compared to ground-based and airborne platforms through their ability to measure local-scale anthropogenic emissions within surface-layer (<100 m) and boundary-layer (<1000 m) altitudes [24–26]. Drone-deployed measurements can also provide vertical emissions profiles, capturing concentration gradients with altitude and supporting improved source characterization and attribution [27]. There are two main small-scale UAV configurations, either fixed-wing or rotary-wing. Fixed-wing UAVs excel in covering extensive geographical areas, offer longer flight endurance, and versatile sensor mounting configurations due to their aerodynamic design and payload [24,27,28]. However, their operational limitations, particularly the inability to hover and the requirement for continuous forward motion, restrict their effectiveness in tracing detailed point-source emissions [27].

Rotary-wing platforms can maintain stable hovering and vertical take-off and landing capabilities compared to fixed-wing platforms. However, they have lower operational velocity and flight duration due to the energy requirements to maintain rotary function [27,29–32]. Another limitation of rotary-wing drones is the additional air flow generated from the propellers, referred to as 'downwash'. Downwash can negatively affect measured concentrations by disrupting air distribution and generating local mixing of air parcels. This can alter the air sampled at the inlet, introducing variability or bias, and affect data quality [33,34]. Rotor downwash largely perturbs the air below the rotor plane; any inlet or sensing volume beneath the aircraft is in the propeller slipstream and will sample accelerated, mixed air, with modest biasing enhancement. Various studies conclude that installing vertical or horizontal tubing inlets away from the platform reduces the magnitude of downwash effects [31,33,35]. Despite these issues, selecting an ideal drone platform and inlet design can enhance measurement accuracy in ambient methane monitoring. The following describes the study region and experimental procedure for applying these techniques.

https://doi.org/10.3390/rs18040549

2. Materials and Methods

2.1. Area of Interest

The Howard University Beltsville Campus (HUBC) Research Center is located 19 km northeast of downtown Washington, DC, spanning 110 acres (44.5 ha) of largely undeveloped land. The center has experimental facilities, including flux and meteorological towers, upper air sounding systems, spectral and broadband radiometers, gas analyzers, and particle sampling equipment (cf. Figure 1). HUBC collaborates on several projects with organizations such as NASA, NOAA, NGA, and the Maryland Department of Environment (MDE).

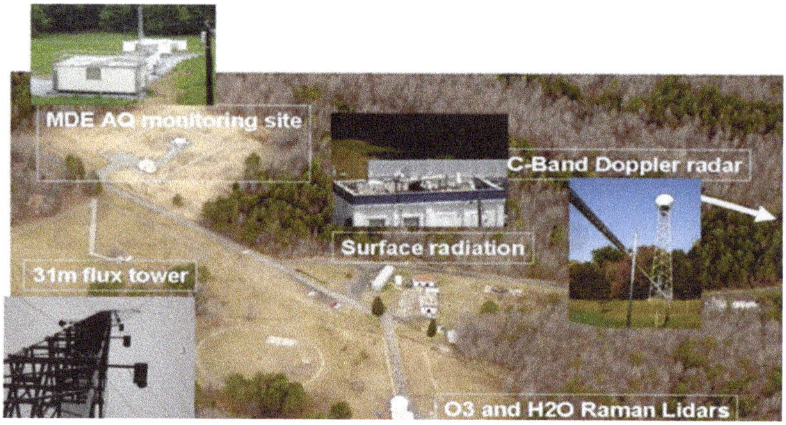

Figure 1. HUBC ground-based testing facility. From the Howard University Beltsville Research Center.

2.2. Aeris Technologies MIRA Strato LDS CH_4 and C_2H_6 Analyzer

The Aeris Technologies MIRA Strato Leak Detection System (LDS) methane and ethane gas analyzer (Aeris Technologies, Inc., Hayward, CA, USA) is a multipass cell and laser absorption spectrometer with built-in GPS capabilities. The Strato LDS operates in the MWIR region, with a sensitivity of <1 ppb/s for CH_4 and 500 ppt/s for C_2H_6, and is compatible with most commercial drones weighing at 2 kg [36]. MWIR absorption spectroscopy uses high-resolution tunable lasers to directly measure light absorption by gas molecules at specific wavelengths [37]. This method is effective because many gases, including methane and ethane, have strong and distinct absorption features in this region (cf. Figure 2), allowing for precise identification and quantification.

The MIRA Strato is equipped to discriminate thermogenic and biogenic CH_4 emissions through its 13-m optical path length and 60 cm^3 (60 mL) multipass cell technology. By folding the laser beam multiple times within the cell, the device maximizes the interaction between light and gas molecules, significantly amplifying the absorption signal without increasing the instrument's physical size [38]. The analyzer internally measures gas concentrations at rates up to 1–5 Hz; these high-frequency readings are averaged and logged at one-second intervals [36], yielding stable time-aligned CH_4 and C_2H_6 readings suitable for drone-based plume detection, as shown in Figure 3.

Configuration and collection of methane measurements proceeded as follows [37]: The MIRA Strato was powered on 10–20 min before deployment to boot, ensuring proper GPS connection, and collected at a one-second duration. Data collection includes:

1. An 'engineering' text file designed as an exhaustive repository of all the measurement parameters and results.
2. Measured spectra in ASCII format (named 'spectra-lite') to diagnose issues or post-process measurements.

https://doi.org/10.3390/rs18040549

3. A KML file that is read directly on Google Earth to map gas concentrations and ethane-to-methane correlations.

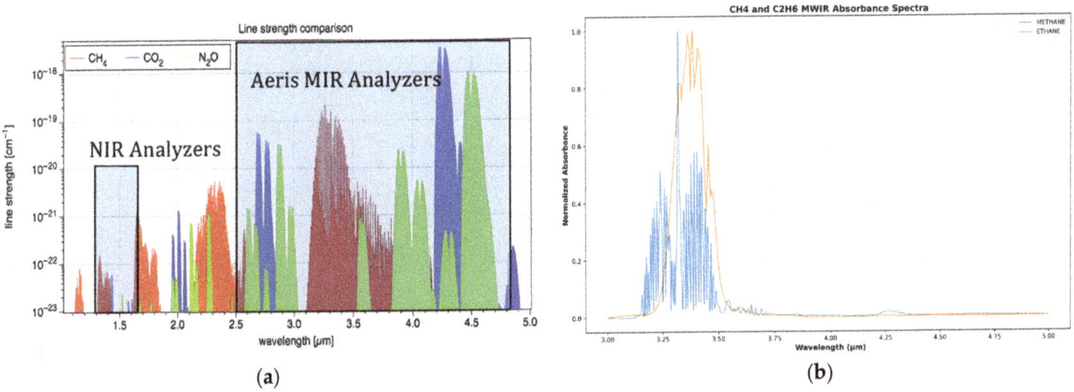

Figure 2. A comparison of methane (CH_4) absorption features. (**a**) Illustration of the spectral collection window of the MIRA Strato analyzer, which operates in MWIR to leverage the strong absorption of ethane (C_2H_6) for biogenic/thermogenic discrimination from aerissensors.com (accessed on 7 May 2025). (**b**) Corresponding normalized MWIR absorbance spectra of CH_4 and C_2H_6 from NIST, demonstrating the spectral region used by the MIRA Strato analyzer.

Metric	***Specification****
Measurement method	Mid-Infrared Laser Absorption Spectroscopy
Species, Sensitivity (σ)	CH_4: <2ppb/s, C_2H_6: 300ppt/s
Drift, typical	1–2% of reading typical, long-term
Temp/Humidity	5–40°C, 10 to 90% RH (non-condensing)
Concentration Range*	10 ppb to 10,000ppm (CH4)
Size	7.5"W x 7.5"D x 3.5"H
Weight	2 kg (4.4 lbs) w/ battery
Power Consumption	17W, 80–90 minute swappable battery
Voltage, current	12–15V DC: 2A, 110–220V AC: 0.5A
Interface/Outputs	Wi-Fi, RS-232, analog output (optional)
Memory	32GB default, scalable
Data Update Rate	1–5Hz Standard

Figure 3. MIRA Strato LDS methane sensor specifications. * Optional ranges, etc., can be configured for specific applications. From aerissensors.com.

2.3. Picarro G2301 CO_2 and CH_4 Analyzer

The Picarro G2301 (Picarro, Inc., Santa Clara, CA, USA) gas concentration analyzer is a Cavity Ring-Down Spectrometer (CRDS) that collects carbon dioxide (CO_2) and methane (CH_4) at parts-per-billion (ppb) and water vapor (H_2O) at percent volume mixing ratio (%) [39]. CRDS is an ultra-sensitive laser absorption technique that uses highly reflective mirrors to trap light pulses in a cavity, resulting in effective path lengths of up to tens of kilometers to detect absorption [40]. Picarro analyzers increase their signal-to-noise ratio using a three-mirror cavity to support continuous traveling light waves [41]. The three-mirror configuration allows for laser light to circulate throughout the cavity, providing a longer path length and amplifying the absorption signal to detect even trace gas levels at the ppb and ppm range [39]. When the laser is turned off, the light within the cavity continues to bounce between the mirrors, gradually decaying due to the reflectivity loss known as 'ring down'. In the presence of a target gas, additional absorption occurs, accelerating the ring-down time. This further increases the analyzer's accuracy by comparing the ring-down

https://doi.org/10.3390/rs18040549

time with and without the target gas. The G2301 has a response time of approximately 2.5 s and offers high measurement precision with a CH_4 sensitivity of ± 1 ppb over 5 s, improving to ± 0.5 ppb over 5 min and ± 0.3 ppb over 60 min [39]. These features make the G2301 well-suited for continuous monitoring and high-precision atmospheric research applications.

The G2301 was used to perform an intercomparing analysis with the MIRA Strato during the study. Both instruments were placed at the MDE Air Quality monitoring station in the northwest portion of the HUBC research center. The two-week analysis from 5–21 September 2023, was conducted to assess the MIRA Strato measurement consistency and accuracy. Instruments were connected via a shared tubing outlet, allowing for simultaneous measurements (cf. Figure 4). A Raspberry Pi single-board computer was connected to the serial port as a backup for power disconnection. The tubing outlet was lastly positioned on the station's roof to collect measurements.

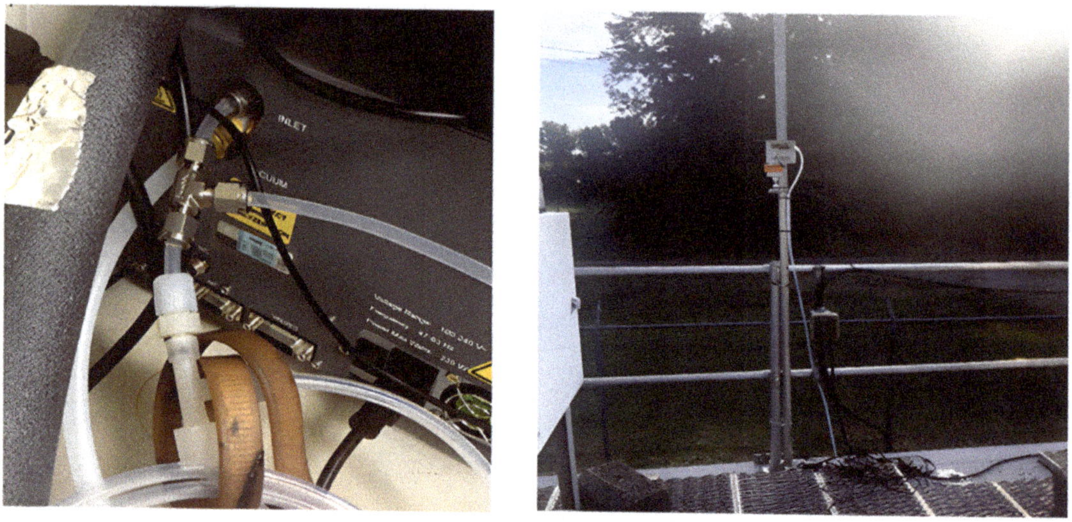

(**a**) (**b**)

Figure 4. Photographic documentation of the MIRA Strato intercomparison with the Picarro G2301 conducted at HUBC in September 2023. (**a**) Inlet tubing configuration connecting the MIRA Strato and Picarro G2301 analyzers to a shared ambient intake line. (**b**) Tubing outlet positioned on the rooftop of the HUBC MDE monitoring site, demonstrating the shared ambient air intake configuration for comparative analysis.

2.4. DJI Matrice 600 Pro Drone

The DJI Matrice 600 Pro is a rotary-wing hexacopter drone weighing 9.5 kg (with six TB47S batteries), with a hovering span of 32 min (without payload), and a 6 kg maximum payload capacity [42,43]. The drone has the A3 Pro flight control system for enhanced stability and precision during atmospheric measurements [43]. During stationary hover, the aircraft uses GPS-based position-hold (DJI P-GPS) with a typical accuracy of ± 0.5 m vertically and ± 1.5 m horizontally under normal satellite reception. While the platform has a maximum ceiling height of 4.5 km, our operations were conducted below 120 m due to drone operating licensing limitations and payload considerations. According to the manufacturer (DJI Matrice 600 Pro user manual), the platform has a maximum wind resistance of 8 m/s and maximum ascent and descent speeds of 5 m/s and 3 m/s, respectively. Payload testing was conducted with a replica model of the Aeris Strato to determine a flight duration of 16–20 min, depending on the windshear. To optimize measurement accuracy and reduce down-washing effects, an 82.5 cm vertical inlet tube was installed

https://doi.org/10.3390/rs18040549

above the propellers with a filter added to the opening to prevent particle contamination (cf. Figure 5). Previous fluid dynamic simulations determined that the Matrice 600 Pro series downwash disturbances decrease at a height of 60 cm above the body [27,29]. While this inlet configuration follows established best practices for mitigating downwash effects reported in literature [27,29], a direct comparative evaluation with and without the inlet extension was not conducted during payload testing. As a result, the effectiveness of the inlet design in reducing downwash-related sampling bias in this study is inferred from prior studies rather than quantitatively verified.

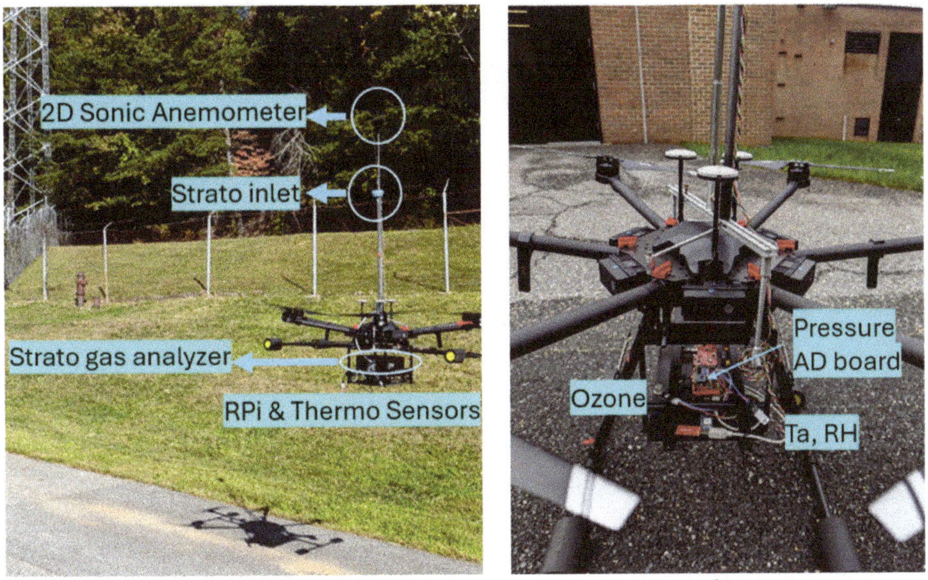

(a) (b)

Figure 5. DJI Matrice 600 Pro drone with inlet tubing during payload testing. (**a**) Placement of the MIRA Strato gas analyzer, inlet, and Raspberry Pi sensor. (**b**) Additional features of the drone include an anemometer, ozone, pressure, and temperature sensors.

Flights were conducted during midday hours, typically starting at 13:20 local time with durations ranging from 17 to 20 min. This timeframe was selected to coincide with the afternoon TROPOMI satellite overpass and peak atmospheric mixing within the boundary layer. Compared to stable nighttime conditions, midday flights allow for greater vertical methane dispersion and reduce the influence of near-surface accumulation, resulting in more representative spatial sampling [44]. Vertical profile measurements were taken at several heights, including 25, 50, 75, 100, and 115 m above ground level, and across different environmental backgrounds such as above HUBC's north building at 6 m structure height, forest canopy at 20–25 m height, and over a meteorological flux tower at 30 m structure height. Stable hovering at the given heights allowed us to record reliable measurements at each level. Flight paths were tracked using the MIRA Strato's KML configuration system, enabling spatial analysis of ethane-to-methane Pearson correlation coefficients in areas likely influenced by biogenic or thermogenic sources (cf. Figure 6). Data points were color-coded: green for correlation coefficients < 0.7, and red for values ≥ 0.7. Elevated correlation zones ($\geq$0.7) may indicate potential areas of high natural gas concentration [36].

https://doi.org/10.3390/rs18040549

(**a**) (**b**)

Figure 6. Example of flight path from deployments, with ethane-to-methane correlations, at HUBC using the MIRA Strato (**a**). Example of flight path (yellow line), derived from the DJI Matrice 600 Pro drone during the campaign, covering various environmental backgrounds within the area of interest, including buildings, forest, and the meteorological flux tower at HUBC (**b**).

A total of 14 flights were completed between October 2023 and September 2024 under varying meteorological conditions, as detailed in Table 1. The measurement strategy incorporated vertical profiling at consistent altitude intervals, weather condition monitoring to ensure safe operation, and real-time data logging through Strato's streaming capability.

Table 1. Flight collection dates, times, duration, altitudes reached, and observations.

Date	Time (LT)	Duration (Approx)	Altitudes Reached (m)	Observations/Notes
11 October 2023	13:20–13:40	20 min	10, 15, 20, 25, 30, 35, 50, 75, 100, 115	Various altitudes reached
12 October 2023	13:20–13:40	20 min	30, 60, 90, 115	Various altitudes reached
19 October 2023	13:20–13:40	20 min	25, 50, 75, 100, 115	Clear sky, various altitudes
23 October 2023	13:20–13:40	20 min	25, 50, 75, 100, 115	Clear sky, various altitudes
25 October 2023	13:20–13:40	20 min	25, 50, 75, 100, 115	Clear sky, various altitudes
27 October 2023	13:20–13:40	20 min	25, 50, 75, 100, 115	Mid-level clouds
2 November 2023	13:20–13:40	20 min	27, 52, 77, 100, 115	Clear sky, various altitudes
6 February 2024	13:20–13:40	20 min	26, 52, 76, 101, 117	Clear sky
26 February 2024	13:22–13:41	19 min	25, 51, 77, 100, 116	Cirrus clouds
25 March 2024	13:20–13:40	20 min	25, 50, 75, 100, 115	Clear sky
29 April 2024	14:22–14:40	18 min	50, 75, 100, 110	Cumulus clouds
27 June 2024	15:10–15:27	17 min	25, 50, 76, 101, 115	Cumulus clouds
14 August 2024	15:00–15:20	20 min	25, 52, 79, 101, 115	Cumulus cloud cover
11 September 2024	14:07–14:27	20 min	28, 51, 77, 100, 115	Some cumulus clouds

https://doi.org/10.3390/rs18040549

For each profile, the UAV hovered at the specified heights and environmental backgrounds for approximately 5 min. The flight path began at the north building, then proceeded to the forest site, and finally to the meteorological flux tower. At each location, ascending and descending measurements were taken. CH_4, C_2H_6, and H_2O concentrations were recorded, with averages, standard deviations, and sample sizes calculated from hovering levels (HL) based on the pressure time series as seen in Figure 7.

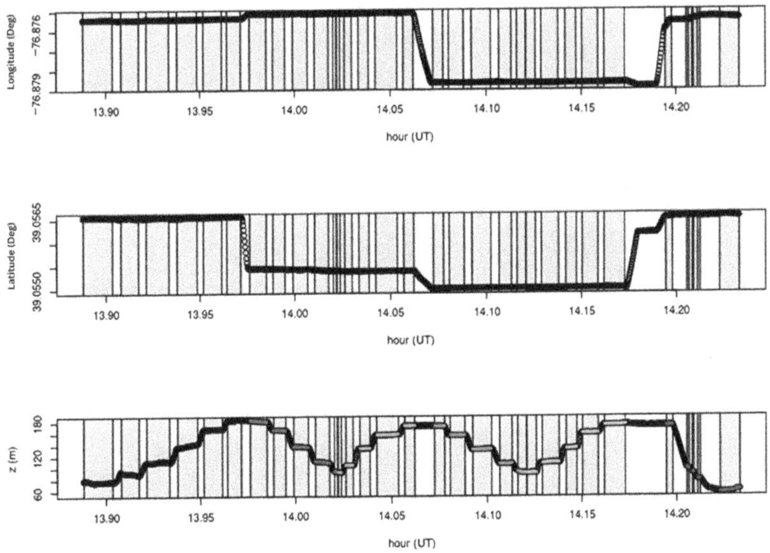

Figure 7. Time series to determine the period for each HL. The top figure shows longitude, the second is latitude, and the last is height derived from the pressure sensor. The shaded areas correspond to the HL, and the colors indicate the profile location: black for ascending building, red for descending forest, green for ascending meteorological tower, blue for ascending forest, cyan for ascending tower, and magenta for descending building.

3. Results

3.1. Intercomparing Analysis

Intercomparing analysis with the Picarro G2301 methane analyzer revealed a correlation with an R^2 value of 0.9845, indicating a near-perfect linear relationship between the two analyzers. Regression analysis showed a slope of 0.9438 and an intercept of 0.0750, demonstrating minimal systematic differences in methane concentration measurements (cf. Figure 8).

3.1.1. Hourly Analysis

Diurnal evaluation determined a high median concentration at HUBC during sunrise between 04:00 and 08:00 LT, reaching approximately 2.2 ppm. The lowest median concentrations occurred during afternoon hours around 13:00–17:00 LT, dropping to about 2.1 ppm (cf. Figure 9). The morning peak is consistent with the expected boundary layer behavior, suggesting surface accumulation with little atmospheric mixing during stable nighttime conditions. On the other hand, the late afternoon drop in concentration is associated with increased vertical mixing and convection, boundary layer growth, and solar heating.

There are several outliers, especially in the late night and early morning, with some spikes as high as 2.5 ppm. The widest interquartile range is also in the morning peak, reflecting greater methane concentration variability. The larger change would likely be due

https://doi.org/10.3390/rs18040549

to the stable nighttime conditions giving way to daytime mixing. During the afternoon (13:00–16:00 LT), concentration ranges are lower and less variable, consistent with improved mixing under unstable weather conditions.

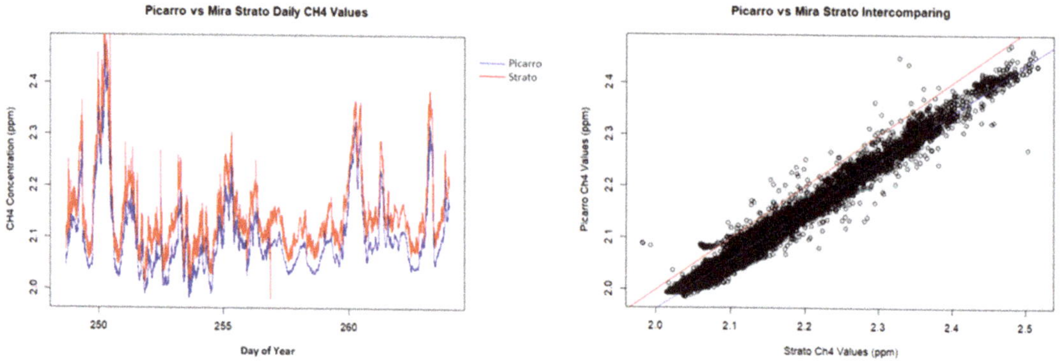

(a) (b)

Figure 8. Intercomparison results: daily values and linear regression. (**a**) Time series of CH_4 concentrations (ppm) showing diurnal variability captured by both instruments, with strong agreement in temporal patterns. (**b**) Linear regression of paired CH_4 observations indicating high correlation between MIRA Strato and Picarro (R^2 = 0.9845, slope = 0.9438).

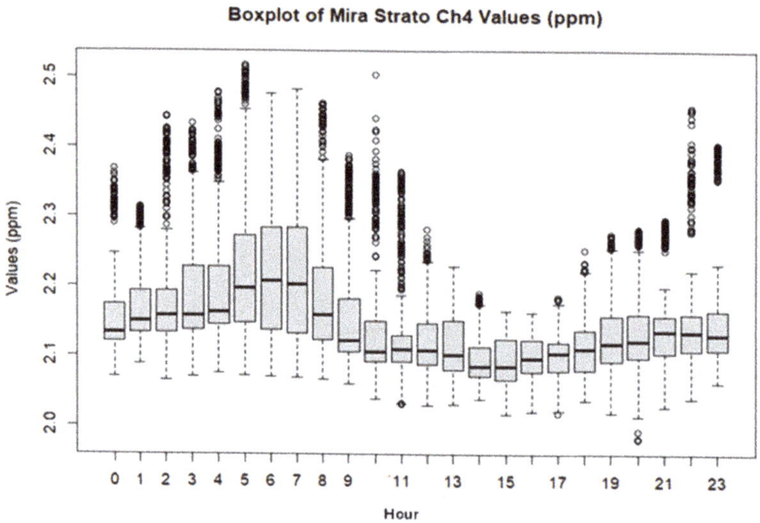

Figure 9. MIRA Strato hourly concentrations during the 2023 intercomparing analysis. Boxplots represent hourly distributions (00:00–23:00 LT), highlighting elevated median CH_4 levels during early morning hours (04:00–08:00 LT) and lower values during afternoon periods (13:00–17:00 LT), consistent with boundary layer dynamics and convective mixing. Whiskers denote 1.5× the interquartile range, and individual points represent outliers, suggesting infrequent enhancement events.

3.1.2. Daily Analysis

Daily variation during the intercomparing analysis showed strong agreement between the MIRA Strato and Picarro G2301 analyzers. The highest median concentrations reached 2.15–2.16 ppm for the Picarro and 2.19–2.20 ppm for the MIRA Strato, with upper quartile values extending to 2.30 ppm and 2.37 ppm, respectively. The lowest median concentrations displayed similar trends at 2.05–2.08 ppm for the Picarro and 2.10–2.08 ppm for the MIRA

https://doi.org/10.3390/rs18040549

Strato. Corresponding upper quantile concentrations ranged from 2.10–2.12 ppm and 2.15–2.18 ppm, respectively. Overall, both instruments tracked daily methane trends closely, with the MIRA Strato reporting slightly higher values in the upper quantile range (cf. Figure 10).

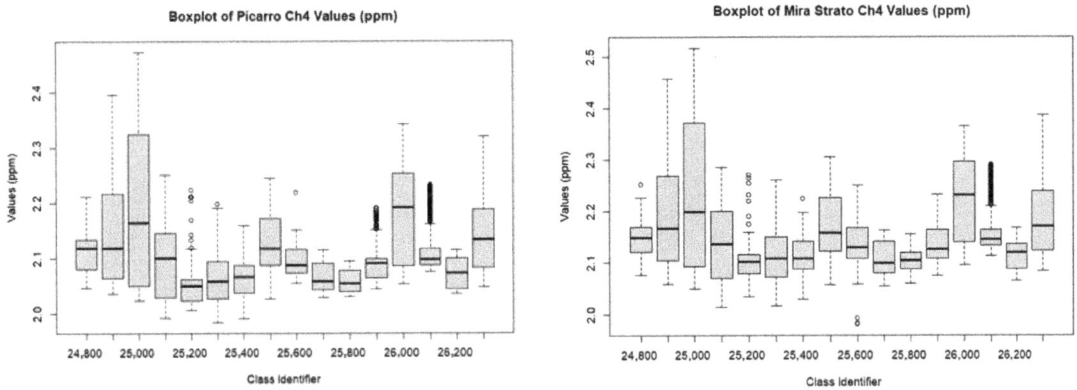

(a) (b)

Figure 10. Boxplot comparing daily methane (CH_4) fluctuation measured by (**a**) the Picarro G2301 and (**b**) the MIRA Strato analyzer during the 2023 intercomparing analysis. The class identifier represents day of year. The plots illustrate similar day-to-day variability and overall agreement in diurnal CH_4 trends, with the MIRA Strato capturing comparable distributional patterns, including median values and dispersion. Whiskers denote 1.5× the interquartile range, and individual points represent outliers, suggesting infrequent enhancement events.

3.2. MIRA Strato Gas Chamber Calibration

Calibration was completed at the NOAA Air Resources Laboratory (ARL) in College Park, MD, with collaboration from NOAA, the University of Maryland, and Howard University. The calibration process involved multiple flow rates, theoretical set points, and statistical analysis to evaluate the MIRA Strato's accuracy in measuring methane (CH_4) and ethane (C_2H_6). The calibration protocol followed NOAA ARL procedures, with gas standards selected to represent typical ambient methane and ethane concentrations and available tank specifications. Gas standards included a Matheson C_2H_6 standard at 4.83 ppm (±2% accuracy), and three CH_4 standards at approximately 1929, 2222.7, and 3152 ppb. Methane standards were introduced directly into the sample line without dilution. Two mass flow controllers (MFCs) were used to regulate the flow rates of C_2H_6 calibration gas and zero air, with the zero-air flow fixed at 2 standard Liters per minute (SLPM) and ethane flow varied from 0 to 10 Standard cubic centimeters per minute (SCCM). Data collection also involved measuring the instrument's response at various voltage set points, ranging from 0 to 5 V, corresponding to the ethane flow range of 0–10 SCCM. Both MFCs were calibrated using two Mesa Labs Definer 220 flow calibrators with ±1% accuracy.

Calibration results for methane, as detailed in Table 2, revealed a strong linear relationship, with a slope of 0.9678, and y-intercept of 45.966, yielding an R^2 value of 0.9986, which indicates high measurement consistency. Similarly, ethane displayed a slope of 1.0111 and a y-intercept of 8.04816, with an R^2 value of 0.9879. The F-statistics and residual sums of squares for both gases were also within acceptable ranges, validating the MIRA Strato's precision across varying concentrations and conditions.

https://doi.org/10.3390/rs18040549

Table 2. MIRA Strato's gas chamber calibration linear regression using gas standards from NOAA ARL. Results indicate high measurement accuracy, with a R^2 value of 0.9986 across varying SRM's.

Gas	Slope	y-Intercept	Standard Error	y-Intercept Standard Error	Coefficient of Determination	Standard Error for the Estimate	F-Statistic	Degrees of Freedom	Regression Sum of Squares	Residual Sum of Squares
Methane	0.97	45.97	0.04	89.15	0.99	32.33	730.56	1	763,611.94	1045.24
Ethane	1.01	8.05	0.03	1.44	0.99	2.95	1116.47	5	9689.27	43.39

3.3. Ambient Methane Profiles from UAV Observations

Across the 14 flights, mean concentrations displayed standard urban ambient levels ranging from 2.05 to 2.19 ppm. Standard deviations remained consistently under 0.02 ppm, pointing toward both precise measurements and stable background methane levels. The highest average concentration was observed at 2.19 ppm on 26 February 2024, while the lowest occurred on 23 October 2023, at 2.05 ppm. Peak values reached 2.23 ppm in late April, and the lowest single reading was 2.02 ppm on 23 October (cf. Figure 11). Throughout all flights, temperatures varied widely, from a low of 12.3 °C in the winter to a high of 32.9 °C in summer flights.

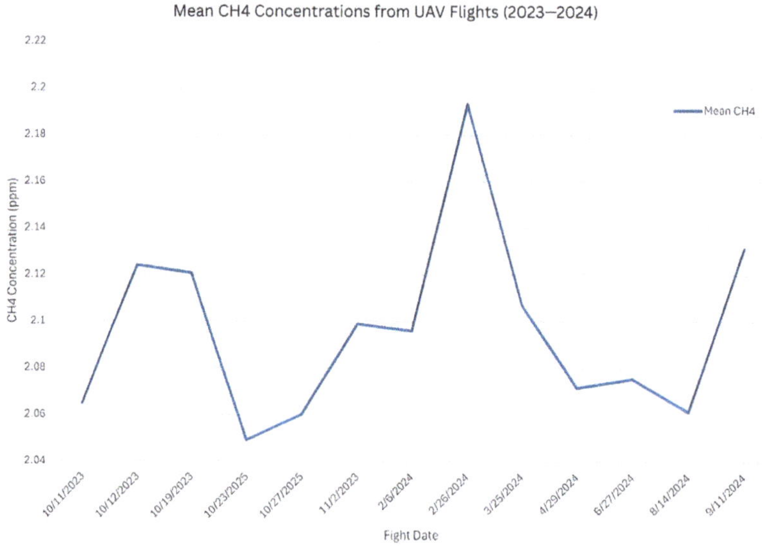

Figure 11. Mean methane (CH_4) concentrations measured using the MIRA Strato analyzer during 14 UAV flights conducted between October 2023 and September 2024.

Flight altitudes ranged between 10 and 117 m, separated into 10-m intervals for vertical profile analysis. Weather conditions, especially temperature inversions, had a noticeable effect on methane patterns. As shown in Figure 12, during the October 19 flight, methane concentrations were highest (2.139 ppm) near the ground (0–10 m) and dropped off with altitude, suggesting that stable air trapped methane close to the surface. In contrast, on 26 February, methane levels actually increased with height—from 2.187 ppm at 0–10 m to 2.199 ppm at 100–110 m—hinting that methane from nearby sources was being advected at that level. Some flights, like the one on 25 October, showed nearly uniform methane concentrations (2.171–2.167 ppm) from 0 to 115 m, indicating well-mixed air. Winter and fall

https://doi.org/10.3390/rs18040549

flights displayed more frequent occurrences of surface-enhanced methane due to possible shallow or stable boundary layers with higher overall concentrations (2.188–2.199 ppm). In contrast, summer flights (June–August 2024) displayed more uniform methane distribution (around 2.060–2.080 ppm) throughout the measured altitudes, a sign that stronger summer mixing was spreading methane more evenly through the atmosphere.

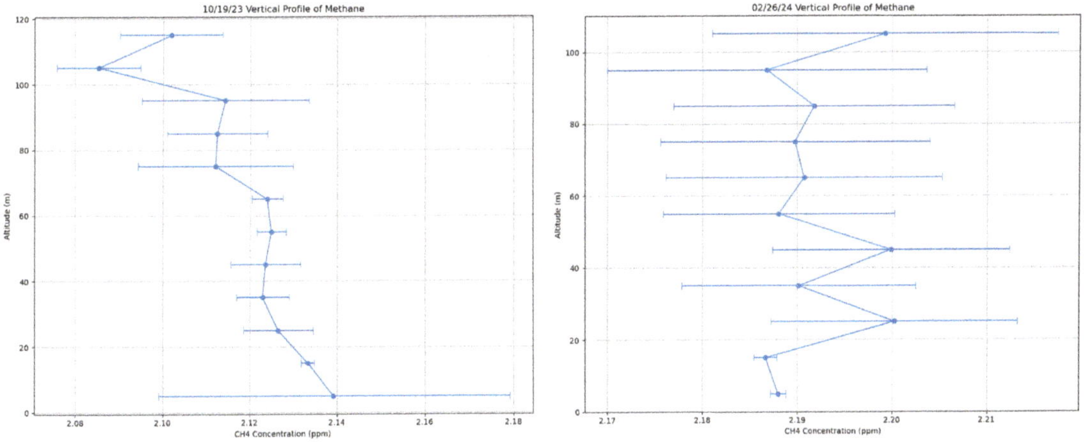

(**a**) (**b**)

Figure 12. Vertical profiles for 19 October 2023, and 26 February 2024. (**a**) 19 October displayed variable methane concentrations near ground (0–10 m), suggesting ground level accumulation; (**b**) 26 February displayed higher fluctuations throughout, with concentration increases with altitude.

3.4. Ethane Co-Tracer UAV Observations and Anomalies

Ethane concentrations as a source apportionment co-tracer, ranged extensively throughout the flight campaign from below detectable because of instrument constraints to as much as greater than 19.41 ppb. A previous study using the Aeris Technologies MIRA Ultra indicates artifacts in ethane readings when sampling dry air with low water vapor content (below 1.05% H_2O) [45]. This is due to MIRA lasers' reliance on water absorption features for wavelength calibration. Under low humidity conditions, the absence of water vapor can cause the laser to lose its frequency lock [45], thus leading to inconsistent measurements (ranging from negative to positive). This is particularly critical in diffuse ambient measurements where single plume concentrations are not available to dominate the signal.

During this study, we believe negative values were attributable to baseline drift and response noise near detection limits. MIRA Strato specifications report a detection threshold of 500 parts per trillion per s (ppt/s) $\approx$ 0.5 ppb for ethane [36], with low concentrations or ambient measurements fluctuating below zero. Excluding these anomalies, concentrations were extremely variable with means between 2.37 and 8.16 ppb on individual flights, and no temporal, seasonal, or spatial patterns were apparent in the ambient measurements.

3.5. Influence of Hovering Environments on Vertical Gas Profiles

Analyzing the vertical distribution of CH_4, C_2H_6, and H_2O concentrations across different hovering environments during UAV flights, based on height parameters defined in Section 2.4, revealed distinct boundary-layer behaviors and the influence of surface interactions. Across all flights, vertical profiles varied modestly by environment but showed consistent overall structure. Methane levels increased during the ascent and descent of HBCU's north building. The building ascending showed a median of 2.098 ppm (mean: 2.114 ± 0.018 ppm), and the building descending showed 2.079 ppm (mean: 2.095 ± 0.023 ppm), indicating a micro-emission influence. In contrast, forest (2.083–2.091 ppm median) and tower coverage (2.077–2.084 ppm median) displayed smoother profiles typical of a convective boundary layer. Notably, ascents over the meteorological tower often reached the upper range of observed methane values, with tower ascending achieving the highest overall mean at 2.100 ppm; a pattern also visible in individual-day profiles such as 26 February 2024 (2.180 ppm; cf. Figure 13), which may indicate localized mixing or nearby plume recirculation around the tower structure.

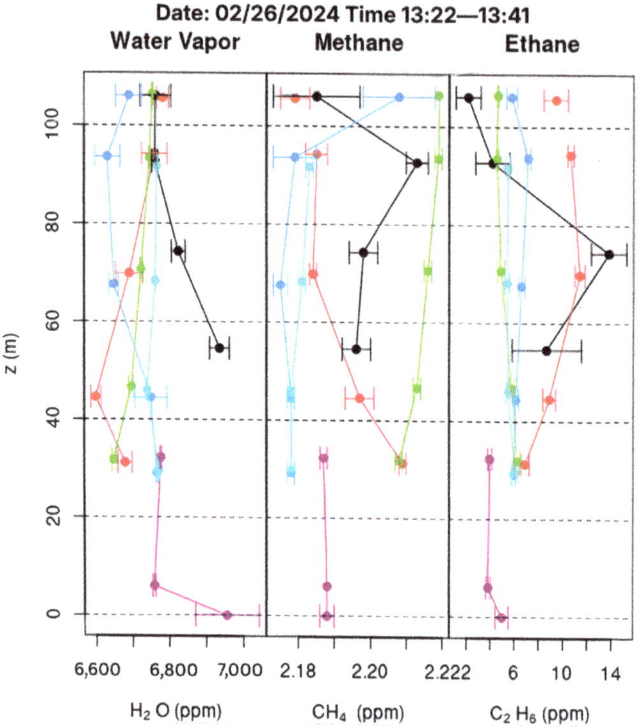

Figure 13. Vertical profiles collected on 26 February 2024, show distinct differences in trace gas structure associated with the hovering environment. Each profile includes ascending building (black), descending building (magenta), ascending forest (blue), descending forest (red), ascending meteorological tower (green), and ascending tower (cyan) flight segments.

Ethane exhibited the strongest environmental sensitivity, with higher near-surface values over the north building (mean: 2.19 ± 2.59 ppb ascending, standard deviation: 6.87 ppb) and lower, more uniform concentrations over the forest (mean: 0.62 ± 1.89 ppb ascending, standard deviation: 5.29 ppb). Unlike methane, ethane frequently exhibited localized perturbations, particularly around the meteorological tower, with measurable concentrations ranging from 0.19 to 8.63 ppb (standard deviation: 2.77 ppb–2.96 ppb across ascending and descending profiles), suggesting the influence of mechanical turbulence or horizontal plume advection. Water vapor showed the clearest gradients, driven primarily by meteorology rather than surface type, with low winter humidity averaging 4561 ppm (range: 3038–6956 ppm) and substantial summer increases averaging 17,272 ppm (range: 15,887–19,097 ppm) near the surface.

Minor systematic differences between ascent and descent traces were observed, particularly in building environments. Near-surface ($z < 50$ m) C_2H_6 concentrations were higher

https://doi.org/10.3390/rs18040549

during building ascent (5.06 ± 5.41 ppb) compared to descent (1.59 ± 2.48 ppb), a difference of 3.47 ppb ($p = 0.12$), suggesting ascending measurements may capture fresher near-surface emissions before vertical mixing homogenizes the signal, though this difference did not reach statistical significance. Descending CH_4 profiles showed no significant ascent-descent difference ($p = 0.47$) despite equipping the UAV with a tubing inlet to minimize downwash. Forest ascent and descent profiles were more consistent, likely due to stronger natural turbulence within the canopy, with the forest maintaining intermediate values between 2.052 and 2.219 ppm (standard deviation: 0.044–0.053 ppm) across measurement days. Coverage over the meteorological tower occasionally displayed mid-altitude deviations during ascent, which may reflect mechanical turbulence rather than true atmospheric gradients, as evidenced by ethane variability at the tower location (standard deviation: 4.22–4.93 ppb; cf. Figure 14).

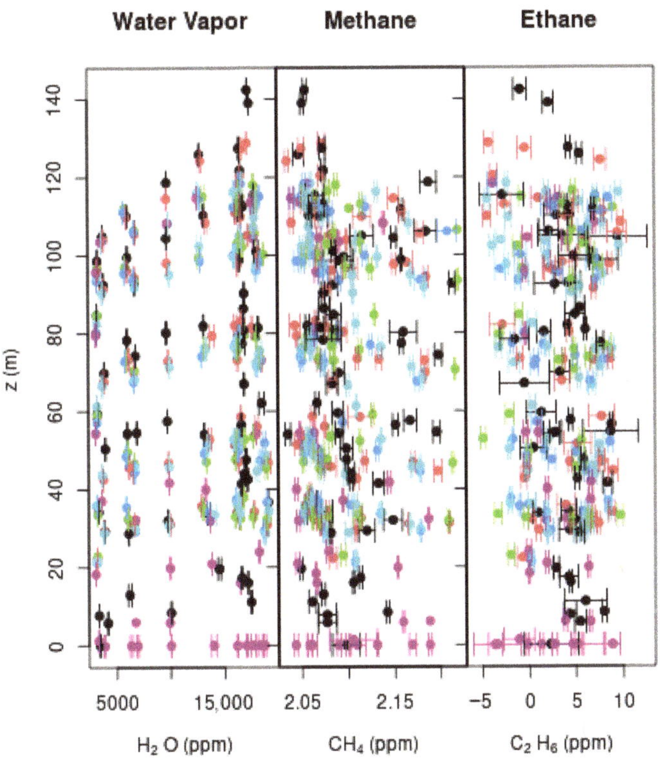

Figure 14. Aggregated vertical profiles of H_2O, CH_4, and C_2H_6 across all UAV flights, grouped by hovering environment. Colors denote ascending building (black), descending building (magenta), ascending forest (blue), descending forest (red), ascending meteorological tower (green), and ascending tower (cyan) flight segments.

3.6. Correlation Analysis

The correlation between methane and environmental variables provided insight into the behavior of methane sources and the atmospheric processes influencing concentrations (cf. Figure 15). Correlations between CH_4 and H_2O were weak and inconsistent, both negative and positive, suggesting a limited influence of humidity on methane variability or possible cross-sensitivity of the instrument. As noted above, CH_4 and C_2H_6 correlations ranged from strong negative values (possibly artifactual) to moderate positive correlations ($r = 0.36$–0.56) suggestive of co-located thermogenic sources, though interpretation remains

https://doi.org/10.3390/rs18040549

limited by established instrument constraints. CH_4-temperature (T_0) correlations varied in strength and direction, most likely reflective of boundary-layer dynamics, diurnal heating, and thermal stratification effects. Among all variables, temperature showed the strongest average correlation with CH_4, followed by C_2H_6 and H_2O. Altitude generally showed a negative correlation with methane, implying surface accumulation. However, positive correlations on certain days (e.g., 29 April) suggest vertical mixing or transport from a nearby source at higher elevations. These patterns reflect the analyzer's sensitivity to temporal meteorological variability and validate its use in source attribution and boundary-layer transport characterization.

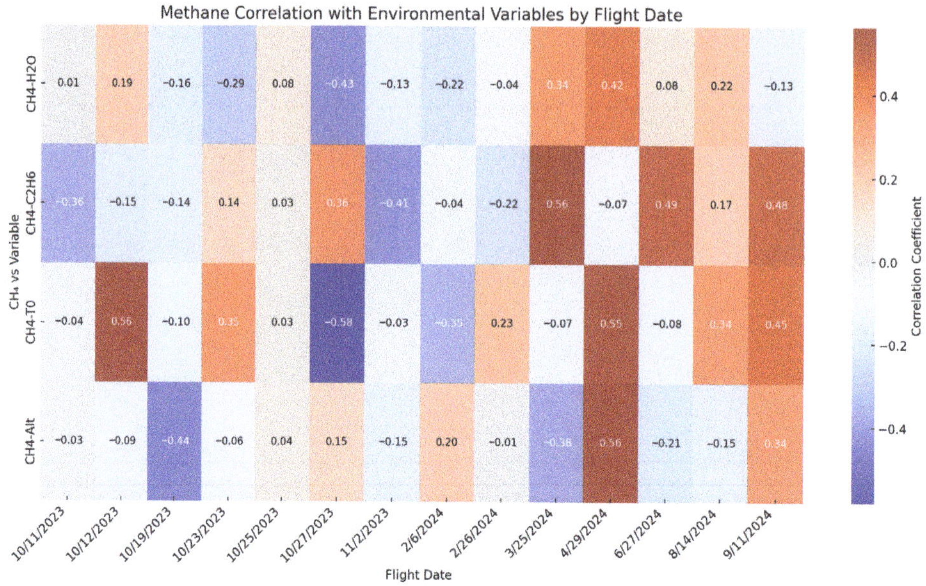

Figure 15. Heatmap of Pearson correlation coefficients between CH_4 concentrations and environmental variables (altitude, temperature, ethane, and water vapor) for each UAV flight. Warmer colors indicate stronger positive correlations, while cooler colors indicate stronger negative correlations.

3.7. *Regression Analysis*

To further evaluate the drivers of ambient methane variability across all UAV flights, we performed multivariate linear regression using temperature, altitude, and ethane as predictors. The regression coefficients differed widely among individual flights, indicating complex and inconsistent relationships without any clear seasonal trends.

3.7.1. Altitude Effects

Altitude coefficients showed high flight-to-flight variability with no consistent seasonal trend. As displayed in Table 3, several flights displayed surface accumulation with significant vertical gradients observed. For example, 19 October (slope = -3.05×10^{-4} ppm/m, $p < 0.001$) and 27 June (slope = -2.70×10^{-5} ppm/m, $p < 0.001$) flights were characterized by negative correlations, indicative of surface-based methane accumulation under stable boundary layer conditions. Few flights showed positive altitude correlation, consistent with convective transport (e.g., 6 February: 9.93×10^{-5} ppm/m, $p < 0.001$; 11 September: 1.50×10^{-4} ppm/m, $p < 0.001$). Two flights showed non-significant altitude effects (25 March: $p = 0.904$; 14 August: $p = 0.351$), suggesting well-mixed atmospheric conditions. Across the 14 flights, seven exhibited surface accumulations, four showed aloft enhancement, and three showed non-significant/well-mixed relationships during typical peak

https://doi.org/10.3390/rs18040549

atmospheric mixing hours of 13:30–15:27 LT. This suggests that local emissions and meteorological variability conditions beyond diurnal mixing alone play a major role in methane's boundary-layer vertical structure at the site.

Table 3. Summary of vertical methane gradient regression results across UAV flights from October 2023 to September 2024. Each row represents the linear regression slope (ppm/m), coefficient of determination (r^2), and p-value for the relationship between CH_4 concentration and altitude.

Date	Slope (ppm/m)	r^2	p-Value	Observations/Notes
11 October 2023	-2.08×10^{-5}	0.0051	0.014	Significant surface accumulation
12 October 2023	-1.03×10^{-4}	0.076	<0.001	Moderate surface accumulation
19 October 2023	-3.05×10^{-4}	0.196	<0.001	Strong surface accumulation
23 October 2023	-3.63×10^{-5}	0.0041	0.026	Weak surface accumulation
25 October 2023	3.72×10^{-5}	0.0078	0.0028	Aloft source
27 October 2023	4.61×10^{-5}	0.047	<0.001	Aloft source
2 November 2023	-6.01×10^{-5}	0.023	<0.001	Moderate surface accumulation
6 February 2024	9.93×10^{-5}	0.041	<0.001	Aloft enhancement
26 February 2024	-3.98×10^{-5}	0.0044	0.036	Slight surface accumulation
25 March 2024	-4.69×10^{-6}	≈0.00008	0.904	No trend, well mixed
29 April 2024	-8.08×10^{-5}	0.0093	0.144	Non-significant, well mixed
27 June 2024	-2.70×10^{-5}	0.029	<0.001	Slight surface accumulation
14 August 2024	7.43×10^{-6}	0.0009	0.351	Well mixed
11 September 2024	1.50×10^{-4}	0.116	<0.001	Aloft enhancement

3.7.2. Temperature Relationships

Temperature coefficients varied from strongly negative to moderately positive among flights, with the most negative coefficient in late spring (29 April: −0.021 ppm/°C, $p < 0.001$) and fall (27 October: −0.008 ppm/°C, $p < 0.001$). Positive temperature relationships were observed in both cool (12 October: +0.008 ppm/°C, $p < 0.001$) and warm seasons (25 March: +0.008 ppm/°C, $p < 0.001$), suggesting that temperature effects are driven by flight-specific atmospheric conditions rather than consistent seasonal processes.

3.7.3. Ethane Coefficients

Despite known instrumental limitations, ethane showed statistically significant coefficients across most flights, though with inconsistent patterns. For fall months, ethane specifically displayed strong negative coefficients (e.g., 11 October: slope = −0.0043 ppm, $p < 0.001$; 2 November: −0.002 ppm, $p < 0.001$), contrary to expectations of enhanced thermogenic emissions during heating season and most likely reflect instrumental artifacts. Positive ethane coefficients were found in warmer months (27 June: 0.0037 ppm, $p < 0.001$; 25 March: 0.0029 ppm, $p < 0.001$), possibly reflecting real thermogenic co-emissions, although interpretation is still constrained by measurement uncertainties (cf. Figure 16).

https://doi.org/10.3390/rs18040549

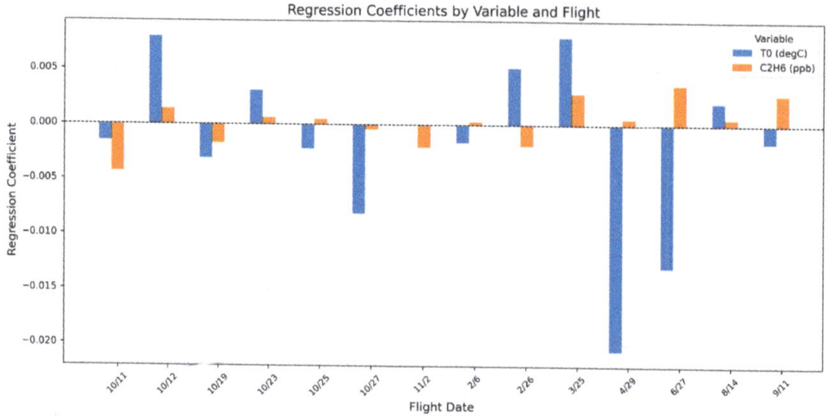

Figure 16. Regression coefficient for temperature (blue) and ethane (orange) across 14 UAV flight dates. Ethane displayed a general positive trend across dates, while temperature varied substantially. The lack of a consistent pattern across dates highlights the influence of localized conditions, variable flight paths, and potential sensor behavior on observed methane concentrations.

3.8. Precision and Temporal Stability of Methane Observations

For each flight, the coefficient of variation (CV) of the CH_4 concentration was determined to assess the MIRA Strato's signal stability and variability under ambient conditions. CV values varied from 0.00397 to 0.01143 (mean CV $\approx$ 0.0067) in all 14 flights, indicating exceptionally low relative variability in measured concentrations. The highest CV was observed on 19 October (CV = 0.01143), coinciding with the strongest CH_4–altitude gradient and negative regression coefficients for all predictors. This implies elevated stratification and sharper vertical CH_4 gradients on that day, to which the analyzer was sensitive enough to detect. In contrast, late summer and fall flights such as 14 August (CV = 0.00398) and 27 October (CV = 0.00397) were less variable, likely reflecting stable atmospheric or instrument conditions (cf. Figure 17). These results highlight the sensor's high precision in the field, with CV values consistently below 1.2%. This is well within acceptable limits for ambient methane analyzers according to instrument validation studies [46] and demonstrates MIRA Strato's stable performance across varying environmental conditions and methane distributions.

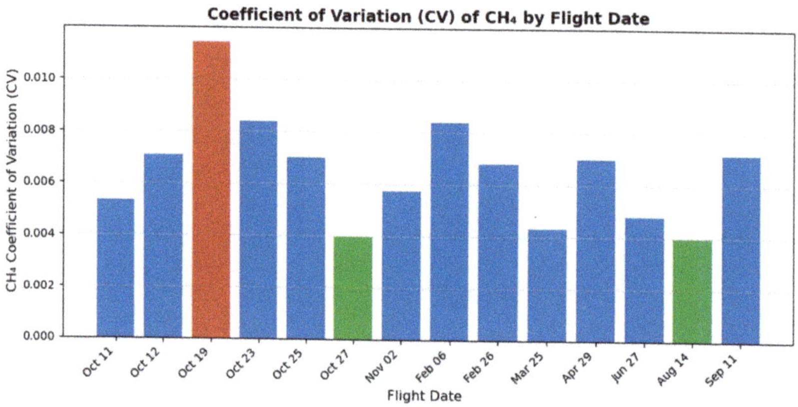

Figure 17. Coefficient of variation (CV) for CH_4 concentration by flight date. Red bar indicates the highest observed CV (19 October 2023); green bars represent comparison flights (27 October 2023, and 14 August 2024); blue bars represent all other flights.

https://doi.org/10.3390/rs18040549

4. Discussion

The UAV-based methane monitoring campaign provided valuable proof-of-concept for microscale methane observations within the atmospheric surface layer. By deploying the Aeris Technologies MIRA Strato analyzer across 14 flights from October 2023 to September 2024 (7 Fall, 3 Winter/Spring, and 4 Summer), we assessed the field stability of the sensor, correlation with environmental drivers, and ability to capture ambient variability, including temporal, spatial, and atmospheric insights into near-surface CH_4 behaviors and relationships.

The MIRA Strato analyzer exhibited strong performance in field and laboratory settings. The high signal-to-noise ratio enabled the detection of subtle ambient variations in CH_4 and co-tracers regardless of in-plume concentrations. Despite some challenges with low-level C_2H_6 detection, the instrument captured consistent spatial patterns and correlation behavior that aligned with atmospheric expectations. The combination of precision, reproducibility, and spatial responsiveness confirms the analyzer's robustness for UAV-deployed ambient methane research.

Although ambient levels throughout the campaign remained around ~2 ppm, concentrations showed seasonal behavior, with elevated concentrations in the fall and early winter (cf. Figure 18). Maximum CH_4 concentrations were observed on 19 October at 2.23 ppm, consistent with expected seasonal shifts in atmospheric mixing. This date also exhibited the most stable conditions and surface accumulation according to the regression analysis. The lowest concentration was observed on 14 August at 2.03 ppm, which also corresponds to the CV value on this date, indicating elevated mixing. The second highest mean concentration (26 February) is not conclusive to seasonal trends but is likely attributable to localized emission influences.

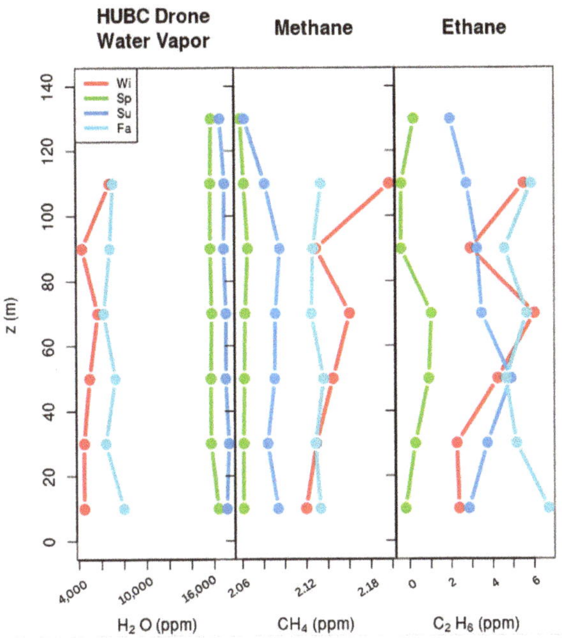

Figure 18. Seasonally averaged vertical profiles of H_2O, CH_4, and C_2H_6 derived from aggregated UAV measurements. Colors represent winter (red), spring (green), summer (blue), and fall (cyan). Methane shows a clear seasonal pattern, with higher concentrations during fall and winter and lower, tightly constrained values in spring and summer, consistent with seasonal emissions.

https://doi.org/10.3390/rs18040549

Correlation analysis further supported broad seasonal changes in methane concentrations due to temperature and time of year, reflecting larger-scale background variability. Satellite observations using TROPOMI trends(cf. Figure 19) corroborate the late-season elevation [44], displaying an increase in mean concentrations from October to December in the region of interest.

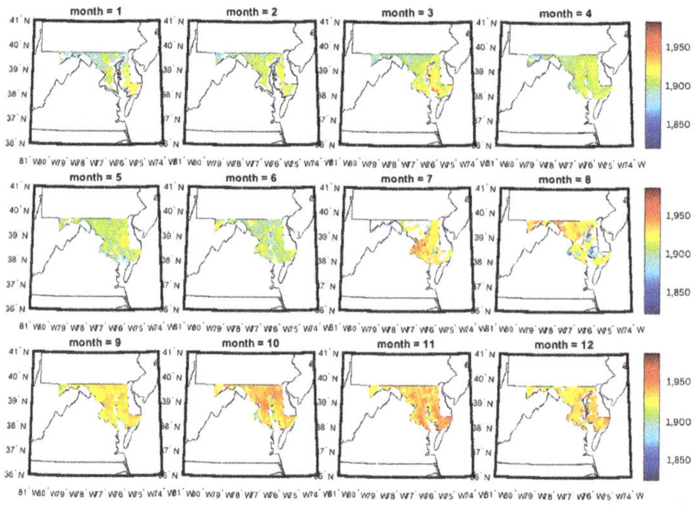

(**a**)

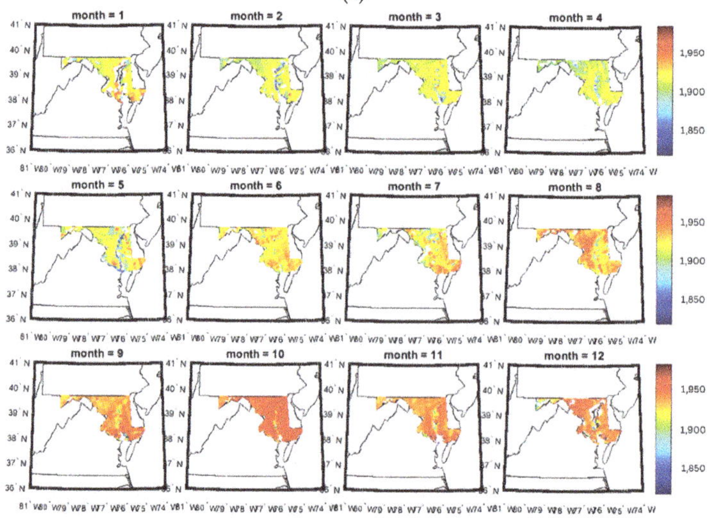

(**b**)

Figure 19. 2023–2024 monthly time series of mean CH_4 concentrations in Maryland, USA, using the TROPOMI satellite for (**a**) 2023 and (**b**) 2024. Monthly trends corroborate ground measurements, displaying an increase in mean concentrations from October to December.

Ethane occasionally recorded negative values, likely reflecting instrument limitations at low ppb concentrations or low-signal bias near detection limits. In order to effectively identify whether a source is thermogenic using the Aeris Technologies instruments in ambient environments, it is recommended to maintain a water vapor mole fraction above 1% in the sample to preserve the integrity of ethane measurements, reduce the laser power

https://doi.org/10.3390/rs18040549

when collecting, or humidify the sample line [45]. This still presents the risk of increased noise from increased water vapor sampling and was not implemented in this study.

Flight-to-flight variation of methane-altitude correlations revealed complex atmospheric stratification patterns, with both negative and positive altitude coefficients observed seasonally. Building on these results, we plan to extend this study by assessing nighttime methane sensitivity using thermometry-based satellite sensors. While the UAV platform and flight patterns were generally consistent, variability in flight path, altitude intervals, and meteorological conditions likely contributed to variability in concentration and correlation results. Hovering environments, such as over buildings, forests, and towers, also introduced small but noticeable changes to the vertical structure, though statistical comparisons indicated these differences remained within the range of natural atmospheric variability, likely reflecting localized turbulence and mechanical mixing near surface features rather than distinct micro-emission signatures. Despite these factors, the Aeris MIRA Strato showed a stable response, low drift, and sensitivity to environmental changes for boundary-layer characterization and ambient methane monitoring.

5. Conclusions

This study demonstrated the capability of the Aeris Technologies MIRA Strato analyzer for UAV-based ambient methane monitoring, capturing seasonal variability, diurnal patterns, and vertical structure within the atmospheric boundary-layer. These results underscore the value of UAV-based MWIR analyzers for ambient monitoring and highlight areas for improvement. Notably, the lack of comparative inlet measurements limited our ability to isolate rotor downwash effects, and detailed environmental characteristics such as building geometry, canopy density, or surrounding source distribution were not captured during profiling. These factors likely contributed to the observed variability across hovering environments. Future research should incorporate controlled inlet design evaluations, wind and turbulence analysis, and expanded site characterization to support more detailed correlation analysis. Overall, this work demonstrates not only the instrument's robustness but also the capability of drone-deployed sensing to resolve fine-scale atmospheric variations without relying on direct plume quantification.

Author Contributions: Formal analysis, investigation, writing—original draft preparation, and visualization D.E. Drone deployment, N.N.K. Drone modification and visualization, R.K.S. Gas chamber calibration, R.K.S., P.S. and X.R. Project Administration, M.W. Writing—review, editing, and visualization, N.R.N. Writing—review and editing, A.F., J.R.V. and J.W. All authors have read and agreed to the published version of the manuscript.

Funding: This research was funded by the U.S Department of Energy (DOE) RENEW (grant no. DE-SC0024489), the National Aeronautics and Space Administration (NASA) Student Airborne Science Activation Program (grant no. 80NSSC21K0435), U.S. Department of Commerce, National Oceanic and Atmospheric Administration (NOAA), through the Educational Partnership Program (grant no. NA22SEC4810015), and the NOAA Cooperative Science Center for Atmospheric Sciences and Meteorology (NCAS-M).

Data Availability Statement: The data and visualization materials supporting the conclusions of this article will be made available by the authors on request.

Acknowledgments: The authors thank the National Geospatial-Intelligence Agency (NGA) and Howard University (HU) Cooperative Research and Development Agreement XHM0476CRFY 20001, Application of Remote Sensing for GEOINT, for providing technical assistance and instrument procurement.

Conflicts of Interest: The authors declare no conflicts of interest.

https://doi.org/10.3390/rs18040549

Disclaimer: The views expressed in this document do not necessarily reflect those of the National Geospatial-Intelligence Agency, the Department of War, or any other department or agency of the US Government.

References

1. EPA Methane Emissions. Available online: www.epa.gov/ghgemissions/overview-greenhouse-gases (accessed on 30 May 2025).
2. Predybaylo, E.; Lelieveld, J.; Pozzer, A.; Gromov, S.; Zimmermann, P.; Osipov, S.; Klingmüller, K.; Steil, B.; Stenchikov, G.; McCabe, M. Surface Temperature and Ozone Responses to the 2030 Global Methane Pledge. *Clim. Change* **2025**, *178*, 78. [CrossRef]
3. Isaksen, I.; Berntsen, T.; Dalsøren, S.; Eleftheratos, K.; Orsolini, Y.; Rogneurd, B.; Stordal, F.; Søvde, O.A.; Zerefos, C.; Holmes, C. Atmospheric Ozone and Methane in a Changing Climate. *Atmosphere* **2014**, *5*, 518–535. [CrossRef]
4. Rosentreter, J.A.; Alcott, L.; Maavara, T.; Sun, X.; Zhou, Y.; Planavsky, N.J.; Raymond, P.A. Revisiting the Global Methane Cycle Through Expert Opinion. *Earths Future* **2024**, *12*, e2023EF004234. [CrossRef]
5. Ming, T.; De Richter, R.; Felzer, B.S.; Li, W. Jump in Tropospheric Methane Concentrations in 2020–2021 and Slowdown in 2022–2024: New Hypotheses on Causation. *Atmosphere* **2025**, *16*, 406. [CrossRef]
6. Zhou, L.; Warner, J.; Nalli, N.R.; Wei, Z.; Oh, Y.; Bruhwiler, L.; Liu, X.; Divakarla, M.; Pryor, K.; Kalluri, S.; et al. Spatiotemporal Variability of Global Atmospheric Methane Observed from Two Decades of Satellite Hyperspectral Infrared Sounders. *Remote Sens.* **2023**, *15*, 2992. [CrossRef]
7. Saunois, M.; Martinez, A.; Poulter, B.; Zhang, Z.; Raymond, P.A.; Regnier, P.; Canadell, J.G.; Jackson, R.B.; Patra, P.K.; Bousquet, P.; et al. Global Methane Budget 2000–2020. *Earth Syst. Sci. Data* **2025**, *17*, 1873–1958. [CrossRef]
8. United Nations Environmental Programme/Climate and Clean Air Coalition. *Global Methane Assessment: 2030 Baseline Report*; United Nations Environment Programme: Nairobi, Kenya, 2022.
9. Stolper, D.A.; Martini, A.M.; Clog, M.; Douglas, P.M.; Shusta, S.S.; Valentine, D.L.; Sessions, A.L.; Eiler, J.M. Distinguishing and Understanding Thermogenic and Biogenic Sources of Methane Using Multiply Substituted Isotopologues. *Geochim. Cosmochim. Acta* **2015**, *161*, 219–247. [CrossRef]
10. Wuebbles, D. Atmospheric Methane and Global Change. *Earth-Sci. Rev.* **2002**, *57*, 177–210. [CrossRef]
11. Stolper, D.A.; Lawson, M.; Davis, C.L.; Ferreira, A.A.; Neto, E.V.S.; Ellis, G.S.; Lewan, M.D.; Martini, A.M.; Tang, Y.; Schoell, M.; et al. Formation Temperatures of Thermogenic and Biogenic Methane. *Science* **2014**, *344*, 1500–1503. [CrossRef] [PubMed]
12. Booth, J.; Rowe, M.; Fischer, K. *Offshore Gas Hydrate Sample Database with an Overview and Preliminary Analysis*; U.S. Geological Survey Open-File Report; USGS Woods Hole Science Center: Woods Hole, MA, USA, 1996.
13. Sherwood, O.A.; Travers, P.D.; Dolan, M.P. Compound-Specific Stable Isotope Analysis of Natural and Produced Hydrocarbon Gases Surrounding Oil and Gas Operations. In *Comprehensive Analytical Chemistry*; Elsevier: Amsterdam, The Netherlands, 2013; Volume 61, pp. 347–372, ISBN 978-0-444-62623-3.
14. Kirschke, S.; Bousquet, P.; Ciais, P.; Saunois, M.; Canadell, J.G.; Dlugokencky, E.J.; Bergamaschi, P.; Bergmann, D.; Blake, D.R.; Bruhwiler, L.; et al. Three Decades of Global Methane Sources and Sinks. *Nat. Geosci.* **2013**, *6*, 813–823. [CrossRef]
15. Li, Q.; Meidan, D.; Hess, P.; Añel, J.A.; Cuevas, C.A.; Doney, S.; Fernandez, R.P.; Van Herpen, M.; Höglund-Isaksson, L.; Johnson, M.S.; et al. Global Environmental Implications of Atmospheric Methane Removal through Chlorine-Mediated Chemistry-Climate Interactions. *Nat. Commun.* **2023**, *14*, 4045. [CrossRef]
16. Gorham, K.A.; Abernethy, S.; Jones, T.R.; Hess, P.; Mahowald, N.M.; Meidan, D.; Johnson, M.S.; Van Herpen, M.M.J.W.; Xu, Y.; Saiz-Lopez, A.; et al. Opinion: A Research Roadmap for Exploring Atmospheric Methane Removal via Iron Salt Aerosol. *Atmos. Chem. Phys.* **2024**, *24*, 5659–5670. [CrossRef]
17. Li, W.; Liu, Y.; Yao, X.; Ming, T.; De Richter, R. Removal of Atmospheric Methane by Increasing Hydroxyl Radicals via a Water Vapor Enhancement Strategy. *Atmosphere* **2024**, *15*, 1046. [CrossRef]
18. Heimann, I.; Griffiths, P.T.; Warwick, N.J.; Abraham, N.L.; Archibald, A.T.; Pyle, J.A. Methane Emissions in a Chemistry-Climate Model: Feedbacks and Climate Response. *J. Adv. Model. Earth Syst.* **2020**, *12*, e2019MS002019. [CrossRef]
19. Stevenson, D.S.; Zhao, A.; Naik, V.; O'Connor, F.M.; Tilmes, S.; Zeng, G.; Murray, L.T.; Collins, W.J.; Griffiths, P.T.; Shim, S.; et al. Trends in Global Tropospheric Hydroxyl Radical and Methane Lifetime since 1850 from AerChemMIP. *Atmospheric Chem. Phys.* **2020**, *20*, 12905–12920. [CrossRef]
20. He, L.; Groom, J.D.; Wilson, E.H.; Fernandez, J.; Konopka, M.C.; Beck, D.A.C.; Lidstrom, M.E. A Methanotrophic Bacterium to Enable Methane Removal for Climate Mitigation. *Proc. Natl. Acad. Sci. USA* **2023**, *120*, e2310046120. [CrossRef] [PubMed]
21. Umweltbundesamt. *Technical Approaches to Removing Methane from the Atmosphere (Methane Removal): Fact Check and Discussion*; Umweltbundesamt: Dessau-Roßlau, Germany, 2024.

https://doi.org/10.3390/rs18040549

22. Committee on Anthropogenic Methane Emissions in the United States: Improving Measurement, Monitoring, Presentation of Results, and Development of Inventories; Board on Atmospheric Sciences and Climate; Board on Agriculture and Natural Resources; Board on Earth Sciences and Resources; Board on Energy and Environmental Systems; Board on Environmental Studies and Toxicology; Division on Earth and Life Studies; National Academies of Sciences, Engineering, and Medicine. *Improving Characterization of Anthropogenic Methane Emissions in the United States*; National Academies Press: Washington, DC, USA, 2018; p. 24987, ISBN 978-0-309-47050-6.
23. Fosco, D.; De Molfetta, M.; Renzulli, P.A.; Notarnicola, B.; Dimauro, G.; Astuto, F.; Spizzirri, U.G. Drone-Based Methane Emission Estimation at Waste, Livestock and Energy Facilities: Experimental Tests in Italy and Challenges of Wind–Concentration Integration with Open-Path Sensors. *J. Environ. Manag.* **2026**, *398*, 128290. [CrossRef] [PubMed]
24. Shaw, J.T.; Shah, A.; Yong, H.; Allen, G. Methods for Quantifying Methane Emissions Using Unmanned Aerial Vehicles: A Review. *Philos. Trans. R. Soc. Math. Phys. Eng. Sci.* **2021**, *379*, 20200450. [CrossRef] [PubMed]
25. Watts, A.C.; Ambrosia, V.G.; Hinkley, E.A. Unmanned Aircraft Systems in Remote Sensing and Scientific Research: Classification and Considerations of Use. *Remote Sens.* **2012**, *4*, 1671–1692. [CrossRef]
26. Yang, S.; Talbot, R.W.; Frish, M.B.; Golston, L.M.; Aubut, N.F.; Zondlo, M.A.; Gretencord, C.; McSpiritt, J. Natural Gas Fugitive Leak Detection Using an Unmanned Aerial Vehicle: Measurement System Description and Mass Balance Approach. *Atmosphere* **2018**, *9*, 383. [CrossRef]
27. Dooley, J.F.; Minschwaner, K.; Dubey, M.K.; El Abbadi, S.H.; Sherwin, E.D.; Meyer, A.G.; Follansbee, E.; Lee, J.E. A New Aerial Approach for Quantifying and Attributing Methane Emissions: Implementation and Validation. *Atmos. Meas. Tech.* **2024**, *17*, 5091–5111. [CrossRef]
28. Hollenbeck, D.; Nunez, G.; Christensen, L.E.; Chen, Y. Wind Measurement and Estimation with Small Unmanned Aerial Systems (sUAS) Using On-Board Mini Ultrasonic Anemometers. In Proceedings of the 2018 International Conference on Unmanned Aircraft Systems (ICUAS), Dallas, TX, USA, 12–15 June 2018; IEEE: Piscataway, NJ, USA, 2018; pp. 285–292.
29. McKinney, K.A.; Wang, D.; Ye, J.; De Fouchier, J.-B.; Guimarães, P.C.; Batista, C.E.; Souza, R.A.F.; Alves, E.G.; Gu, D.; Guenther, A.B.; et al. A Sampler for Atmospheric Volatile Organic Compounds by Copter Unmanned Aerial Vehicles. *Atmos. Meas. Tech.* **2019**, *12*, 3123–3135. [CrossRef]
30. Villa, T.; Gonzalez, F.; Miljievic, B.; Ristovski, Z.; Morawska, L. An Overview of Small Unmanned Aerial Vehicles for Air Quality Measurements: Present Applications and Future Prospectives. *Sensors* **2016**, *16*, 1072. [CrossRef] [PubMed]
31. Burgués, J.; Marco, S. Environmental Chemical Sensing Using Small Drones: A Review. *Sci. Total Environ.* **2020**, *748*, 141172. [CrossRef] [PubMed]
32. Acharya, B.S.; Bhandari, M.; Bandini, F.; Pizarro, A.; Perks, M.; Joshi, D.R.; Wang, S.; Dogwiler, T.; Ray, R.L.; Kharel, G.; et al. Unmanned Aerial Vehicles in Hydrology and Water Management: Applications, Challenges, and Perspectives. *Water Resour. Res.* **2021**, *57*, e2021WR029925. [CrossRef]
33. Jońca, J.; Pawnuk, M.; Bezyk, Y.; Arsen, A.; Sówka, I. Drone-Assisted Monitoring of Atmospheric Pollution—A Comprehensive Review. *Sustainability* **2022**, *14*, 11516. [CrossRef]
34. Kaliszewski, M.; Włodarski, M.; Młyńczak, J.; Jankiewicz, B.; Auer, L.; Bartosewicz, B.; Liszewska, M.; Budner, B.; Szala, M.; Schneider, B.; et al. The Multi-Gas Sensor for Remote UAV and UGV Missions—Development and Tests. *Sensors* **2021**, *21*, 7608. [CrossRef] [PubMed]
35. Madokoro, H.; Kiguchi, O.; Nagayoshi, T.; Chiba, T.; Inoue, M.; Chiyonobu, S.; Nix, S.; Woo, H.; Sato, K. Development of Drone-Mounted Multiple Sensing System with Advanced Mobility for In Situ Atmospheric Measurement: A Case Study Focusing on PM2.5 Local Distribution. *Sensors* **2021**, *21*, 4881. [CrossRef]
36. Aeris Technologies. *MIRA Strato LDS: UAV/Drone Natural Gas Leak Detection System w/GPS*; Aeris Technologies: Hayward, CA, USA, 2021; p. 2. Available online: https://aerissensors.com/wp-content/uploads/2021/12/MIRA-Strato-LDS_Drone_211209.pdf (accessed on 7 May 2025).
37. Paschotta, R. Spectroscopy. *RP Photonics Encycl.* Available online: https://www.rp-photonics.com/spectroscopy.html (accessed on 8 April 2025).
38. Paschotta, R. Multipass Gas Cells. *RP Photonics Encycl.* Available online: https://www.rp-photonics.com/multipass_gas_cells.html (accessed on 23 April 2025).
39. Picarro, Inc. *G2401 Gas Concentration Analyzer.* Available online: https://www.picarro.com/environmental/products/g2401_gas_concentration_analyzer (accessed on 30 April 2025).
40. Wheeler, M.D.; Newman, S.M.; Orr-Ewing, A.J.; Ashfold, M.N.R. Cavity Ring-down Spectroscopy. *J. Chem. Soc. Faraday Trans.* **1998**, *94*, 337–351. [CrossRef]
41. Borhan, M.S.; Khanaum, M.M. Sensors and Methods for Measuring Greenhouse Gas Emissions from Different Components of Livestock Production Facilities. *J. Geosci. Environ. Prot.* **2022**, *10*, 242–272. [CrossRef]
42. Bauer, P.; Nagy, M. Flight-Data-Based High-Fidelity System Identification of DJI M600 Pro Hexacopter. *Aerospace* **2024**, *11*, 79. [CrossRef]

https://doi.org/10.3390/rs18040549

43. *DJI Matrice 600 Pro User Manual PDF*. Available online: https://dl.djicdn.com/downloads/m600%20pro/20180417/Matrice_600_Pro_User_Manual_v1.0_EN.pdf (accessed on 7 June 2025).
44. Lampert, A.; Pätzold, F.; Asmussen, M.O.; Lobitz, L.; Krüger, T.; Rausch, T.; Sachs, T.; Wille, C.; Sotomayor Zakharov, D.; Gaus, D.; et al. Studying Boundary Layer Methane Isotopy and Vertical Mixing Processes at a Rewetted Peatland Site Using an Unmanned Aircraft System. *Atmos. Meas. Tech.* **2020**, *13*, 1937–1952. [CrossRef]
45. Commane, R.; Hallward-Driemeier, A.; Murray, L.T. Intercomparison of Commercial Analyzers for Atmospheric Ethane and Methane Observations. *Atmos. Meas. Tech.* **2023**, *16*, 1431–1441. [CrossRef]
46. Levrault, C.M.; Difford, G.F.; Steinheim, G.; Groot Koerkamp, P.W.G.; Ogink, N.W.M. Validation of the Methane Production Measurement Accuracy and Ranking Capacity of Portable Accumulation Chambers for Use with Small Ruminants. *Biosyst. Eng.* **2023**, *236*, 201–211. [CrossRef]

https://doi.org/10.3390/rs18040549

remote sensing

MDPI

Article

Low-Cost Real-Time Remote Sensing and Geolocation of Moving Targets via Monocular Bearing-Only Micro UAVs

Peng Sun [1,2,3], Shiji Tong [1,2,3], Kaiyu Qin [1,2,3,*], Zhenbing Luo [4], Boxian Lin [1,2,3] and Mengji Shi [1,2,3]

1 School of Aeronautics and Astronautics, University of Electronic Science and Technology of China, Chengdu 611731, China; pengsun@std.uestc.edu.cn (P.S.); shijitong@std.uestc.edu.cn (S.T.); linbx@uestc.edu.cn (B.L.); maangat@uestc.edu.cn (M.S.)
2 Aircraft Swarm Intelligent Sensing and Cooperative Control Key Laboratory of Sichuan Province, Chengdu 611731, China
3 National Laboratory on Adaptive Optics, Chengdu 610209, China
4 College of Aerospace Science and Engineering, National University of Defense Technology, Changsha 410073, China; luozhenbing@163.com
* Correspondence: kyqin@uestc.edu.cn

Highlights

What are the main findings?

- A bearing-only localization framework is proposed for micro UAVs operating over uneven terrain, combining a pseudo-linear Kalman filter (PLKF) with a sliding-window nonlinear least squares optimization to achieve real-time 3D positioning and motion prediction.
- An observability-enhanced flight trajectory planning method is designed based on the Fisher Information Matrix (FIM), which improves localization accuracy in flight experiments, with an average gain of up to 4.34 m.

What are the implications of the main findings?

- The method addresses the limitations of monocular cameras with scale ambiguity and lack of depth measurement, providing a practical solution for low-cost UAVs in remote sensing and moving target localization.
- The results suggest potential applications in emergency response, target monitoring, and traffic security, demonstrating the feasibility of high-accuracy localization on resource-constrained UAV platforms.

Abstract

Low-cost and real-time remote sensing of moving targets is increasingly required in civilian applications. Micro unmanned aerial vehicles (UAVs) provide a promising platform for such missions because of their small size and flexible deployment, but they are constrained by payload capacity and energy budget. Consequently, they typically carry lightweight monocular cameras only. These cameras cannot directly measure distance and suffer from scale ambiguity, which makes accurate geolocation difficult. This paper tackles geolocation and short-term trajectory prediction of moving targets over uneven terrain using bearing-only measurements from a monocular camera. We present a two-stage estimation framework in which a pseudo-linear Kalman filter (PLKF) provides real-time state estimates, while a sliding-window nonlinear least-squares (NLS) back end refines them. Future target positions are obtained by extrapolating the estimated trajectory. To improve localization accuracy, we analyze the relationship between the UAV path and the Cramér–Rao lower bound (CRLB) using the Fisher Information Matrix (FIM) and derive an observability-enhanced trajectory planning method. Real-flight experiments validate the framework, showing that accurate geolocation can be achieved in real time using only low-cost monocular bearing measurements.

Keywords: target geolocation; low-cost; motion state estimation; UAV; bearing-only; trajectory optimization

1. Introduction

1.1. Background and Significance

Target localization is a key problem in aerial remote sensing, supporting applications such as high-resolution geographic mapping [1], target surveillance and tracking [2,3], land-resource monitoring [4], moving-target enclosure [5], and search-and-rescue missions [6]. With the miniaturization and cost reduction of unmanned aerial vehicles (UAVs), micro UAVs equipped with lightweight visual sensors have become valuable platforms for real-time data acquisition in both civilian and defense contexts. In these scenarios, geolocation depends on estimating the relative state between the UAV and ground targets from onboard observations, often under strict payload and energy constraints, complex terrain, and dynamic environments.

Although many tracking and localization techniques have been developed in remote sensing [7,8], aerospace [9,10], and robotics [11,12], bearing-only localization from monocular imagery remains particularly challenging due to scale ambiguity, lack of direct range measurements, and strong dependence on relative motion. To address these challenges, this paper (i) employs bearing-only measurements from a monocular gimbal camera for moving-target localization, (ii) quantifies how relative motion and UAV trajectories affect observability and estimation accuracy, and (iii) develops an observability-enhanced trajectory planning strategy for real-time geolocation performance.

1.2. Related Works

Previous studies have aimed to improve UAV-based visual detection and geographic localization capabilities [13–18]. Li et al. [13] proposed a detection algorithm based on the Swin Transformer network to improve image quality and target detection accuracy. Meng et al. [14] enhanced remote sensing photogrammetry accuracy through anomaly detection. Yang et al. [15] discussed UAV localization and control under occluded camera scenarios in remote sensing contexts. Savkin et al. [16] proposed a distributed navigation algorithm to monitor targets moving along curved roads. Guo et al. [17] developed a novel adaptive extended Kalman filter method for UAVs equipped with electro-optical pods to enhance localization accuracy. Zhou et al. [18] applied related methods to large fixed-wing UAVs operating at high altitudes.

However, micro UAVs are constrained by payload and power supply, limiting their capability to carry high-precision sensors such as electro-optical pods, thus typically equipped only with monocular gimbal cameras. Due to the inherent scale uncertainty and lack of direct distance measurement, accurately estimating the 3D pose of targets remains challenging. Current research on monocular camera-based localization mainly employs prior-based [19,20] and learning-based approaches [21,22]. Zhou et al. [19] used a flat terrain assumption and UAV altimeter measurements to approximate relative target heights, which is unsuitable for non-flat terrains. Pizani et al. [20] utilized Digital Elevation Models (DEMs) to provide height priors, but this requires large-scale maps unsuitable for micro UAVs. Wang et al. [21] and Yang et al. [22] recovered depth from monocular images through learning methods, but these involve complex training and high computational demands. Zhang et al. [23] proposed an approach without prior information, applicable only to stationary ground targets, with the difficulty of estimation increasing significantly at greater distances.

To address these challenges, we propose a bearing-only localization method that avoids direct depth measurements or scale priors, relying solely on angular measurements collected during continuous relative motion. This concept has proven viable in radar and maritime applications. Farina et al. [24] initially used radar bearings to estimate target 3D positions. Li et al. [25] applied angular measurements to localize maritime vessels. In aerospace and remote sensing, bearing-only methods have been widely utilized [26–28]. Chu et al. [26] applied the technique in spacecraft rendezvous and docking, while Wang et al. [27] and Li et al. [28] used it for UAV localization and aerial target navigation.

Because bearing-only measurements provide direction but not range, accurate estimation requires relative motion and multiple views. This creates two main challenges for UAV remote sensing. First, solving the problem hinges on estimating the target's state between consecutive frames when both the UAV and the target are moving. In practice, incremental Kalman filtering like EKF and the PLKF [29–31] is commonly used, but fast maneuvers can accumulate inter-frame errors, and the nonlinear measurement model increases computational load, making high-precision real-time estimation difficult. Second, not all relative motions lead to the same 3D accuracy under noise; this is an issue of observability. Typical analyses use bearing rigidity, rank conditions, or information-theoretic tools. For example, Ferdowsi [29] studied 2D observability via matrix conditions, while Jiang [32] and Woffinden [33] analyzed the 3D single-target case under Gaussian noise using FIM/CRLB, showing that different motion patterns yield different minimum error bounds.

Beyond these foundations, several recent works focus explicitly on bearing-only localization and observability enhancement. Wang et al. [27] derive a 3D bearing-only helical homing law and reveal weak observability under proportional navigation guidance (PNG), motivating guidance that explicitly increases observability. Zheng et al. [34] propose a cooperative STT-R estimator that exploits bearing rates in a distributed RLS framework to improve observability and reduce velocity lag. In multi-robot settings, Lu et al. [35] present a concurrent-learning-based relative localization method that relaxes persistent excitation and enhances inter-robot observability during shape formation. For air-to-air scenarios, Liu et al. [36] combine detection, KF-based tracking, and interval replanning to maintain fixed-distance pursuit for 3D reconstruction of a non-cooperative UAV.

Compared with these studies, our work considers a single micro-UAV with bearing-only sensing over non-flat terrain and couples a PLKF→SW-NLS estimator with an FIM-driven, observability-enhancing trajectory planner. The entire pipeline runs onboard and is validated in real flight.

1.3. Contributions

To solve the challenge of localizing targets on non-flat terrains using bearing-only measurements from micro UAVs, this study proposes a comprehensive localization and trajectory planning framework addressing cumulative inter-frame errors and observability-enhanced trajectory design. The primary contributions are as follows.

(1) Bearing-Only 3D Localization Framework for Uneven Terrain: To address the limitations of prior studies that typically assume flat ground [19] or rely on DEM-based elevation priors [20], this paper develops a real-time localization framework that operates effectively over uneven terrain. It enables a micro-UAV equipped only with a monocular gimbal camera to perform 3D localization and prediction of moving targets. The framework has been validated through real flight experiments, demonstrating high-accuracy geolocation under low-cost conditions and providing a systematic solution for micro-UAV remote sensing.

(2) Two-Stage Fusion Algorithm with PLKF Front End and Sliding-Window Nonlinear Least-Squares Back End: We propose a two-stage structure that combines a PLKF based

front end with a sliding-window optimization back end. The front end provides inter-frame priors and projects bearing residuals, while the back end corrects accumulated errors through a SW-NLS formulation. This design significantly improves estimation accuracy while maintaining real-time performance, making it suitable for onboard deployment on micro-UAVs. Compared with the traditional EKF method, the front end achieves a computation speed improvement of approximately 8 ms per frame and offers better numerical stability while maintaining comparable accuracy.

(3) FIM-Based Observability Analysis and Closed-Loop Trajectory Planning Method: By quantitatively analyzing observability under Gaussian noise using the Fisher Information Matrix (FIM), this work reveals how the relative motion between the UAV and the target influences localization accuracy. Based on this analysis, we develop an observability-enhancing trajectory planning method that yields a closed-form, easily implementable local-orbit design rule, executed in discrete form. Compared with [28], our method continuously improves observability during flight along the planned trajectory, forming a closed-loop information flow that integrates state estimation, observability analysis, trajectory optimization, and enhanced estimation. Real flight results show that the proposed trajectory improves localization accuracy by more than 4.34 m compared with direct pursuit and manual flight, demonstrating both the effectiveness and engineering practicality of the method.

1.4. Organization

The remainder of the paper is organized as follows. Section 2 describes the system overview and problem formulation. Section 3 details the PLKF-based motion estimation, sliding-window optimization, observability analysis, and trajectory planning. Section 4 presents real-flight experiments and comparative results. Section 5 concludes the paper.

2. Problem Formulation

2.1. System Overview

Considering that micro-UAVs equipped with inexpensive monocular cameras cannot directly measure the depth of targets during remote sensing tasks over uneven terrain, this paper proposes a real-time remote sensing target localization framework using bearing-only measurements. The overall system structure is shown in Figure 1, and its workflow and functionality are described as follows:

Firstly, the UAV collects ground images in real-time using the onboard monocular camera. Targets are detected using the YOLOv8 algorithm, providing real-time target coordinates in the image frame. After successful detection, image coordinates are converted into a unit direction vector R_C in the camera frame using intrinsic parameters. Using extrinsic camera parameters and the UAV's real-time pose data, this vector is further transformed into the world coordinate frame, yielding the direction measurement R_W.

Next, the UAV continuously observes the target from different positions and perspectives, acquiring multiple angle measurements. These measurements are sequentially stored, forming an observation data sequence for target localization. To obtain initial estimates required for optimization, the system employs the Pseudo-Linear Kalman Filter (PLKF) method, which estimates the target's state contains position and velocity in real-time using a motion model each time new measurement data arrives.

Subsequently, a backend optimization procedure refines localization accuracy based on these measurements and initial estimates. Historical measurements beyond a certain threshold have negligible effects and add computational burden; hence, the system utilizes a fixed-length sliding window. Only the relative positions within this sliding window undergo nonlinear least-squares optimization, with earlier data serving as known prior positions. This approach maintains continuity and enhances computational efficiency.

Then, Bernstein polynomials fit the initially obtained target coordinates, generating a smooth and continuous trajectory curve. Recognizing that actual target motion is constrained by velocity and acceleration limits, a sliding-window mechanism is employed to extrapolate short-term trajectory predictions, providing necessary information for subsequent UAV trajectory optimization.

Finally, based on Fisher Information Matrix (FIM), the paper quantitatively analyzes how different UAV-target relative trajectories affect localization precision under Gaussian noise. Leveraging insights from this observability analysis, an observability-enhanced UAV trajectory planning method is proposed. This method significantly improves localization accuracy, is easy to implement, and does not require strictly tracking precise predefined paths. The integration of bearing-only localization and observability-enhanced trajectory planning effectively enhances the efficiency and accuracy of remote sensing tasks over uneven terrain. The overall workflow is shown in Figure 1.

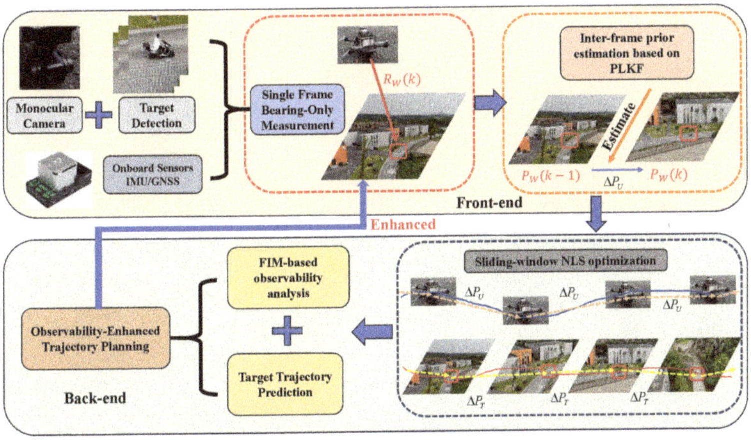

Figure 1. System overview and module interconnection. The monocular detector produces image coordinates that are converted to a world-frame bearing R_W (Equations (1)–(4)). The PLKF front end provides inter-frame priors; the sliding-window NLS back end optimizes target positions using projector-whitened bearing residuals and the PLKF prior (Equations (19)–(26)). A short-horizon trajectory $\hat{B}(t)$ is fit for prediction, and an observability-aware planner evaluates the FIM M and isotropy J to generate an observability-enhancing path (Equations (38)–(41)). The controller tracks the waypoint onboard, and the updated UAV pose produces new bearings, closing the estimation–planning loop. Our goal is accurate real-time geolocation on non- flat terrain using monocular, bearing-only sensing.

2.2. Measurement Model

To clearly formulate the mathematical modeling of remote sensing target localization, we define four coordinate systems: the world coordinate system W, the UAV body coordinate system U, the target body coordinate system T, and the camera coordinate system C onboard the UAV. The world coordinate system W is fixed to the ground and serves as an inertial reference frame. The origin of the UAV body coordinate system U is located at the UAV's center of mass, with its x-axis pointing towards the nose, y-axis pointing to the left wing, and z-axis pointing upwards, forming a right-handed coordinate system. Similarly, the target coordinate system T has its origin at the target's center of mass, with axes defined identically to the UAV body coordinate system. The camera coordinate system C is centered at the camera's optical center, with the z-axis along the optical axis, and the

x- and y-axes forming a right-handed system on the imaging plane. Figure 2 shows the relationships among these coordinate frames.

The geographical localization of targets based on visual measurements can be achieved by combining the camera imaging model and rigid body transformations. Consider a ground target P, whose projection onto the camera imaging plane is denoted as P_T. The position of the target P in the world coordinate system can be represented as:

$$P_W = P_C + \lambda \cdot R_c^w \cdot R_C, \tag{1}$$

where P_C denotes the camera's optical center position in the world coordinate system W, R_c^w is the rotation matrix from the camera coordinate system C to the world coordinate system W, R_C is the unit direction vector from the camera to the target in the camera coordinate system C, and λ represents the scale factor, i.e., the actual distance from the camera optical center to the target point.

Using the pinhole camera model, the unit direction vector R_C can be expressed in terms of pixel coordinates (u, v) on the image plane and camera focal length f as:

$$R_C = \frac{1}{\sqrt{u^2 + v^2 + f^2}} (u, v, f)^{\mathrm{T}}. \tag{2}$$

Substituting this into the target position equation, we obtain the target position in the world coordinate system:

$$(x_w, y_w, z_w)^{\mathrm{T}} = (x_c, y_c, z_c)^{\mathrm{T}} + \lambda \cdot R_c^w \cdot \frac{1}{\sqrt{u^2 + v^2 + f^2}} (u, v, f)^{\mathrm{T}}. \tag{3}$$

Since monocular vision cannot directly measure the actual distance λ from the camera to the target (due to scale ambiguity), we rearrange this expression to derive the explicit formula for the scale factor:

$$\lambda = \frac{(z_w - z_c)\sqrt{u^2 + v^2 + f^2}}{\mathbf{e}_z R_c^w (u, v, f)^{\mathrm{T}}}. \tag{4}$$

Here, $\mathbf{e}_z = (0, 0, 1)$ denotes the direction of the z-axis in the world coordinate system. Substituting the scale factor back into the position expression, we obtain the target's horizontal position:

$$(x_w, y_w)^{\mathrm{T}} = (x_c, y_c)^{\mathrm{T}} + (z_w - z_c) \frac{\sqrt{u^2 + v^2 + f^2}}{\mathbf{e}_z R_c^w (u, v, f)^{\mathrm{T}}} \begin{bmatrix} 1 & 0 & 0 \\ 0 & 1 & 0 \end{bmatrix} R_c^w (u, v, f)^{\mathrm{T}}. \tag{5}$$

The rotation matrix R_c^w is determined by the following chain relation:

$$R_c^w = R_u^w R_c^u,$$

where R_c^u represents the extrinsic rotation matrix between the camera and the UAV, a time-varying parameter due to gimbal movements, calculated in real-time from camera-gimbal calibration and measured gimbal rotation angles. R_u^w is the UAV's attitude rotation matrix in the world coordinate system, measured by the UAV's attitude and position sensors. The camera focal length f is obtained through camera calibration, and (u, v) are the detected target image coordinates. The camera optical center position is determined by the UAV's onboard GPS device and camera extrinsic parameters.

Equation (5) provides a clear geometric method to derive the target's three-dimensional coordinates from two-dimensional image data. However, as each image measurement provides only two independent equations, monocular visual localization inherently suffers from depth ambiguity. Therefore, additional constraints or prior information are required to determine the complete 3D coordinates. A common assumption in UAV localization tasks is the flat ground assumption, assuming negligible ground elevation changes. In practical scenarios, the relative altitude between UAV and target $h = z_U - z_T$ can initially be approximated using UAV GPS altitude information. This approximation provides a feasible initial guess for optimization. Once the initial estimate of the target position is established, subsequent optimizations iteratively refine the target location using the previously estimated position, a method proven feasible in [37].

For accurate 3D target position estimation, besides obtaining direction vectors R_W at each measurement, the UAV state information (position and velocity) at discrete times t_k must also be acquired:

$$\mathcal{S}_U^{6D} = \{(P_U(k), V_U(k)) | k = 1, 2, ..., n\}, \tag{6}$$

with

$$P_U(k) \in \mathbb{R}^3, \quad V_U(k) \in \mathbb{R}^3.$$

Thus, at each measurement time t_k, the system records direction vector $R_W(k)$, UAV position $P_U(k)$, and UAV velocity $V_U(k)$. These measurements form a discrete time-sequence:

$$\mathcal{S}_U = \{(R_W(k), P_U(k), V_U(k)) | k = 1, 2, ..., n\}. \tag{7}$$

Given that measurement data accumulate over time and early data's influence diminishes, we adopt a dynamic sliding-window approach, maintaining a window of fixed length L:

$$\mathcal{S}_U^L = \{(R_W(k), P_U(k), V_U(k)) | k = t - L + 1, ..., t\}. \tag{8}$$

Additionally, reasonable prior estimations of target motion between two measurements are necessary to achieve stable nonlinear least-squares optimization. The prior estimation of target states between observations is derived using individual measurement data and motion models, detailed further in Section 3. Figure 2 illustrates the process of bearing-only measurements of the target by the UAV during its motion.

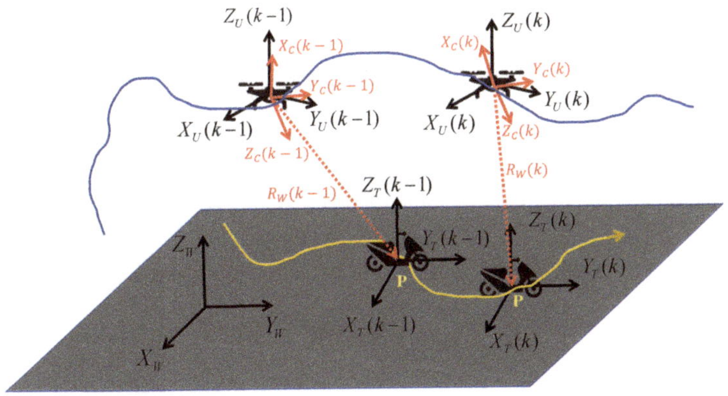

Figure 2. Schematic illustration of bearing-only measurements taken by the UAV for a moving target at multiple discrete time instants, where the PLKF estimates the inter-frame prior, and the NLS optimizes the relative motion among multiple frames. The blue line represents the UAV flight trajectory, and the yellow line represents the target's trajectory.

3. Methodology

3.1. Inter-Frame Target Motion Estimation Based on Bearing-Only Measurements

The target motion between two consecutive frames from time $t(k-1)$ to $t(k)$ can be estimated using the angle measurements obtained at time $t(k)$. In this section, the relative target motion is modeled using orthogonal projection vectors and pseudo-linear Kalman filtering (PLKF). We systematically describe the mathematical modeling, filtering steps, and characteristics of the inter-frame motion estimation.

Consider the measurement at time $t(k)$. The UAV state, comprising position and velocity in the world coordinate system W, is denoted as:

$$S_U^{6D}(k) = (P_U(k), V_U(k)),$$

and the target state as:

$$S_T^{6D}(k) = (P_T(k), V_T(k)).$$

The relative state between the UAV and the target, which is to be estimated, is expressed as:

$$S(k) = \left[(P_U(k) - P_T(k))^\top,\ (V_U(k) - V_T(k))^\top\right]^\top \in \mathbb{R}^6. \tag{9}$$

The direction vector from the UAV to the target, measured by the monocular gimbal camera, is defined as:

$$R_W(k) = \frac{P_T(k) - P_U(k)}{\|P_T(k) - P_U(k)\|} \in \mathbb{R}^3, \quad \|R_W(k)\| = 1. \tag{10}$$

Due to Gaussian measurement noise $\mu \sim \mathcal{N}(0, \Sigma)$, where $\Sigma = \sigma^2 I_3$, the noisy measurement is expressed as:

$$\tilde{R}_W(k) = R_W(k) + \mu. \tag{11}$$

Given the visual detection frequency of approximately 15–30 Hz, both UAV and target motions can be reasonably approximated as constant velocity within the small interval $\Delta t = t(k) - t(k-1)$. The kinematic state equations for the target and UAV are:

$$S_T^{6D}(k) = A\, S_T^{6D}(k-1) + Bw(k), \tag{12}$$

and

$$S_U^{6D}(k) = A\, S_U^{6D}(k-1) + Bv(k), \tag{13}$$

where

$$A = \begin{bmatrix} I_3 & \Delta t\, I_3 \\ 0_3 & I_3 \end{bmatrix}, \quad B = \begin{bmatrix} \frac{1}{2}\Delta t^2 I_3 \\ \Delta t I_3 \end{bmatrix}, \tag{14}$$

with

$$w(k) \sim \mathcal{N}(0, \Sigma_1), \quad v(k) \sim \mathcal{N}(0, \Sigma_2). \tag{15}$$

Considering UAV maneuvers, the combined relative motion model can be expressed as:

$$\hat{S}(k) = A(\hat{S}(k-1) + S_U^{6D}(k-1)) - S_U^{6D}(k). \tag{16}$$

Directly expressing the angle measurement $\tilde{R}_W(k)$ as a nonlinear function complicates subsequent computations. Hence, we introduce an orthogonal projection operator for pseudo-linearization. For any vector $\mathbf{v} \in \mathbb{R}^3$, the orthogonal projection operator P_v is defined as:

$$P_v = I_3 - \frac{vv^\top}{\|v\|^2}, \quad v \neq 0, \tag{17}$$

which projects vectors onto a plane perpendicular to v and satisfies $P_v^2 = P_v = P_v^\top$. Multiplying both sides of $\tilde{R}_W(k) = R_W(k) + \mu$ by $\lambda_k P_{\tilde{R}(k)}$, where $\lambda_k = \|P_T(k) - P_U(k)\|$, yields:

$$0_3 = P_{\tilde{R}(k)}(P_T(k) - P_U(k)) + \lambda(k) P_{\tilde{R}(k)}\mu. \tag{18}$$

Rewriting this equation using relative state $S(k)$, we obtain a pseudo-linear measurement equation:

$$0_3 = H(k)S(k) + V_k\mu, \quad H(k) = \left[P_{\tilde{R}(k)}, 0_3\right], \quad V_k = \lambda(k)P_{\tilde{R}(k)}. \tag{19}$$

Note that $H(k)$ and $V(k)$ are time-varying and $H(k)$ is rank-deficient. In practice, the predicted distance $\hat{\lambda}(k) = \|\hat{S}(k)^-_{1:3}\|$ substitutes the unknown $\lambda(k)$.

Based on this model and the measurement noise $\mu(k) \sim \mathcal{N}(0, \Sigma_3)$, with $\Sigma_3 = \sigma_3^2 I_3$, the measurement covariance becomes:

$$T_k = V_k \Sigma_3 V_k^\top. \tag{20}$$

The PLKF algorithm consists of prediction and update steps. The prediction step is:

$$\hat{S}(k)^- = A(\hat{S}_{k-1} + S_U^{6D}(k-1)) - S_U^{6D}(k), \quad P(k)^- = AP(k-1)A^\top + BW(k)B^\top. \tag{21}$$

The Kalman gain is:

$$K(k) = P(k)^- H(k)^\top \left(H(k)P(k)^- H(k)^\top + T(k)\right)^\dagger, \tag{22}$$

where $(\cdot)^\dagger$ denotes the Moore-Penrose pseudo-inverse used due to rank deficiency of $H(k)$. As the measurement vector is zero, the state update equations are:

$$\hat{S}(k) = \hat{S}(k)^- - K(k)H(k)\hat{S}(k)^-, \quad P(k) = (I_6 - K(k)H(k))P(k)^-. \tag{23}$$

Through pseudo-linearization, measurement information is implicitly integrated into matrices $H(k)$ and $T(k)$, providing a linear formulation that facilitates prediction and updates. The PLKF method avoids nonlinear trigonometric expansions of angular measurements, enhancing numerical stability and computational efficiency, making it highly suitable for high-frequency visual measurements under constrained onboard computing resources.

3.2. Target Position Optimization Based on Nonlinear Least Squares

To mitigate the cumulative errors caused by the prolonged operation of pseudo-linear Kalman filtering (PLKF) and to fully exploit the geometric constraints from multi-frame bearing-only measurements, this paper designs a batch optimization sliding-window backend for discrete target position estimation on top of the PLKF frontend.

Let the sliding window length be L and the current frame index be k. The target positions to be estimated within the window are defined as:

$$\mathbf{S} = \left[P_T(k-L+1)^\top, P_T(k-L+2)^\top, \ldots, P_T(k)^\top\right]^\top \in \mathbb{R}^{3\times L}.$$

For any frame $j \in [k-L+1, k]$ within the window, the whitened residual of direction measurements is defined as:

$$\tilde{r}_j = \frac{1}{\sigma} P_{\tilde{R}(j)} \left(P_T(j) - P_U(j) \right) \in \mathbb{R}^3, \quad P_{\tilde{R}(j)} = I_3 - \tilde{R}_W(j) \tilde{R}_W^\top(j), \tag{24}$$

where σ denotes the standard deviation of the bearing measurement. As $P_{\tilde{R}} P_{\tilde{R}}^\top = P_{\tilde{R}}$, the above residual is normalized to have a unit covariance matrix, constraining only two degrees of freedom orthogonal to the measurement direction. The unobserved depth dimension is supplemented by the PLKF prior.

PLKF provides an estimated relative position $\hat{\mathbf{r}}(k) = (\hat{S}(k))_{1:3}$ and associated covariance $\mathbf{\Lambda}_k = P_{k,r}$ for the latest frame k. The normalized prior residual is defined as:

$$r^{\mathrm{pri}} = \mathbf{L}_k^{-\frac{1}{2}} \left(P_T(k) - \hat{r}(k) \right), \quad \mathbf{L}_k^{-\frac{1}{2}} \mathbf{L}_k^{-\frac{1}{2}\top} = \mathbf{\Lambda}_k^{-1}, \tag{25}$$

where $\mathbf{L}_k^{-\frac{1}{2}}$ is derived via Cholesky decomposition of the prior covariance matrix $\mathbf{\Lambda}_k$.

Combining these residual terms, the total optimization cost function for the sliding-window backend is defined as:

$$J(\mathbf{S}) = \sum_{j=k-L+1}^{k} |\tilde{r}_j|_2^2 + |r^{\mathrm{pri}}|_2^2. \tag{26}$$

The first term minimizes the orthogonal plane error between the target positions and measurement rays, achieving geometric intersection across multiple frames. The second term provides stable depth anchoring from single-frame scale estimates provided by the PLKF frontend. Their weights are adaptively adjusted via their respective covariance matrices to prevent redundant information weighting.

Applying increments $\delta\mathbf{S}$ to the current estimate $\mathbf{S}^{(0)}$, the residuals can be linearized as:

$$\tilde{r}_j(\mathbf{S}) \approx \tilde{r}_j + \mathbf{J}_j \delta\mathbf{S}, \quad \mathbf{J}_j = [\mathbf{0}, \ldots, \sigma^{-1} P_{\tilde{R}(j)}, \ldots, \mathbf{0}], \tag{27}$$

$$r^{\mathrm{pri}}(\mathbf{S}) \approx r^{\mathrm{pri}} + \mathbf{J}_{\mathrm{pri}} \delta\mathbf{S}, \quad \mathbf{J}_{\mathrm{pri}} = [\mathbf{0}, \ldots, \mathbf{L}_k^{-\frac{1}{2}}]. \tag{28}$$

The corresponding normal equation is:

$$\mathbf{H}\delta\mathbf{S} = -\mathbf{b}, \quad \mathbf{H} = \mathbf{J}^\top \mathbf{J}, \quad \mathbf{b} = \mathbf{J}^\top \mathbf{r}, \tag{29}$$

where $\mathbf{J}$ stacks each frame's $\mathbf{J}_j$ and prior $\mathbf{J}_{\mathrm{pri}}$, and $\mathbf{H}$ is a sparse block matrix that can be efficiently solved using sparse $LDL^\top$ decomposition.

When the window slides, $\mathbf{H}$ and $\mathbf{b}$ are partitioned by $[\mathbf{T}_{k-L+1}, \mathbf{S}_{\mathrm{rem}}]$, marginalizing out the oldest state via Schur complement:

$$\tilde{\mathbf{H}} = \mathbf{H}_{rr} - \mathbf{H}_{rm} \mathbf{H}_{mm}^{-1} \mathbf{H}_{mr}, \quad \tilde{\mathbf{b}} = \mathbf{b}_r - \mathbf{H}_{rm} \mathbf{H}_{mm}^{-1} \mathbf{b}_m. \tag{30}$$

The resulting $(\tilde{\mathbf{H}}, \tilde{\mathbf{b}})$ forms the new Gaussian prior for the next optimization loop, retaining full information and maintaining a 3D variable dimension.

Practically, the prior residuals generated when historical frames enter the window, as given by:

$$r_i^{\mathrm{pri}} = \mathbf{L}_i^{-1/2} (P_T(i) - \hat{\mathbf{r}}_i), \quad i < k, \tag{31}$$

are continuously incorporated into the current Gaussian prior block through Schur complements as the window advances. This process ensures each window retains scale information from L frames and preserves the sparsity and conditioning of the least squares system.

The total cost function defined in (9) is solved using the Levenberg–Marquardt (LM) algorithm. LM linearizes $J(\mathbf{S})$ at each iteration, solves the increment $\delta\mathbf{S}$ after damping $\mathbf{H}$ by adding $\lambda\mathbf{I}$, and iteratively updates $\mathbf{S}$ until convergence.

For this optimization, predicting the target's future trajectory is necessary. Considering bounded continuity of velocity and acceleration, the trajectory is extrapolated using optimized past observations. Specifically, a time-decayed quadratic regularized least squares fitting is applied to the FIFO observation sequence $\mathbf{Q}_{\text{target}} = \{(\mathbf{p}_{t_i}, t_i)\}_{i=1}^{L}$, minimizing:

$$J_{\text{pre}} = \sum_{i=1}^{L} w_{t_i} \|\hat{B}(t_i) - \mathbf{p}_{t_i}\|_2^2 + w_p L \int_{t_1}^{t_p} \|\hat{B}^{(2)}(t)\|_2^2 \, dt, \quad w_{t_i} = \tanh\left(\frac{k_t}{t_L - t_i}\right), \tag{32}$$

The fitting cost J_{pre} consists of a time-decayed data term and a curvature regularization term. The first term minimizes the error between the fitted curve and the historical trajectory points, while the second term penalizes curvature to ensure the smoothness of the predicted trajectory and to avoid sharp fluctuations. $\hat{B}(t) \in \mathbb{R}^3$ is an nth-order Bézier curve with control points c_i, w_{t_i} downweights older samples, $w_p > 0$ controls smoothness, and t_p denotes the end of the prediction horizon. The velocity and acceleration bounds in Equation (33) are applied to c_i through standard Bézier derivatives. subject to velocity and acceleration constraints:

$$\pm n(c^i - c^{i-1}) \le \pm v_{mp}, \quad \pm n(n-1)\frac{(c^i - 2c^{i-1} + c^{i-2})}{s_t} \le \pm a_{mp}. \tag{33}$$

Solving this constrained quadratic programming (QP) yields an n-th order Bézier curve $\hat{B}(t)$, producing a smooth, continuously differentiable trajectory over both observed and predicted intervals.

Equation (32) is updated in a sliding-window manner with each incoming visual detection. Since the visual detection frequency exceeds 15 Hz, the predicted target trajectory is rapidly updated as well, allowing the planner to keep up with the target's frequent and irregular movements.

3.3. Observability-Enhanced Trajectory Planning

Building upon the PLKF-NLS-based bearing-only target localization framework, this section addresses another important issue: analyzing how the drone's relative motion impacts the theoretical lower bound of target estimation accuracy under bearing measurements with Gaussian noise. Specifically, the analysis is carried out using the close relationship between the Fisher Information Matrix (FIM) and the Cramér–Rao lower bound (CRLB), where the CRLB defines the minimum achievable variance for any unbiased estimator. Even within our nonlinear optimization framework, the lowest achievable error bound is constrained by the CRLB. The FIM serves as a crucial tool for characterizing estimation precision, with its inverse providing a theoretical lower bound for the covariance of estimation errors. Based on this, we quantitatively analyze how different UAV motion patterns influence the lower bound of the target position estimation error and propose an observability-enhanced trajectory planning method for the UAV.

In this process, the camera measurement $R_C(u, v, f)$ is rotated into the world frame through $R_c^w = R_u^w R_c^u$, making it dependent on the UAV's attitude but independent of its translation. The UAV position appears in the measurement model only through the relative displacement $P_T - P_U$ in the back-end residual and through the range r_j used for weighting in the FIM (Equations (38)–(41)). Consequently, attitude errors act as angular noise on $R_W(k)$, as reflected in Equations (11) and (24).

The target positions to be estimated within the sliding window are denoted as:

$$\mathbf{S} = \left[P_T(k-L+1)^\top, P_T(k-L+2)^\top, \ldots, P_T(k)^\top\right]^\top \in \mathbb{R}^{3\times L}, \tag{34}$$

with the corresponding noisy bearing measurements:

$$\mathbf{R} = \left[\tilde{R}_W(k-L+1), \tilde{R}_W(k-L+2), \ldots, \tilde{R}_W(k)\right] \in \mathbb{R}^{3\times L}, \tag{35}$$

where each noisy measurement $\tilde{R}_W(j)$ is the true bearing vector plus Gaussian noise:

$$\tilde{R}_W(j) = R_W(j) + \mu_j, \quad \mu_j \sim \mathcal{N}(0, \sigma^2 I_3), j = k-L+1, \ldots, k. \tag{36}$$

According to the Cramér–Rao lower bound, the estimation error covariance matrix for any unbiased estimator $\hat{\mathbf{S}}(\mathbf{R})$ satisfies:

$$\mathbf{P} = \mathbb{E}[(\hat{\mathbf{S}}_3 - \mathbf{S}_3)(\hat{\mathbf{S}_3} - \mathbf{S}_3)^\top] \succeq \mathbf{M}^{-1}, \tag{37}$$

where the FIM is defined as:

$$\mathbf{M} = -\mathbb{E}\left[\frac{\partial^2 \log p_{\mathbf{R}|\mathbf{S}_3}}{\partial \mathbf{S}_3^2}\right]. \tag{38}$$

Under the assumptions of independent measurements with zero-mean Gaussian noise, the above FIM can be expressed as the sum of measurement information across frames within the sliding window:

$$\mathbf{M} = \sum_{j=k-L+1}^{k} \left[\frac{\partial R_W(j)}{\partial P_T(j)}\right]^\top \Sigma_j^{-1} \left[\frac{\partial R_W(j)}{\partial P_T(j)}\right] = \sum_{j=k-L+1}^{k} \frac{1}{\sigma_j^2 r_j^2}\left(I_3 - R_W(j) R_W(j)^\top\right), \tag{39}$$

where the unit bearing vector is defined as $R_W(j) = (P_T(j) - P_U(j))/r_j$, and the range is $r_j = \|P_T(j) - P_U(j)\|$. Particularly, the matrix $(I - R_W R_W^\top)$ is a projection operator that projects any vector onto the plane perpendicular to the bearing vector R_W. It is idempotent, satisfying $(I - R_W R_W^\top)^2 = I - R_W R_W^\top$, thus simplifying the form of the FIM.

In optimization, the determinant of the FIM, $\det \mathbf{M}$, is inversely proportional to the volume of the probability ellipsoid for estimation error, and thus commonly used as an indicator of observational information content. However, directly optimizing $\det \mathbf{M}$ becomes highly complex in three-dimensional space for a sliding window length $L > 2$. Therefore, to obtain a more analytically convenient indicator, this section introduces the isotropy metric of the FIM:

$$J = \|\mathbf{M} - \bar{\lambda} I_3\|_F^2, \qquad \bar{\lambda} = \tfrac{1}{3}\operatorname{tr}\mathbf{M}, \tag{40}$$

which measures the difference among the eigenvalues of the FIM. Specifically, $J = 0$ indicates an ideal isotropic observational state. Expanding the above expression yields a more explicit formulation:

$$J = \|G\|_F^2 - \frac{1}{3}S_3^2, \quad S = \sum_{j=k-L+1}^{k} \frac{1}{\sigma_j^2 r_j^2}, \quad G = \sum_{j=k-L+1}^{k} \frac{1}{\sigma_j^2 r_j^2} R_W(j) R_W(j)^\top. \tag{41}$$

From the above formulation, it is clear that two key factors influence the indicator J: the first involves the distance term $1/r_j^4$, and the second is related to the inner product between different measurement bearing vectors $(R_W(i)^\top R_W(j))^2$. Evidently, shorter measurement distances and more orthogonal consecutive measurement bearings lead to a smaller J value and thus stronger observability.

Hence, we formulate the trajectory planning as a constrained optimization problem: minimizing the indicator J subject to safety distance r_d and UAV dynamic constraints:

$$\min_{\{P_U(j)\}} J \quad \text{s.t.} \begin{cases} r_j \geq r_{\min}, \\ \|P_U(j+1) - P_U(j)\| \leq v_{\max}\Delta t, \\ \|V_U(j+1) - V_U(j)\| \leq a_{\max}\Delta t. \end{cases} \tag{42}$$

The FIM analysis indicates that maximizing the information content requires two conditions: (i) keeping the relative distance between UAV and target near the minimal allowed distance r_d; (ii) making the bearing vectors as orthogonal as possible within the sliding window, thereby minimizing their coupling terms $(R_W(i)^\top R_W(j))^2$. Ideally, this corresponds to the UAV orbiting around the target at a constant angular velocity, forming a spherical trajectory with radius r_d. Due to UAV speed and vision sampling constraints, achieving perfectly orthogonal consecutive measurements is impractical; thus, ensuring spatially well-distributed measurements with significant angular differences is sufficient.

However, as the target is on the ground, the UAV can only orbit horizontally at fixed altitude z, implying a horizontal observation distance r_h. Through detailed analysis, we derive the optimal horizontal distance and angular velocity that maximize observability under speed limit $v_{\max}$:

$$r_h^\star = z, \quad \omega^\star = \frac{v_{\max}}{z}, \quad v^\star = v_{\max}. \tag{43}$$

Since the target also has a certain motion velocity, a margin should be reserved in practice.

Finally, explicitly incorporating the target's motion speed, the discrete trajectory at each control interval Δt can be given by:

$$\begin{aligned} P_U(k) = {} & P_U(k-1) + \dot{\hat{B}}(t_k)\Delta t \\ & + r_h\big[\cos(\theta_0 + (k-1)\Delta\theta)\mathbf{N}(t_k) + \sin(\theta_0 + (k-1)\Delta\theta)(\mathbf{T}(t_k) \times \mathbf{N}(t_k))\big] \\ & - r_h\big[\cos(\theta_0 + (k-2)\Delta\theta)\mathbf{N}(t_{k-1}) + \sin(\theta_0 + (k-2)\Delta\theta)(\mathbf{T}(t_{k-1}) \times \mathbf{N}(t_{k-1}))\big], \end{aligned} \tag{44}$$

where $\Delta\theta = \omega\Delta t$ and the angular velocity constraint must satisfy:

$$|\dot{\hat{B}}(t_k)| + \omega r_h \leq v_{\max}, \quad \omega^2 r_h \leq a_{\max}. \tag{45}$$

At each perception update, we refit $\hat{B}(t)$ over the latest window according to Equation (32) and recompute $T(k)$ and $N(k)$. Then, under the constraints of Equation (44), Equation (43) is executed once as a discrete update, followed by re-planning in the next cycle. This receding-horizon execution preserves the FIM-guided geometric properties (Equations (41) and (42)) while maintaining a fast and adaptive response to unknown and rapidly changing target motion.

This discrete expression explicitly reflects the combined effect of target motion and UAV orbiting motion, ensuring optimal observability while strictly adhering to dynamic constraints.

In contrast, position errors mainly alter the geometric structure that governs observability—specifically, the distances and the angles between adjacent bearing rays. As long as the trajectory shape remains approximately circular while dynamically orbiting the target, the variation of $R_W(k)$ relative to the planned trajectory will remain small.

4. Results

In this section, experiments were conducted in real-world environments to validate the effectiveness of the proposed method. Specifically, Section 4.1 describes our experimental site, the UAV platform used, and the experimental details. Section 4.2 analyzes the experimental results. Section 4.3 first investigates the performance of the NLS-based back-end optimization and compares the performance of the PLKF estimator with that of the EKF. Finally, it comprehensively analyzes the influence of trajectories on localization performance.

4.1. Flight Experiment Details

To validate the effectiveness of the proposed method, we used a typical remote-sensing scenario involving target localization using a micro UAV. Specifically, a motorbike moving in uneven terrain was selected as the target for tracking and localization. This scenario is prevalent in civilian applications, effectively demonstrating the general applicability of the proposed method in practical remote sensing tasks. The experimental site was a 200 m × 200 m area with varied terrain, where the elevation difference between the highest and lowest points reached 15 m. The airspace was open and provided reliable GNSS signal reception for the UAV at all locations. Additionally, the experiment was conducted under clear weather conditions, with wind speeds ranging between 1 m/s and 3 m/s, thus posing no significant impact on UAV flight stability. The target vehicle was driven randomly on uneven roads within the site and was equipped with a GNSS receiver to provide accurate ground-truth trajectory data for algorithm performance comparison. The experimental site is illustrated in Figure 3.

Figure 3. Our experimental site includes a road with left and right turns and a slope. The target's motion path is roughly indicated by the red arrows: the target moves straight first, then turns left, followed by a right turn, and finally climbs up to the top of the slope.

The experiment utilized a custom-designed micro UAV platform as shown in Figure 4. The UAV has a wheelbase of 35 cm, weighs only 1.3 kg, and has a maximum flight speed of 15 m/s, making it suitable for single-person portability and deployment. It is equipped with an onboard computing unit, an autonomous navigation system, and a vision system. Specifically, the onboard computing unit is an Intel NUC12 i7-1260P, the flight controller is a CUAV x7, and the GNSS receiver is a CUAV NEO3 pro. Navigation data is jointly obtained from the inertial measurement unit (IMU) of the flight controller and the GNSS receiver. The propulsion system comprises high-performance brushless motors and a 22.2 V, 2600 mAh lithium-polymer battery. The vision system includes a camera mounted on a three-axis gimbal and onboard target detection and tracking algorithms. The target detection algorithm, developed based on YOLOv8, identifies and locates the target vehicle in real-time within the camera's field of view. The automatic gimbal tracking algorithm dynamically adjusts the gimbal orientation based on the target's position in the image, ensuring the target remains centered and continuously records the gimbal's three-axis angles. The camera has a resolution of 1920 × 1080 pixels and a frame rate of 25 Hz. These subsystems collectively enable the real-time fusion of geospatial localization, UAV status, and visual data. The software architecture of the entire system is based on ROS Noetic, and the nonlinear least-squares optimization problem is solved using the Ceres library. The trajectory planning generated waypoints are tracked using the model predictive control (MPC) method from [38], with control commands sent to the flight controller as angular velocities and throttle percentages.

In practical environments, wind enters Equations (12)–(15) as a bounded process disturbance. Its primary effect is geometric, causing slight deviations in the range r_j and the bearing angles that determine the FIM. A cascaded PID controller running at 100 Hz is employed to counteract such position errors, and each motor provides a maximum thrust exceeding 3 kg, offering substantial control redundancy for wind resistance.

During the experiment, lasting over 200 s, the target vehicle continuously moved along the road while the UAV maintained a safe flight altitude of 25 m, ensuring both safe distance and continuous effective observation of the target.

Figure 4. The micro UAV platform used in the experiment.

4.2. UAV Target Localization Performance Analysis

Figures 5 and 6 together provide a comprehensive overview of the UAV-based target-tracking and localization experiment. Figure 5 presents the entire mission trajectories. the blue curve is the UAV flight path, the orange curve is the ground-truth target trajectory obtained from GNSS, and the yellow curve is the target trajectory estimated from bearing-only observations. Figure 6b–f illustrate the key stages from takeoff and target search to stable tracking and localization. We set the instant when the UAV completes takeoff and begins searching for the target (Figure 6) as time zero. At $t \approx 15$ s, the UAV reaches the vicinity of the target and switches to an observability-enhancing trajectory for stable tracking and localization, as shown in Figure 6b. During the experiment the target turns left at $t \approx 70$ s (Figure 6c), turns right at $t \approx 120$ s (Figure 6e), then climbs the slope and stops at the summit at $t \approx 220$ s, as shown in Figure 6f.

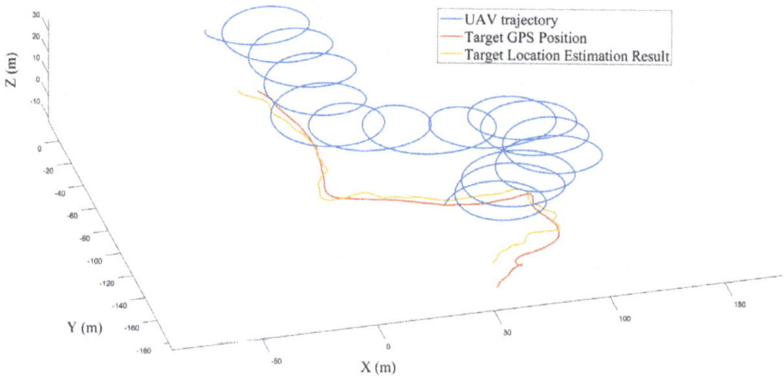

Figure 5. 3D visualization of the target and UAV trajectories. The red curve represents the ground truth target trajectory obtained from the GNSS receiver, the yellow curve shows the estimated target positions, and the blue curve depicts the UAV flight trajectory.

Figure 6. (**a**) The UAV completes takeoff and reaches the designated altitude. (**b**) The UAV detects the target and begins localization. (**c**) The target turns left. (**d**) The target moves onto another road. (**e**) The target turns right and starts climbing the slope. (**f**) The target finishes the climb and reaches the destination.

After take-off, the UAV first pans its gimbal camera to search for the target; once the target appears in the field of view, the camera continuously tracks it. In the initial localization frames, the UAV assumes a flat ground prior and flies almost straight toward the target to quickly reduce the range and iteratively refine the target position. At this stage, no observability-enhancing trajectory is used, so the localization accuracy is relatively low. As the UAV approaches, the horizontal distance r decreases, the bearing angle rate increases, and the uncertainty in the target position estimate decreases rapidly (Figure 7a). When $t \approx 15$ s and r reach the desired value of $r = 25$ m, the UAV flies around the target at maximum linear speed until the end of the experiment ($t \approx 220$ s), producing high-precision estimates of the current and past positions of the target.

Figure 7b shows the RMSE between the estimated and true target positions throughout the experiment. The RMSE starts at 7.85 m, drops to 5.53 m as the UAV closes in, and further converges to 3.25 m once the observability-enhancing trajectory begins. A sharp left turn at $t \approx 70$ s briefly raises the RMSE to 6.13 m, but it quickly recovers to 2.93 m. A similar spike to 4.51 m occurs at the right turn at $t \approx 120$ s, then decreases to 3.88 m. The uphill motion slightly increases the RMSE to 4.03 m. During the whole process, the RMSE is 3.79m. Throughout the trial, the gimbal rapidly responds to target speed changes and the trajectory-prediction module adjusts the flight path in real time, allowing the estimation error to reconverge quickly. These results demonstrate that the proposed method maintains robust localization and tracking performance even when the target undergoes significant dynamic maneuvers.

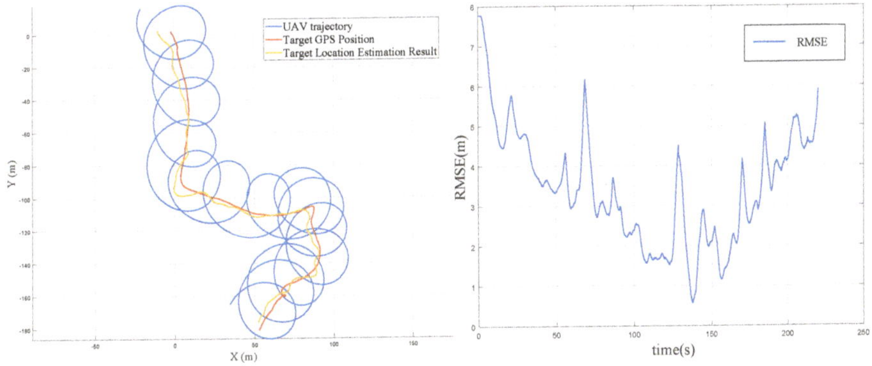

(**a**) (**b**)

Figure 7. (**a**) Blue lines denote the UAV's 2D flight trajectory, orange the target's GPS trajectory, and yellow the estimated target positions. (**b**) Estimation error over the entire flight.

4.3. Comparative Experiments

In this section, we first examine how back-end optimization influences overall system performance. We then disable the back-end and compare the performance gains achieved by the EKF [39] and PLKF front-end filtering algorithms. Finally, we analyze the impact of various flight-path geometries on localization accuracy, thereby providing a comprehensive validation of the effectiveness and advantages of the proposed method.

The specific experimental setup is as follows. A real flight dataset based on the observability-enhanced trajectory proposed in Section 3.3 was selected. During the flight, the UAV employed the proposed PLKF front end and Sliding-Window Nonlinear Least-Squares (SW-NLS) back end to perform real-time target localization and trajectory optimization. After the flight, the recorded data were replayed offline: first, the onboard real-time PLKF estimation results were extracted to produce the outcomes shown in Figure 8c,d; then, the same GNSS/IMU data, gimbal attitude angles, and image detection results were input into an independently implemented EKF estimator to compute and record its runtime and localization error, resulting in Figure 8a,b. This setup ensures a fair comparison between the two algorithms under identical input and environmental conditions.

Figure 8 illustrates the position-estimation results obtained after removing the back-end module and relying solely on the EKF or PLKF front end. Specifically, Figure 8a,b show the localization trajectory and RMSE obtained with the EKF front end, whereas Figure 8c,d present the corresponding results obtained with the PLKF front end. Under these conditions, the RMSE increased dramatically to 10.10 m for the EKF front-end and to 8.56 m for the PLKF front-end. Although PLKF consistently outperformed EKF in both computational speed and accuracy, it still could not achieve high localization accuracy when used independently. These results highlight that our proposed method, which combines the fast and effective PLKF front-end with the high-accuracy NLS back-end, achieves an optimal balance between computational speed and localization accuracy. Besides, the PLKF front-end had an inter-frame solving time ranging from 6 ms to 9 ms, while the EKF front-end required 14 ms to 18 ms, clearly demonstrating the computational speed advantage of PLKF.

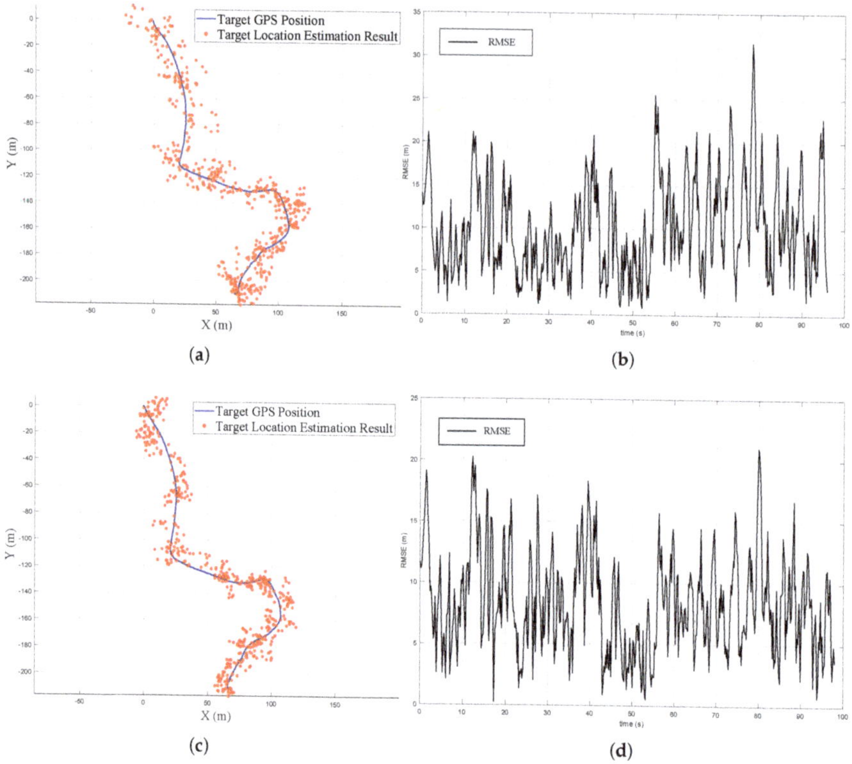

Figure 8. Target position estimates obtained with the EKF and PLKF front ends. Because the back-end module is disabled, the estimates appear as scattered points. (**a**) Target positions estimated with the EKF front end. (**b**) Estimation error (RMSE) for the EKF front end. (**c**) Target positions estimated with the PLKF front end. (**d**) Estimation error (RMSE) for the PLKF front end.

To evaluate the effectiveness of the proposed observability-enhanced trajectory, we further compared it with two commonly used UAV surveillance trajectories: the parallel-following trajectory and the manually operated trajectory. In the experiments, the ground target moved along the same route and within the same speed range as shown in Figures 5 and 6, while the UAV maintained a flight altitude of approximately 25 m. Across all three experiments, the sensor configuration, detection algorithm, and localization method were kept identical.

First, we tested a commonly used parallel-following trajectory in remote sensing tasks; second, a professional operator manually piloted the UAV freely to track the target. In both sets of experiments, the UAV maintained a flight altitude of approximately 25 m. We recorded trajectory data and localization errors during these experiments and compared them against our observability-enhanced trajectory. Figure 9a–d show the four different flight trajectories and their localization errors relative to the actual target positions. The parallel-following trajectory resulted in an RMSE of 12.04 m, indicating a near-total localization failure. The RMSE for the manually operated trajectory was 8.13 m, respectively, significantly higher compared to our observability-enhanced trajectory, which achieved an RMSE of only 3.79 m. These results comprehensively demonstrate the effectiveness and superiority of the observability-enhanced trajectory proposed in this study in significantly improving remote sensing localization accuracy.

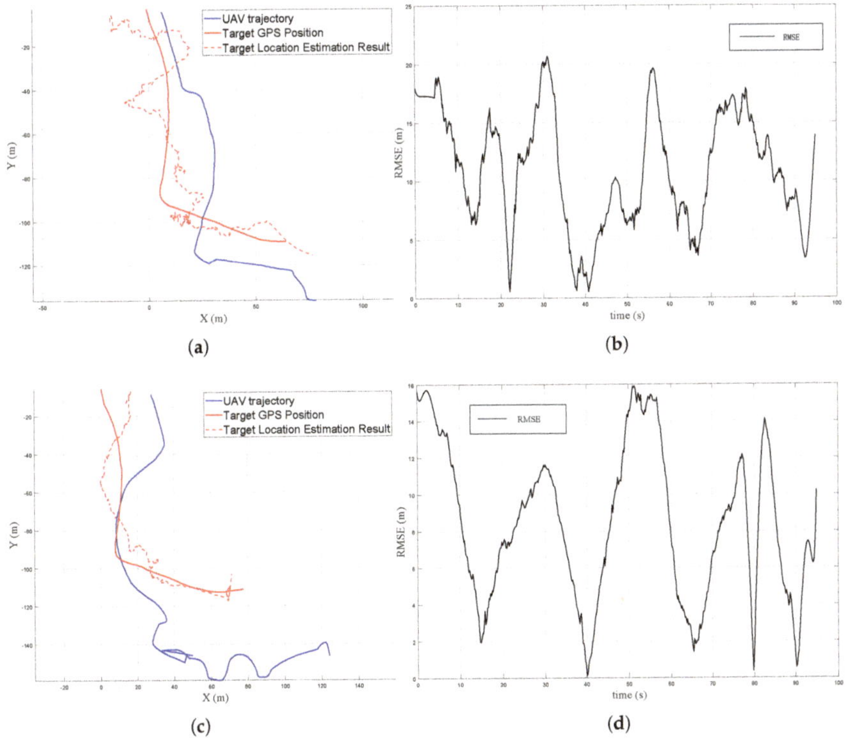

Figure 9. (**a**) Parallel-escort flight trajectory. (**b**) Estimation error for the parallel escort case. (**c**) Trajectory flown manually by an expert pilot. (**d**) Estimation error for the manual flight.

5. Conclusions

In this study, we systematically proposed and validated a UAV-based remote sensing target localization method using bearing-only measurements. The method comprises a localization framework integrating a front-end filtering algorithm with back-end optimization, along with an observability-enhanced trajectory planning strategy. Comparative experimental results demonstrated that combining the PLKF front-end with the sliding-window NLS back-end achieves an optimal balance between computational speed and localization accuracy. Specifically, the PLKF front-end significantly outperformed the EKF in computational efficiency, reducing inter-frame solving time by approximately two-thirds while maintaining comparable localization accuracy when paired with the NLS back-end.

Further experimental evaluations using various flight trajectories clearly highlighted the advantages of our proposed observability-enhanced trajectory. Compared to common tracking methods, such as parallel-following and manual operations, our approach substantially improved localization accuracy, significantly reducing RMSE. Specifically, the RMSE achieved with the observability-enhanced trajectory was only 3.79 m.

In conclusion, the localization method combining the fast PLKF front-end, high-precision NLS optimization, and observability-enhanced trajectory planning provides a robust and efficient solution for remote sensing target localization tasks, demonstrating strong potential for practical applications in various UAV-based sensing scenarios.

Author Contributions: Methodology, P.S.; software, P.S. and S.T.; validation, P.S. and S.T.; supervision, K.Q. and Z.L.; writing—original draft preparation, P.S.; writing—review and editing, B.L. and M.S. All authors have read and agreed to the published version of the manuscript.

Funding: This work was supported by the Natural Science Foundation of Sichuan Province (2024NSFSC0021), the Sichuan Science and Technology Programs (MZGC20240139), and the Fundamental Research Funds for the Central Universities (ZYGX2024K028, ZYGX2025K028).

Data Availability Statement: The original contributions of this study are included within this article. Further inquiries regarding this work can be directed to the corresponding author.

Conflicts of Interest: The authors declare no conflicts of interest.

References

1. Cheng, Z.; Gong, W.; Jaboyedoff, M.; Chen, J.; Derron, M.-H.; Zhao, F. Landslide Identification in UAV Images Through Recognition of Landslide Boundaries and Ground Surface Cracks. *Remote Sens.* **2025**, *17*, 1900. [CrossRef]
2. de Moraes, R.S.; de Freitas, E.P. Multi-UAV Based Crowd Monitoring System. *IEEE Trans. Aerosp. Electron. Syst.* **2020**, *56*, 1332–1345. [CrossRef]
3. Li, W.; Yan, S.; Shi, L.; Yue, J.; Shi, M.; Lin, B.; Qin, K. Multiagent Consensus Tracking Control over Asynchronous Cooperation–Competition Networks. *IEEE Trans. Cybern.* **2025**, 55, 4347–4360. [CrossRef] [PubMed]
4. Cao, Y.; Huang, X. A Coarse-to-Fine Weakly Supervised Learning Method for Green Plastic Cover Segmentation Using High-Resolution Remote Sensing Images. *ISPRS J. Photogramm. Remote Sens.* **2022**, *188*, 157–176. [CrossRef]
5. Li, W.; Yue, J.; Shi, M.; Lin, B.; Qin, K. Neural Network-Based Dynamic Target Enclosing Control for Uncertain Nonlinear Multi-Agent Systems over Signed Networks. *Neural Netw.* **2025**, *184*, 107057. [CrossRef] [PubMed]
6. Lyu, M.; Zhao, Y.; Huang, C.; Huang, H. Unmanned Aerial Vehicles for Search and Rescue: A Survey. *Remote Sens.* **2023**, *15*, 3266. [CrossRef]
7. Liu, C.; Peng, S.; Li, S.; Qiu, H.; Xia, Y.; Li, Z.; Zhao, L. A Novel EAGLe Framework for Robust UAV-View Geo-Localization. *IEEE Trans. Geosci. Remote Sens.* **2025**, *63*, 5629917. [CrossRef]
8. Zhou, S.; Fan, X.; Wang, Z.; Wang, W.; Zhang, Y. High-Order Temporal Context-Aware Aerial Tracking with Heterogeneous Visual Experts. *Remote Sens.* **2025**, *17*, 2237. [CrossRef]
9. Kim, J.; Lee, D.; Cho, K.; Kim, J.; Han, D. Two-Stage Trajectory Planning for Stable Image Acquisition of a Fixed Wing UAV. *IEEE Trans. Aerosp. Electron. Syst.* **2014**, *50*, 2405–2415. [CrossRef]
10. Sun, N.; Zhao, J.; Shi, Q.; Liu, C.; Liu, P. Moving Target Tracking by Unmanned Aerial Vehicle: A Survey and Taxonomy. *IEEE Trans. Ind. Inform.* **2024**, *20*, 7056–7068. [CrossRef]
11. Zhang, T.; Li, C.; Zhao, K.; Shen, H.; Pang, T. ROAR: A Robust Autonomous Aerial Tracking System for Challenging Scenarios. *IEEE Robot. Autom. Lett.* **2025**, *10*, 7571–7578. [CrossRef]
12. Tian, R.; Ji, R.; Bai, C.; Guo, J. A Vision-Based Ground Moving Target Tracking System for Quadrotor UAVs. In Proceedings of the 2023 IEEE International Conference on Unmanned Systems (ICUS), Hefei, China, 20–22 October 2023; pp. 1750–1754. [CrossRef]
13. Li, R.; Zhao, X. LSwinSR: UAV Imagery Super-Resolution Based on Linear Swin Transformer. *IEEE Trans. Geosci. Remote Sens.* **2024**, *62*, 5641013. [CrossRef]
14. Meng, C.; Yang, H.; Jiang, C.; Hu, Q.; Li, D. Improving UAV Remote Sensing Photogrammetry Accuracy Under Navigation Interference Using Anomaly Detection and Data Fusion. *Remote Sens.* **2025**, *17*, 2176. [CrossRef]
15. Yang, H.; Dong, H.; Zhao, X. Safety-Critical Control Allocation for Obstacle Avoidance of Quadrotor Aerial Photography. *IEEE Control Syst. Lett.* **2024**, *8*, 1973–1978. [CrossRef]
16. Savkin, A.V.; Huang, H. Navigation of a UAV Network for Optimal Surveillance of a Group of Ground Targets Moving Along a Road. *IEEE Trans. Intell. Transp. Syst.* **2022**, *23*, 9281–9285. [CrossRef]
17. Guo, Z.; Zhao, F.; Sun, Y.; Chen, X.; Wu, R. Research on Target Localization in Airborne Electro-Optical Stabilized Platforms Based on Adaptive Extended Kalman Filtering. *Measurement* **2024**, *234*, 114794. [CrossRef]
18. Zhou, X.; Jia, W.; He, R.; Sun, W. High-Precision Localization Tracking and Motion State Estimation of Ground-Based Moving Target Utilizing Unmanned Aerial Vehicle High-Altitude Reconnaissance. *Remote Sens* **2025**, *17*, 735. [CrossRef]

19. Zhou, Y.; Tang, D.; Zhou, H.; Xiang, X. Moving Target Geolocation and Trajectory Prediction Using a Fixed-Wing UAV in Cluttered Environments. *Remote Sens.* **2025**, *17*, 969. [CrossRef]
20. Pizani Domiciano, M.A.; Shiguemori, E.H.; Vieira Dias, L.A.; de Cunha, A.M. Particle Collision Algorithm Applied to Automatic Estimation of Digital Elevation Model From Images Captured by UAV. *IEEE Geosci. Remote Sens. Lett.* **2018**, *15*, 1630–1634. [CrossRef]
21. Wang, Y.; Li, H.; Li, X.; Wang, Z.; Zhang, B. UAV Image Target Localization Method Based on Outlier Filter and Frame Buffer. *Chin. J. Aeronaut.* **2024**, *37*, 375–390. [CrossRef]
22. Yang, Z.; Xie, F.; Zhou, J.; Yao, Y.; Hu, C.; Zhou, B. AIGDet: Altitude-Information-Guided Vehicle Target Detection in UAV-Based Images. *IEEE Sens. J.* **2024**, *24*, 22672–22684. [CrossRef]
23. Zhang, L.; Deng, F.; Chen, J.; Bi, Y.; Phang, S.K.; Chen, X.; Chen, B.M. Vision-Based Target Three-Dimensional Geolocation Using Unmanned Aerial Vehicles. *IEEE Trans. Ind. Electron.* **2018**, *65*, 8052–8061. [CrossRef]
24. Farina, A. Target Tracking with Bearings–Only Measurements. *Signal Process.* **1999**, *78*, 61–78. [CrossRef]
25. Li, X.; Wang, Y.; Qi, B.; Hao, Y.; Li, S. Optimal Maneuver Strategy for an Autonomous Underwater Vehicle with Bearing-Only Measurements. *Ocean Eng.* **2023**, *278*, 114350. [CrossRef]
26. Chu, X.; Liang, Z.; Li, Y. Trajectory Optimization for Rendezvous with Bearing-Only Tracking. *Acta Astronaut.* **2020**, *171*, 311–322. [CrossRef]
27. Wang, Y.; Wu, Z.; Piao, H.; He, S. Three-Dimensional Bearing-Only Helical Homing Guidance. *IEEE Trans. Aerosp. Electron. Syst.* **2024**, *60*, 4794–4807. [CrossRef]
28. Li, J.; Ning, Z.; He, S.; Lee, C.-H.; Zhao, S. Three-Dimensional Bearing-Only Target Following via Observability-Enhanced Helical Guidance. *IEEE Trans. Robot.* **2023**, *39*, 1509–1526. [CrossRef]
29. Ferdowsi, M.H. Observability Conditions for Target States with Bearing-Only Measurements in Three-Dimensional Case. In Proceedings of the 2006 IEEE Conference on Computer Aided Control System Design, International Conference on Control Applications, and International Symposium on Intelligent Control (CACSD-CCA-ISIC), Munich, Germany, 4–6 October 2006; pp. 1444–1449. [CrossRef]
30. Song, T.; Speyer, J. A Stochastic Analysis of a Modified Gain Extended Kalman Filter with Applications to Estimation with Bearings-Only Measurements. *IEEE Trans. Autom. Control* **1985**, *30*, 940–949. [CrossRef]
31. Zhao, S.; Zelazo, D. Bearing Rigidity Theory and Its Applications for Control and Estimation of Network Systems: Life Beyond Distance Rigidity. *IEEE Control Syst. Mag.* **2019**, *39*, 66–83. [CrossRef]
32. Jiang, H.; Cai, Y.; Yu, Z. Observability Metrics for Single-Target Tracking with Bearings-Only Measurements. *IEEE Trans. Syst. Man Cybern. Syst.* **2022**, *52*, 1065–1077. [CrossRef]
33. Woffinden, D.C.; Geller, D.K. Observability Criteria for Angles-Only Navigation. *IEEE Trans. Aerosp. Electron. Syst.* **2009**, *45*, 1194–1208. [CrossRef]
34. Zheng, C.; Guo, H.; Zhao, S. A Cooperative Bearing-Rate Approach for Observability-Enhanced Target Motion Estimation. In Proceedings of the 2025 IEEE International Conference on Robotics and Automation (ICRA), Atlanta, GA, USA, 19–23 May 2025; pp. 12579–12585. [CrossRef]
35. Lu, J.; Ze, K.; Yue, S.; Liu, K.; Wang, W.; Sun, G. Concurrent-Learning Based Relative Localization in Shape Formation of Robot Swarms. *IEEE Trans. Autom. Sci. Eng.* **2025**, *22*, 11188–11204. [CrossRef]
36. Liu, Y.; Xi, L.; Sun, Z.; Zhang, L.; Dong, W.; Lu, M.; Chen, C.; Deng, F. Air-to-Air Detection and Tracking of Non-Cooperative UAVs for 3D Reconstruction. In Proceedings of the 2024 IEEE 18th International Conference on Control & Automation (ICCA), Reykjavík, Iceland, 18–21 June 2024; pp. 936–941. [CrossRef]
37. He, S.; Wang, J.; Lin, D. Three-Dimensional Bias-Compensation Pseudomeasurement Kalman Filter for Bearing-Only Measurement. *J. Guid. Control Dyn.* **2018**, *41*, 2675–2683. [CrossRef]
38. Torrente, G.; Kaufmann, E.; Föhn, P.; Scaramuzza, D. Data-Driven MPC for Quadrotors. *IEEE Robot. Autom. Lett.* **2021**, *6*, 3769–3776. [CrossRef]
39. Kim, J.; Suh, T.; Ryu, J. Bearings-Only Target Motion Analysis of a Highly Manoeuvring Target. *IET Radar Sonar Navig.* **2017**, *11*, 1011–1019. [CrossRef]

remote sensing

MDPI

Article

Optimizing Camera Settings and Unmanned Aerial Vehicle Flight Methods for Imagery-Based 3D Reconstruction: Applications in Outcrop and Underground Rock Faces

Junsu Leem [1], Seyedahmad Mehrishal [2], Il-Seok Kang [1], Dong-Ho Yoon [1], Yulong Shao [2], Jae-Joon Song [2,*] and Jinha Jung [3]

[1] Department of Energy Systems Engineering, Seoul National University, Seoul 08826, Republic of Korea; sooyalim@snu.ac.kr (J.L.); xenopsis@snu.ac.kr (I.-S.K.); 4energy@snu.ac.kr (D.-H.Y.)
[2] Research Institute of Energy and Resources, Seoul National University, Seoul 08826, Republic of Korea; ahmad@snu.ac.kr (S.M.); snu_yulong.shao@snu.ac.kr (Y.S.)
[3] Lyles School of Civil and Construction Engineering, Purdue University, West Lafayette, IN 47907, USA; jinha@purdue.edu
* Correspondence: songjj@snu.ac.kr

Abstract: The structure from motion (SfM) and multiview stereo (MVS) techniques have proven effective in generating high-quality 3D point clouds, particularly when integrated with unmanned aerial vehicles (UAVs). However, the impact of image quality—a critical factor for SfM–MVS techniques—has received limited attention. This study proposes a method for optimizing camera settings and UAV flight methods to minimize point cloud errors under illumination and time constraints. The effectiveness of the optimized settings was validated by comparing point clouds generated under these conditions with those obtained using arbitrary settings. The evaluation involved measuring point-to-point error levels for an indoor target and analyzing the standard deviation of cloud-to-mesh (C2M) and multiscale model-to-model cloud comparison (M3C2) distances across six joint planes of a rock mass outcrop in Seoul, Republic of Korea. The results showed that optimal settings improved accuracy without requiring additional lighting or extended survey time. Furthermore, we assessed the performance of SfM–MVS under optimized settings in an underground tunnel in Yeoju-si, Republic of Korea, comparing the resulting 3D models with those generated using Light Detection and Ranging (LiDAR). Despite challenging lighting conditions and time constraints, the results suggest that SfM–MVS with optimized settings has the potential to produce 3D models with higher accuracy and resolution at a lower cost than LiDAR in such environments.

Keywords: structure-from-motion; multiview stereo; camera; UAV; LiDAR; illumination; time; optimization; rock

1. Introduction

Structure from motion (SfM) and multiview stereo (MVS) photogrammetry have been widely recognized for their effectiveness in generating high-resolution and accurate 3D point clouds or 2.5D digital elevation models (DEMs) at a relatively low cost [1,2]. These techniques can be further enhanced by integration with mobile camera platforms, such as unmanned aerial vehicles (UAVs), enabling rapid data acquisition even in remote or previously inaccessible areas [1–9]. The applications of SfM–MVS are diverse, including surface roughness estimation [1,6–8], topographic change detection [6,10,11], vegetation monitoring [3,4], and rock mass characterization [12–14]. However, despite its versatility, the SfM–MVS technique involves inherent complexities, making it challenging to ensure the quality of the generated 3D models [1].

Enhancing the quality of 3D models fundamentally depends on acquiring high-quality images, as they serve as the primary input for SfM–MVS processing. Ensuring proper image exposure is critical, as overexposed or underexposed images can hinder effective feature extraction and matching in the SfM–MVS workflow [4,15]. Achieving optimal image brightness requires a careful balance between blur and noise, which can be controlled through appropriate camera settings. An ideal solution is reducing ISO values and exposure time while increasing the F-number [16], but this is often impractical under actual survey conditions, where camera settings must be determined based on environmental factors. One common approach is to use the camera's auto mode, which allows the internal algorithm to automatically calculate scene exposure and adjust camera settings for optimal image quality. While this method is convenient, the resulting images do not always guarantee high-quality 3D reconstructions. Alternatively, a manual configuration that maintains an F-number of 3.5, an ISO value of 200, and a shutter speed faster than 1/60 s to ensure adequate brightness has been shown to produce high-quality 3D reconstruction results [17]. This strategy is particularly effective under favorable lighting conditions, such as bright daylight. However, these settings may not be feasible in low-light environments.

While appropriate camera settings are essential for image quality, additional challenges arise when the camera is mounted on a UAV. In particular, mounting a camera on a UAV introduces several challenges, including image quality degradation due to blurring and vibrations [18,19]. While gimbaled cameras can help mitigate vibration-related issues [1], additional measures, such as restricting shutter speed, are required to minimize motion blur. Camera settings that maintain motion blur below 1.5 px and diffraction within the pixel pitch are recommended [20]; however, adhering to these standards can be particularly challenging under low-light conditions [21], such as nighttime photography [22] or underground environments [23,24]. When still photogrammetry is employed, high dynamic range (HDR) images can significantly improve the quality of 3D reconstructions [25]. However, during mobile or UAV-based photogrammetry, HDR images cannot be captured properly due to the UAV's motion.

While UAV flight methods are closely related to the selection of camera settings, previous studies have largely overlooked the importance of their relationship. UAV flight methods are often derived based on the UAV's mission planning tools. However, these derived flight methods may be infeasible according to the field or survey conditions. For instance, in underground tunnel construction, only 3–10 min are typically allocated for rock face surveys. Surveys generally require a specific target resolution or ground sampling distance (GSD), which subsequently limits the permissible height or distance from the target [26]. For urban mapping and forestry applications, altitudes between 100 and 120 m provide a good balance between resolution and coverage [27,28]. Nevertheless, such limitations restrict the range of feasible UAV flight paths, which in turn affects image quality.

Moreover, existing studies on UAV-based 3D reconstruction have largely focused on isolated parameter optimizations, such as auto-exposure settings or UAV mission planning tools. Optimizing camera settings for a given illumination condition and optimizing flight methods for specific flight constraints (e.g., time) may each be optimal within their respective domains. However, these fragmented approaches often result in sub-optimal configurations, as they fail to account for the interactions between camera settings and UAV flight methods. Therefore, it is crucial to perform a simultaneous optimization of both camera settings and flight methods to achieve a globally optimal configuration.

According to a survey conducted by the Korea Institute of Construction Technology [29] on satisfaction with 3D modeling using SfM–MVS and Light Detection and Ranging (LiDAR), a major drawback identified was the lack of clear guidelines for workers on how to effectively capture images for SfM–MVS. Additionally, concerns were raised regarding the reliability of the point cloud quality generated by SfM–MVS, particularly in tunnel construction sites where inadequate lighting can significantly impact results. Consequently, despite the higher capital costs, potential for shadow zones, lack of texture, and longer acquisition times associated with LiDAR, workers tended to favor this technique due to its perceived reliability under challenging conditions.

In this study, we developed a method to optimize photography settings, encompassing both camera settings and UAV flight methods simultaneously, to achieve the best image quality that minimizes errors in 3D point clouds under given lighting and time constraints. Utilizing the error prediction model based on image error propagation outlined in our previous research [30], Section 2 explains the methodology for determining optimal photography settings and their applications. Section 3 presents a comparative analysis of point cloud errors generated using the optimized settings versus arbitrary settings in controlled laboratory environments and under field conditions, specifically at an outcrop rock mass on Mt. Gwanak, Seoul, Republic of Korea. Additionally, we evaluated the performance of SfM–MVS using the optimal photography settings and compared it with LiDAR data from an underground tunnel face in Yeoju-si, Republic of Korea. Section 4 discusses the results, demonstrating the effectiveness of the proposed photography settings and highlighting their potential benefits for improving SfM–MVS applications. Section 5 concludes by discussing the practicality of determining and implementing these optimal photography settings, emphasizing their significance in enhancing the accuracy and reliability of SfM–MVS-derived 3D models across various operational environments.

2. Materials and Methods

2.1. Theoretical Background of Error Prediction

Three methodologies for assessing errors in SfM–MVS have been documented: point-to-point (PP), point-to-raster (PR), and raster-to-raster (RR) [6]. PP-type errors involve a direct comparison between two point clouds, while PR-type errors are assessed by comparing a digital elevation model (DEM) with reference points obtained from a total station (TS) or a differential global positioning system (dGPS). In contrast, RR-type errors measure discrepancies between two DEMs. Common metrics for quantifying these errors include root mean squared error (RMSE), mean error (ME), and mean absolute error (MAE) [1]. In this study, the error metric was based on PP-type RMSE, as it circumvents issues related to sampling errors in raster data [10] and uncertainties in reference points obtained using TS and dGPS [1,6,7], while also facilitating statistical manageability. Furthermore, we defined "optimal photography settings (OPS)" as the camera settings that minimize PP-type RMSE in the 3D point cloud generated under given imaging conditions.

The term "photography settings" encompasses both camera settings, including shutter speed (t), F-number (N), ISO (S), and pixel resolution (p), and factors related to the motion of the camera platform, such as the distance from the target (D) and the speed of the platform or UAV (v). For this study, the acronym "UAV" was used to denote all types of mobile camera platforms, with v representing the UAV's flight speed.

Our previous study argued that the selection of photography settings can have a significant impact on the error levels in both the captured images and the resulting 3D point clouds, with these relationships exhibiting complex interactions [30]. The relationship between image errors (ε_{2D}) and errors in SfM–MVS-generated point clouds (ε_{3D}) is described as

$$\epsilon_{3D} = s\epsilon_{2D} \cong \frac{D}{f}\epsilon_{2D}, \tag{1}$$

where the scaling factor, s, is approximately equivalent to the ratio of D divided by the focal length (f). Equation (1) implies that image errors are propagated and magnified in the resulting point cloud. To further quantify image errors (Equation (A1)), three primary sources of distortion were considered: motion blur (Equations (A2) and (A3)) out-of-focus blur (Equations (A4) and (A5)), and image noise (Equations (A6) and (A7)). These distortions were modeled based on the camera settings and UAV flight parameters, with detailed mathematical derivations and probabilistic representations provided in Appendix A.

To mitigate systematic errors resulting from inappropriate image brightness, we assumed that the captured images were adequately saturated by employing camera settings that satisfy Equation (2) under given illumination conditions [31].

$$\frac{ESt}{N^2} = 250 \text{ lx} \cdot \text{s} \quad (2)$$

where E stands for the level of illumination.

For UAV image capture, it was assumed that images were taken while the UAV traveled at a constant v, at a constant D, employing a parallel-axis image acquisition scheme to ensure consistent viewing angles from various shooting positions [7]. Additional assumptions made included the predominance of photon shot noise as the main source of image noise [20] and a reduction in pixel resolution through pixel binning to mitigate image noise [32].

This analysis did not account for systematic errors arising from invalid bundle adjustment (BA), lens distortion, inappropriate use of ground control points (GCPs), insufficient image overlap, uncalibrated camera settings, and texture deficiency. In general, BA is deemed reliable due to the robustness of feature detection algorithms [33,34] and outlier removal techniques [35–37].

Lens distortion is considered to be effectively eliminated and has a minimal impact on the accuracy of point cloud errors, as modern SfM software is capable of precise calibration [11,38]. While incorporating additional GCPs can help reduce errors, the marginal benefits tend to plateau when using four or more GCPs [8,39–41]. Similarly, increasing the number of images can improve accuracy to a certain extent [42–44]; however, beyond a certain threshold, an excessively high image count may lead to computational inefficiencies [7,38]. A practical alternative is extracting a sufficient number of frames from a video taken with a mobile camera, instead of relying solely on individual photos [45]. While video-based approaches benefit from higher frame counts and greater image overlap, capturing high-resolution still images is generally a more effective strategy for maximizing point cloud accuracy, as still images typically offer significantly higher resolution than video frames. This approach is most effective when the required image overlap can be maintained; otherwise, video extraction may still be a viable alternative. However, a discussion on the ideal overlap for these images is beyond the scope of this study.

Camera calibrations can be refined during the BA process, but inaccuracies in calibration may lead to systematic errors affecting the final 3D model [11]. It is advised to ensure camera calibration is accurately fixed [9,11,16], particularly in cases of weak camera geometry, such as when employing a parallel-axis image acquisition scheme [38].

Adequate texture is crucial for successful feature extraction and dense matching; images with insufficient texture may compromise these processes [3,46]. This issue can be addressed by ensuring images have sufficient texture at a given scale [5] or by applying artificial texture to the target surface [16,47].

Errors in SfM–MVS can be pre-determined by identifying key parameters, including illumination, camera specifications, camera settings, and UAV flight methods at the time of image capture [30]. However, the challenge lies in selecting the optimal combination of these parameters from a vast range of possible configurations. To address this, Section 2.2 outlines two primary constraints—illumination and time—that narrow down the selection of potential parameter combinations. Subsequently, Section 2.3 presents a methodology for determining the OPS within these constraints.

2.2. Constraints: Illumination and Time

In this study, we present example figures and applications using Bentley's ContextCapture software v10.17.0.39 (https://bentley.com/software/itwin-capture-modeler/, accessed on 2 March 2025) for the SfM–MVS processing, along with a DJI Mavic 2 Pro drone camera (DJI Technology Co., Ltd., Shenzhen, China; refer to Table 1) for image acquisition. For these specific tools, we used the M value of 13 px (Equation (A3)), and the Q value of 2.62×10^{-5} $\mathrm{lx}^{0.5}\mathrm{s}^{0.5}\mathrm{m/px}$ (Equation (A7)) [30].

Table 1. Specifications of DJI Mavic 2 Pro (https://www.dji.com/mavic-2, accessed on 2 March 2025).

Sensor	1″ CMOS Effective pixels: 20 M
Lens	Field of view (FOV): approximately 77° Focal length: 28 mm (35 mm equivalent) Aperture: $f/2.8$–$f/11$ Shooting range: 1 m–∞
ISO range	100–12,800 (for still images) 100–6400 (for videos)
Shutter speed	8–1/8000 s
Still image resolution	5.4 K: 5472 × 3648 @ 24/25/30 fps
Video resolution	4 K: 3840 × 2160 @ 24/25/30 fps 2.7 K: 2688 × 1512 @ 24/25/30/48/50/60 fps FHD: 1920 × 1080 @ 24/25/30/48/50/60/120 fps

2.2.1. Illumination

Equations (1) and (A1)–(A7) define nine parameters for calculating SfM–MVS errors: camera settings (N, S, t, p), UAV flight methods (D, v), camera specifications (sensor size d, f), and illumination (E). While camera settings and UAV flight methods can be adjusted as needed, camera specifications and illumination are often treated as uncontrollable parameters. Modifying camera specifications is generally avoided [9,11,16]; they are effectively treated as fixed parameters. Therefore, illumination remains the only uncontrollable parameter that can vary, thereby acting as a principal constraint that restricts the selection of camera settings and UAV flight methods.

Figure 1 illustrates the optimization of camera settings under an illumination constraint of E = 100 lx when flying the UAV under D = 3 m and v = 0.8 m/s conditions. While every camera settings combination satisfies the appropriate image brightness (Equation (2)), error levels significantly fluctuate with different camera settings. Among these combinations, the optimal camera settings, yielding a minimum error of 2.05 mm under the given conditions, are N = 2.8, t = 1/160 s, S = 3200, and p = 1920 px. If the illumination conditions or UAV flight methods change, the optimal camera settings and the corresponding lowest achievable error can be determined using the procedure outlined in this study. As mentioned earlier, the illumination conditions are typically uncontrollable, meaning that UAV flight methods must be determined under an additional constraint, time, which is introduced in the next section.

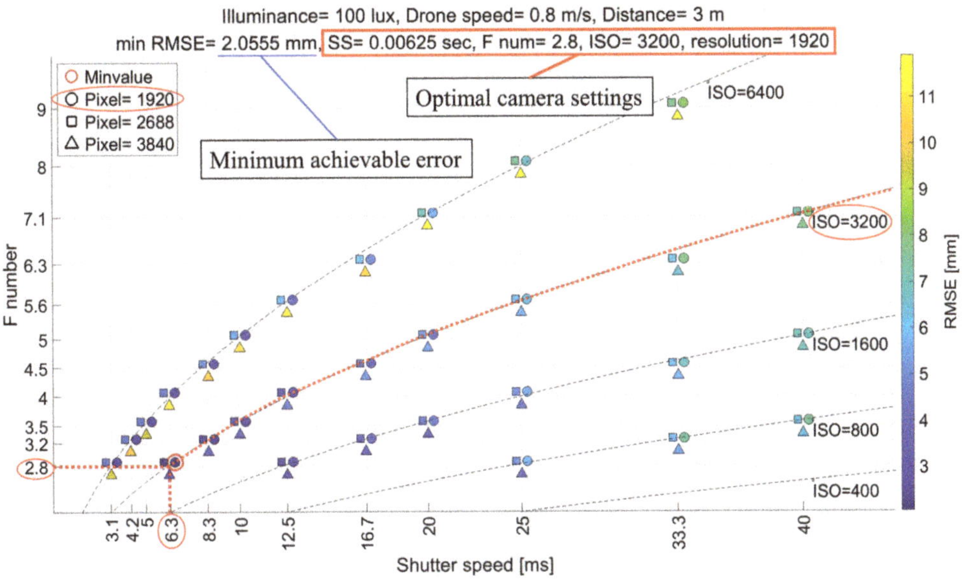

Figure 1. Optimization of the camera settings under specific *E*–*D*–*v* conditions.

2.2.2. Time

Survey sites often have strict time constraints for image acquisition, which may be influenced by factors such as limited survey periods, UAV battery capacity, and changes in outdoor illumination over time. To ensure that the entire survey is completed within the allocated timeframe, the UAV flight method—including flying speed and distance from the target—must be selected while considering photography time limitations.

The faster the UAV speed (Figure 2a) and the greater the distance from the target (Figure 2b), the larger the area covered per unit of time. When the UAV navigates in a zig-zag pattern parallel to the target [27], increasing the speed reduces the time required for horizontal traversal. Similarly, as the distance from the object increases, the captured scene encompasses a larger area, thereby reducing the need for multiple horizontal passes to complete the survey. The field of view (FOV) affects the area captured at a given distance; however, in the absence of a zoom lens, the camera's FOV remains fixed. Disregarding the acceleration and deceleration phases near trajectory boundaries and the time needed for vertical traversal, and presuming that the target area is considerably large, the area photographed per unit of time (A) is directly proportional to both v and D from the target, as expressed in Equation (3). The precise proportionality constant depends on the camera's FOV, which is approximately 0.78 for the DJI Mavic 2 Pro camera.

$$A \propto vD \tag{3}$$

In Figure 1, the optimal camera settings were identified for UAV flight methods of $D = 3$ m and $v = 0.8$ m/s, resulting in a minimum error of 2.1 mm. However, exploring alternative UAV flight methods under different *D*–*v* conditions, while maintaining the same area covered per unit of time, could potentially yield lower errors. This highlights the importance of examining various UAV flight methods under identical time constraints.

Figure 3 illustrates the optimization of UAV flight methods. In Figure 3a, the color assigned to each point on the *D*–*v* chart illustrates the minimum error achievable for each *D*–*v* condition at a given illuminance level (see Figure 1). For instance, a cyan color represents the minimum achievable error of 1.47 mm at $D = 3$ m, $v = 0.3$ m/s, and $E = 100$ lx. The optimal UAV flight method is determined by identifying the points with minimum RMSE values, which are marked as red dots in Figure 3b–e. These points are selected across all D–v conditions that correspond to equivalent A values, as indicated by the dashed lines in the figures.

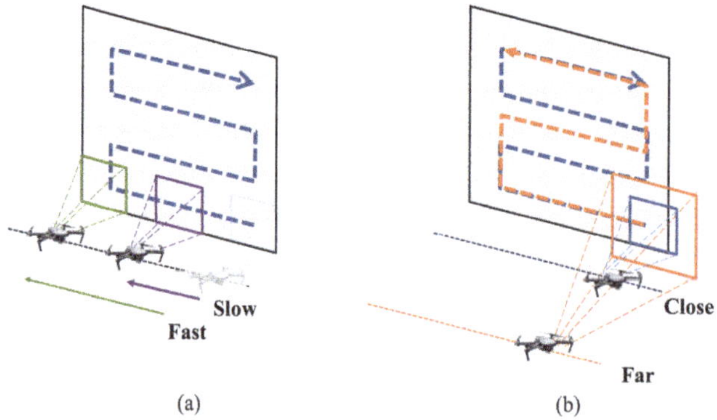

Figure 2. Impact of UAV flight methods on the area photographed per unit of time: (**a**) UAV speed; (**b**) distance from the object.

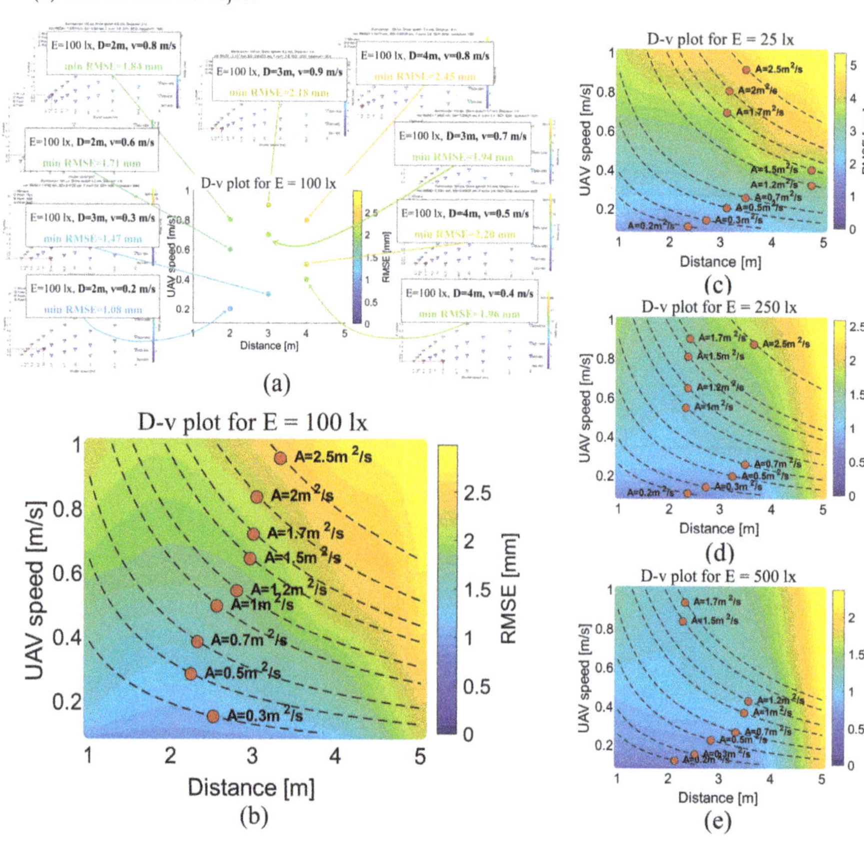

Figure 3. Optimization of UAV flight methods under E–A constraints. (**a**) The procedure for generating a full D–v chart. (**b**) Optimization for $E = 100$ lx; each dashed line indicates points with identical A, with the red dot on each line indicating the point of minimum error for that specific A constraint. (**c**) Optimization for $E = 25$ lx. (**d**) Optimization for $E = 250$ lx. (**e**) Optimization for $E = 500$ lx.

The relationship between the optimal (D, v) and A is complex, as it is affected by variations in camera settings at each (D, v) point. In Figure 3b, the most favorable D–v condition for A = 0.5 m^2/s, D = 2 m, and v = 0.3 m/s leads to a minimum error of 1.25 mm with camera settings of t = 1/100 s, N = 3.5, S = 3200, and p = 1920 px. In contrast, for A = 1.5 m^2/s with D = 3 m and v = 0.7 m/s, the minimum error is 1.97 mm using camera settings of t = 1/160 s, N = 2.8, S = 3200, and p = 1920 px. Furthermore, the optimal (D, v) also varies depending on illuminance levels. For example, at E = 100 lx, the optimal UAV flight method for A = 0.5 m^2/s is D = 2 m and v = 0.3 m/s (as shown in Figure 3b), whereas at E = 250 lx, it shifts to D = 3 m and v = 0.2 m/s for the same A (as shown in Figure 3d).

2.3. Step-by-Step Process of the OPS Determination

OPS is defined as the combination of the optimal camera settings and the optimal UAV flight methods that minimize error under given illumination and time constraints. The process of determining the OPS consists of four main steps (Figure 4): (1) Measure the illumination level (E) using a lux meter. In the absence of a lux meter, auto-exposure settings can serve as an alternative method to estimate relative brightness levels, although they may lack the absolute accuracy of dedicated illuminance sensors. (2) Calculate the required area to be photographed per unit of time (A); (3) determine the optimal UAV flight methods (D, v); (4) determine the optimal camera settings (N, t, S, p). While Steps 1 and 2 can be performed in any order, Step 3 requires the completion of both Steps 1 and 2, and Step 4 needs Steps 1, 2, and 3 to be completed beforehand. For a measured illuminance of 100 lx, with the requirement to photograph the entire target area of 100 m^2 within 4 min, the OPS were identified as follows: flying a drone at D = 2 m with v = 0.3 m/s, combined with camera settings of N = 3.5, t = 1/100 s, S = 3200, and p = 1920 px.

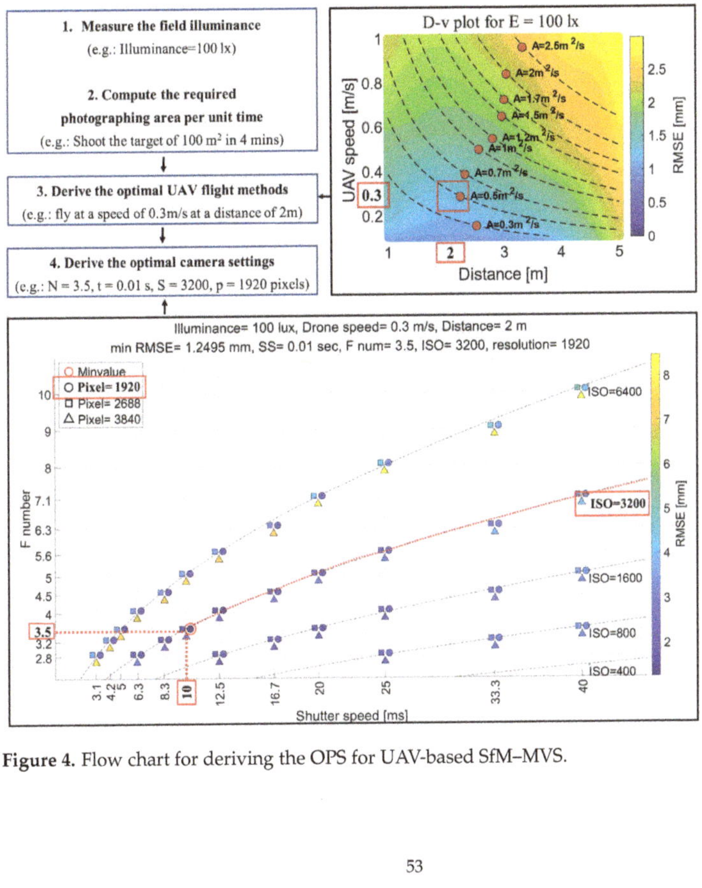

Figure 4. Flow chart for deriving the OPS for UAV-based SfM–MVS.

A point of potential confusion is that the *D–v* chart, utilized in Step 3, is derived from analyzing the minimum camera setting values under various *E–D–v* conditions, as detailed in Step 4. Nevertheless, determining the optimal UAV flight methods must precede the selection of optimal camera settings, as the latter is influenced by the former. It is important to note that the optimal UAV flight methods identified in Step 3 are only effective when combined with the optimal camera settings identified in Step 4. This dependency highlights the intricate relationship between UAV flight methods and camera settings in minimizing error.

3. Application of the OPS

This section presents examples of OPS applications and evaluates their effectiveness. Point clouds generated employing the OPS were compared with those obtained employing arbitrary photography settings (APS). The APS were not chosen freely but were selected from camera settings that satisfy Equation (2) to ensure adequate image brightness, and from UAV flight methods that were constrained by the same time limitations (Equation (3)). Furthermore, the performance of SfM–MVS utilizing the OPS was compared to LiDAR data acquired from an underground tunnel construction site. This comparison aimed to demonstrate the potential of the OPS in enhancing the performance of SfM–MVS under challenging environmental conditions.

3.1. Comparison Between the OPS and APS Under Laboratory Conditions

An experimental setup was designed in a laboratory to evaluate the error differences between the OPS and APS. A textured reference target was established on a vertical flat wall consisting of three planes, each with 60 cm edge lengths and inclined approximately 27 degrees from the wall (Figure 5). Three LED lamps were placed to ensure uniform diffuse illumination across the target, and their intensity was adjustable to control illuminance levels.

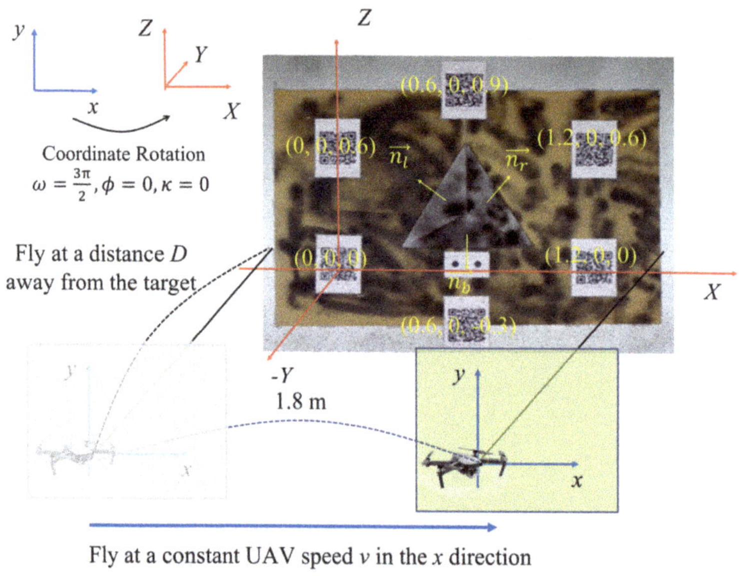

Figure 5. Flow chart for deriving the OPS for UAV-based SfM–MVS. Experimental setup and axis settings. The normal vector for the left plane is represented as n_l = (0.423, 0.876, −0.217), for the right plane as n_r = (0.438, 0.869, −0.229), and for the bottom plane as n_b = (−0.001, 0.857, 0.516). ω, ϕ, and κ represent the rotation angles around the *x*-, *y*-, and *z*-axes, respectively, between the 2D camera coordinate system and the 3D object coordinate system.

To minimize systematic errors, six QR codes were utilized as GCPs [8,39]. These QR codes were generated using the "ContextCapture Target Creator" feature within Bentley's ContextCapture software (v10.17.0.39). This feature allows the creation of QR codes embedded with predefined 3D coordinates (x, y, z) in a local coordinate system, specified in meter units. In this study, the QR codes were arranged in a hexagonal layout (Figure 5), and their physical placement was performed using a ruler, ensuring an accuracy within 1 mm. When the images containing these QR codes were processed in ContextCapture, the software automatically recognized the center of each QR code as the predefined coordinates. The detected coordinates exhibited an RMS 3D error of 4.9 mm (ranging from 4.5 to 5.2 mm). This error was primarily attributed to limitations in ruler precision during placement and variations in GSD caused by differences in image resolution and camera distance.

The imagery was captured using a DJI Mavic 2 Pro drone camera, with the onboard gimbal helping to dampen vibrations. A parallel-axis image acquisition scheme was employed, ensuring the camera's viewing angle remained consistently orthogonal to the target [7]. Instead of taking individual photographs, a video was recorded at 24 fps in a single take, and frames were later extracted from the video. During the video capture, the UAV maintained a constant horizontal flight speed, ensuring that the extracted frames were sequential with uniform baseline distances. To prevent systematic errors due to insufficient image overlap, 20 frames were selected [7,38], with the total distance from the center of the first frame to the center of the last frame measuring 1.8 m.

The experiment involved capturing the target using the OPS and two APS under specific constraints of illuminance (E = 100 lx) and a combined factor of distance and velocity ($D \times v = 1.8\ \mathrm{m}^2/\mathrm{s}$), which, considering the FOV of the camera, corresponded to $A \cong 1.4\ \mathrm{m}^2/\mathrm{s}$. Following the proposed methodology (Figure 4), the OPS were determined, and as previously mentioned, the APS were selected based on camera settings and UAV flight methods that satisfied these constraints. Details of these photography settings are provided in Table 2.

Table 2. The photography settings utilized for the laboratory experiment. Each setting was repeated 5 times, resulting in a total of 15 shots.

Settings Index	N	t (s)	S	p (px)	D (m)	v (m/s)
OPS	2.8	1/160	3200	1920	3	0.6
APS#1	2.8	1/320	6400	1920	6	0.3
APS#2	2.8	1/320	6400	1920	9	0.2

For point cloud generation, Bentley's ContextCapture software was utilized, and the camera calibration was fixed to specified values during the SfM–MVS procedure [16]. To avoid potential smoothing effects that could occur during 3D mesh production [45], the final outputs were produced in 3D point cloud format (LAS format) rather than 3D mesh models. The generated 3D point clouds were then registered to a predefined object coordinate system using the open-source 3D point-cloud processing software CloudCompare v2.12 alpha, ensuring that the data from different settings could be accurately compared and analyzed.

Due to the challenges in accurately identifying the reference point locations of distorted points [1], direct measurement of point-to-point errors is difficult. Instead, we measured the distances from the point cloud to the planes of the reference target (denoted as $\varepsilon_{\mathrm{plane}}$), which are straightforward to measure, and converted these values to PP-type RMSE ($\varepsilon_{\mathrm{point}}$). The conversion from $\varepsilon_{\mathrm{plane}}$ to $\varepsilon_{\mathrm{point}}$ is based on Equation (4) and can be applied when the reference target has a planar shape [30].

$$\epsilon_{\text{point}} = \epsilon_{\text{plane}} \times \sqrt{\frac{\sum_k (dx_k \hat{r_1} + dy_k \hat{r_2})^2}{\sum_k \{dx_k(\hat{n} \cdot \hat{r_1}) + dy_k(\hat{n} \cdot \hat{r_2})\}^2}}, \tag{4}$$

where dx_k and dy_k represent errors on the image plane, while r_1 and r_2 are the column vectors of the rotation matrix that converts 2D image coordinates to 3D object coordinates. Additionally, n denotes the normal vector of the reference target plane (Figure 5). The comparison results of $\varepsilon_{\text{point}}$ for the OPS and two APS are presented in Section 4.1.1.

3.2. Comparison Between the OPS and APS Under Field Conditions

The effectiveness of the OPS was validated through a field experiment conducted at Mt. Gwanak, Seoul, Republic of Korea. The site features a well-exposed rock mass outcrop (Figure 6) composed of Jurassic-age granite bedrock. The joint planes, primarily oriented in the southeast and northeast directions, consist of smooth, planar surfaces characteristic of tensile fracturing. Additionally, previous studies utilizing LiDAR surveys at the same site have been reported [48].

The site features a rock mass outcrop approximately 3 × 2.5 m in size. We designated six joint planes as targets, each measuring 15 × 15 cm, with their indices displayed in Figure 6. Instead of creating a separate 3D model for each joint plane, we surveyed the entire site and then extracted the six joint planes from the complete 3D model. The reference 3D model was captured using a high-precision handheld laser scanner, the SCANTECH KSCAN Magic (SCANTECH (HANGZHOU) CO., LTD., Hangzhou, China; Table 3), which offers an accuracy of 0.05 mm and a GSD of 0.05 mm. To ensure high accuracy, calibration markers were attached to the target surface, which partially obstructed some sections of the rock face. The entire site was first scanned with a GSD of 1 mm. Subsequently, only the region of interest—the six joint planes and their vicinities—was rescanned at a finer GSD of 0.1 mm to minimize the scanning time. However, despite the scanning area covering only 7.5 m^2, the scanning process required seven hours, making it impractical for routine on-site surveys.

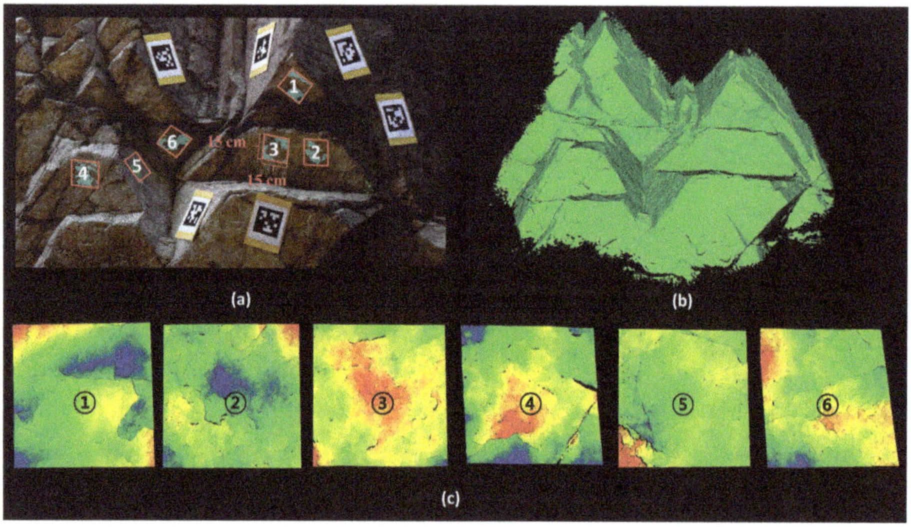

Figure 6. (**a**) An image of the outcrop rock mass located at Mt. Gwanak, Seoul, Republic of Korea. (**b**) Reference model scanned with the SCANTECH KSCAN Magic laser scanner with a GSD of 1 mm. (**c**) Reference models of each joint plane, rescanned with a finer GSD of 0.1 mm. Joint planes are numbered from 1 to 6 for indexing purposes.

Table 3. SCANTECH KSCAN-Magic laser scanner specification (https://www.3d-scantech.com/product/kscan-magic-composite-3d-scanner/, accessed on 2 March 2025).

Accuracy	≥0.01 mm
Scanning rate	≥1,350,000 measurements/s
Scanning area	≥1440 × 860 mm
Resolution	≥0.01 mm
Stand-off distance	300 mm
Depth of field	925 mm

For the SfM–MVS process, we attached six April codes uniformly around the target region and used them as GCPs to minimize systematic errors [8,39]. The distances between the April codes were measured and input into ContextCapture software, which automatically recognized the codes and provided absolute scale information. The target joint planes were captured under two different illuminance constraints (E = 50, 1500 lx) and one time constraint (A = 0.3 m^2/s). These targets were located in an open-air environment, where illuminance was indirectly controlled by adjusting the time or weather conditions of the survey. For instance, 50 lx corresponds to low brightness and similar post-sunset conditions, while 1500 lx represents intermediate brightness, typical of a cloudy day. Brighter conditions, such as those on a sunny day, were deliberately avoided, as non-diffuse illumination (e.g., shadowed areas) could introduce systematic errors. The time constraint (A = 0.3 m^2/s) reflects realistic conditions found in tunnel construction site surveys, where rock faces measuring 50–200 m^2 must be surveyed within 3–10 min.

We used a camera mounted on the DJI Mavic 2 Pro drone and employed a parallel-axis image acquisition scheme. A single video was recorded per shot while the drone maintained a constant horizontal flight speed. For 50 lx, we captured four videos, with each OPS and one APS repeated twice. For 1500 lx, we recorded eight videos, repeating the OPS and three APS twice each. Details of these photography settings are provided in Table 4. From each video, 50–60 frames were extracted and processed using ContextCapture software to generate 3D point clouds of the rock mass.

Table 4. The photography settings utilized for the outcrop rock mass survey include settings #1 and #3 as the OPS, and the other settings as the APS. Each setting was repeated twice, resulting in a total of 12 shots.

Settings Index	E (lx)	N	t (s)	S	p (px)	D (m)	v (m/s)	Optimal
1	50	3.5	1/50	3200	1920	2	0.2	O
2	50	3.5	1/50	3200	1920	4	0.1	X
3	1500	5.6	1/200	800	3840	2	0.2	O
4	1500	5.6	1/40	200	3840	4	0.1	X
5	1500	4	1/200	400	3840	4	0.1	X
6	1500	6.3	1/40	200	3840	2	0.2	X

The error correction method used for the indoor target (detailed in Equation (4)) was feasible because the target consisted of flat planes. However, this method could not be applied to the non-flat geometry of the target joint planes. Consequently, alternative quantitative indicators relevant to error assessment were measured [49]: The cloud-to-mesh (C2M) distance and the multiscale model-to-model cloud comparison (M3C2) distance [50].

In this study, point clouds generated using SfM–MVS were compared against a reference 3D model produced by laser scanning. The standard deviation (std) of the C2M and M3C2 distances (stdC2M and stdM3C2, respectively) served as the quality indicator for the point clouds, with calculations performed using CloudCompare software (Figure 7). Approximately 30 visually corresponding feature points between the reference model and the point clouds were identified initially. The models were then coarsely aligned based on these feature points, followed by a more precise alignment using the iterative closest point (ICP) algorithm. From the 3D models, we extracted six joint planes, each measuring 15 × 15 cm. The "compute cloud/mesh distance" function in the software was used to calculate the stdC2M for each joint plane under each photography setting (Figure 7d,e). Similarly, the "M3C2 distance" function in the software was used to calculate the stdM3C2 with the default parameter settings utilized. The comparison results for the stdC2M and stdM3C2 between the OPS and APS are presented in Section 4.1.2.

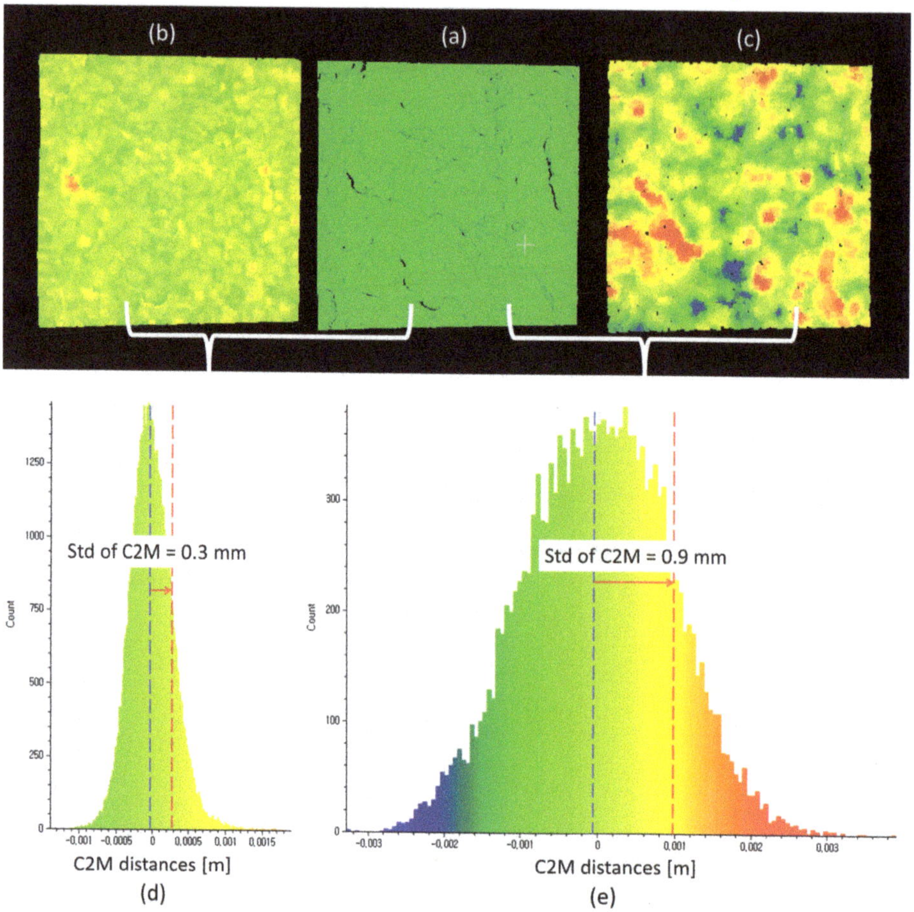

Figure 7. Example of calculating the stdC2M distances. Colors represent the C2M distances to the reference model. The same procedure was conducted for calculating the stdM3C2 distances: (**a**) Reference model of joint plane #2; (**b**) point cloud generated using SfM–MVS with settings #3; (**c**) point cloud generated using SfM–MVS with settings #4; (**d**) histogram of C2M distances for settings #3, with a resulting stdC2M of 0.3 mm; (**e**) histogram of C2M distances for settings #4, with a resulting stdC2M of 0.9 mm.

3.3. Comparison with LiDAR Data

Underground tunnel construction sites feature a low-light environment, and the surveying time is often limited due to ongoing construction activities. Moreover, the Global Navigation Satellite System (GNSS) is typically unavailable in tunnels, further complicating the surveying process. To evaluate the SfM–MVS method under such challenging conditions, we implemented SfM–MVS using the OPS and conducted a comparative analysis with LiDAR. This comparison was conducted at an underground tunnel construction site in Yeoju-si, Republic of Korea, where a surveying team used a LiDAR instrument to scan the tunnel face. As shown in Figure 8, the target rock face area was approximately 70 m^2. With the help of field lighting (Figure 8a), the illuminance in the area was approximately 25 lx.

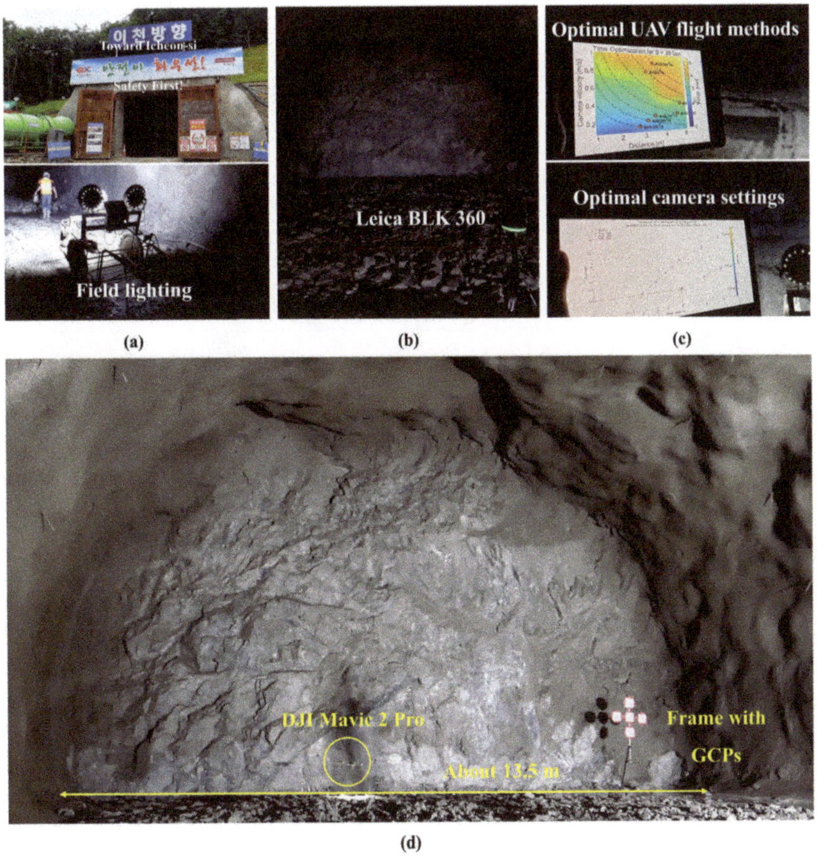

Figure 8. Field application of SfM–MVS using the OPS at an underground tunnel construction site in Yeoju-si, Republic of Korea: (**a**) Entrance of the tunnel and field lighting inside; (**b**) the Leica BLK 360 LiDAR used by the site surveying team; (**c**) derivation of the OPS for the site; (**d**) the target rock face and DJI Mavic 2 Pro used in the survey. A frame with GCPs was placed in front of the tunnel face.

The surveying team utilized a Leica BLK360 imaging laser scanner (Leica Geosystems AG, Heerbrugg, Switzerland; Figure 8b), with its specifications detailed in Table 5. It is important to distinguish between the ranging accuracy, which refers to errors in the direction of the laser beam, and 3D point accuracy. The error term in this study corresponds specifically to the 3D point accuracy listed in Table 5. The LiDAR instrument was positioned

10 m away from the tunnel face, and the scanning process took approximately 5 min, which is a typical duration for surveying in tunnel construction sites. The laser-scanned data were then post-processed using Leica Cyclone REGISTER 360 software v2022.1.0 to remove outlier points and refine the dataset.

Since the field survey team operated the LiDAR instrument for 5 min, the required time constraint was calculated as A = 70 m^2/4 min, allowing for a tolerance period to accommodate the approach and retreat of the drone. Under the constraints of E = 25 lx and A = 0.3 m^2/s, the OPS were determined as follows (Figure 8c): N = 2.8, t = 1/40 s, S = 3200, p = 1920 px, D = 3 m, and v = 0.2 m/s. The overall shooting method remained consistent with the previous sections. However, one notable difference from the previous sections was the deployment and distribution of GCPs. Although it was not feasible to place GCPs evenly along the tunnel face due to safety concerns and restricted accessibility, an independent frame was positioned in front of the tunnel face, equipped with five GCP QR codes (Figure 8d). These QR-coded GCPs were utilized to scale the reconstructed models and minimize systematic errors.

Table 5. Leica BLK360 imaging laser scanner specifications (https://leica-geosystems.com/products/laser-scanners/scanners/blk360 accessed on 2 March 2025).

Field of view	360° (horizontal)/300° (vertical)
Range	0.6–60 m
Point measurement rate	≤360,000 pts/s
Ranging accuracy	4 mm @ 10 m/7 mm @ 20 m
Measurement speed	<3 min for complete full dome scan
3D point accuracy	6 mm @ 10 m/8 mm @ 20 m

Additionally, the DJI Mavic 2 Pro drone camera could not capture the entire height from the floor to the ceiling (approximately 6.4 m) within its FOV when positioned 3 m away. As a result, the drone had to fly a zig-zag path (refer to Figure 2), which required acceleration and deceleration at both ends. Nonetheless, the majority of the shooting was conducted at a uniform drone speed, and only the frames captured during this uniform flight speed were utilized for analysis.

Unlike the validation described in Section 3.1 or Section 3.2, obtaining a reference model for the tunnel face was not feasible due to time limitations. Instead, we compared the point clouds generated by both methods using quantitative indices such as point resolution and idealistic accuracy, and qualitative indices like joint expression capability. The results of these comparisons are presented in Section 4.2.

4. Results

Section 2 outlined the procedure for deriving the OPS under illuminance and time constraints. To highlight the importance of the OPS, we compared the quality of point clouds obtained using the OPS with those obtained using the APS (Section 3.1), with detailed results presented in Section 4.1. Furthermore, to demonstrate the potential of the SfM–MVS method when utilizing the OPS, we conducted a comparative analysis of point cloud quality between SfM–MVS and LiDAR in an underground tunnel site (Section 3.2), where both lighting conditions and surveying time were limited. The results of this comparison are detailed in Section 4.2.

4.1. Effectiveness of the OPS

In Section 3.1, we evaluated the effectiveness of the OPS by comparing the quality of the point clouds obtained using the OPS and APS. First, the error levels in the point cloud of an indoor reference target, which consisted of flat planes, were assessed, and then the stdC2M distances for the joint planes on a rock mass outcrop in Mt. Gwanak were compared.

4.1.1. Point Cloud Errors for the Indoor Reference Target

We generated 3D point clouds of an indoor reference target using the OPS and two APS under E = 100 lx and A = 1.4 m^2/s, with five repetitions for each setting. The camera orientation and lens correction parameters were well calibrated, resulting in an average reprojection error of less than 1 px during the SfM procedure [39]—OPS = 0.67–0.7 px, APS#1 = 0.68–0.69 px, and APS#2 = 0.65–0.66 px.

Figure 9 illustrates the point cloud error levels for the reference. A significant difference in error levels was observed between the OPS and APS. The mean error level for the OPS (red dots in Figure 9) was 2.59 mm, while the APS results showed notably higher errors: 7.58 mm for APS#1 (cyan squares in Figure 9) and 8.67 mm for APS#2 (green diamonds in Figure 9). The improvement in error levels when using the OPS was expected, as it represents the minimum error achievable with photography settings derived from the error pre-determination model developed by Leem et al. [30], which has demonstrated high predictive performance (note the alignment between the predicted and observed RMSEs in Figure 9). Furthermore, severe distortion in the representation of flat planes was observed in the APS-generated point clouds, as shown in the side views of Figure 9. Notably, without modifying the existing illuminance and time constraints, substantially better point cloud quality could be achieved simply by adjusting the camera settings and UAV flight methods.

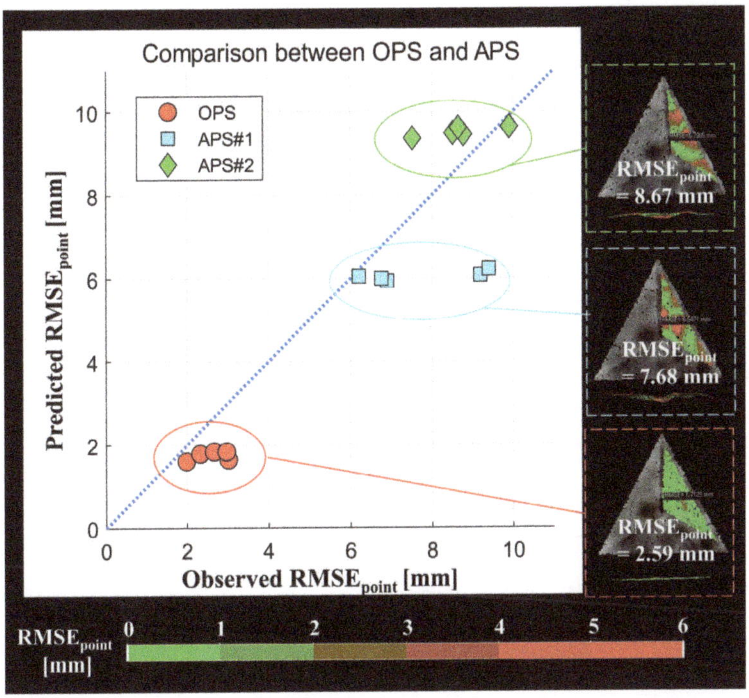

Figure 9. Comparison between the OPS and APS under E = 100 lx and A = 1.4 m^2/s ($D \times v$ = 1.8 m^2/s) constraints. The side views on the right are vertically exaggerated to help the reader's understanding.

4.1.2. StdC2M Distances for the Joint Planes on an Outdoor Rock Mass

We generated 3D models of an outdoor rock mass using both laser scanning and SfM–MVS. The laser-scanned model served as the reference, and we measured the stdC2M and the stdM3C2 between the reference model and those generated using SfM–MVS. In total, 12 shots were taken, including four for the OPS and eight for the APS. Additionally, the camera orientation and lens correction parameters for the SfM–MVS were well calibrated, with reprojection errors ranging from a minimum of 0.43 px to a maximum of 0.67 px. Since the mean values of C2M or M3C2 were significantly lower than the std values (maximum mean of C2M: 0.038 mm, minimum stdC2M: 3.01 mm; maximum mean of M3C2: 0.057 mm, minimum stdM3C2: 3.04 mm), it can be inferred that the SfM–MVS-generated models were well registered with the reference model.

The stdC2M and stdM3C2 values of the OPS and APS are listed in Tables 6 and 7, respectively, with their comparison illustrated in Figure 10. In the figure, red dots represent the std values for the OPS, while black dots indicate the std values for the APS. It is evident that the red dots consistently lie below the black dots under both 50 and 1500 lx conditions. With the exception of two cases for stdC2M (joint planes #4 and #5, E = 1500 lx), where the APS achieved the second lowest stdC2M, the lowest and second lowest stdC2M and stdM3C2 values were obtained using the OPS in all other cases. This result suggests that the OPS can generate 3D models that more closely resemble the reference than those created using the APS, thereby indicating that the OPS produced more accurate 3D models.

Table 6. StdC2M for each photography setting across joint planes #1 to #6, listed in millimeter units. The lowest and second-lowest stdC2M values for each plane are highlighted in bold.

Settings	Planes	#1	#2	#3	#4	#5	#6	Optimal
50 lx	#1	**1.43** **1.09**	**0.91** **0.92**	**0.72** **0.68**	**1.01** **0.98**	**0.81** **0.70**	**1.38** **1.09**	O
	#2	2.16 1.85	1.51 1.97	1.27 1.43	1.49 1.89	1.23 1.49	3.18 3.17	X
1500 lx	#3	**0.67** **0.55**	**0.42** **0.30**	**0.37** **0.43**	**0.65** 0.71	0.65 **0.43**	**0.71** **0.58**	O
	#4	1.94 1.54	0.94 1.01	1.02 1.00	2.16 **0.66**	0.96 **0.60**	1.37 1.10	X
	#5	0.81 1.06	1.26 0.66	1.50 0.66	0.78 0.76	0.66 1.82	1.49 0.82	X
	#6	1.12 0.86	0.90 0.60	1.09 0.73	0.17 1.05	2.25 1.53	0.25 0.13	X

Table 7. StdM3C2 for each photography setting across joint planes #1 to #6, listed in millimeter units. The lowest and second-lowest stdM3C2 values for each plane are highlighted in bold.

Settings	Planes	#1	#2	#3	#4	#5	#6	Optimal
50 lx	#1	**1.92** **1.53**	**1.05** **1.04**	**0.85** **0.78**	**1.26** **1.22**	**0.98** **0.83**	**1.66** **1.34**	O
	#2	2.74 2.52	1.70 2.17	1.56 1.74	1.66 2.22	1.55 1.88	3.57 3.71	X
1500 lx	#3	**0.95** **0.82**	**0.44** **0.30**	**0.38** **0.45**	**0.79** **0.76**	**0.71** **0.45**	**0.78** **0.63**	O
	#4	2.39 2.01	1.04 1.11	1.18 1.14	2.44 0.87	1.12 0.72	1.60 1.29	X
	#5	1.21 1.45	1.37 0.72	1.67 0.74	1.00 0.94	0.77 2.14	1.68 0.95	X
	#6	1.50 1.19	0.98 0.63	1.22 0.79	1.37 1.26	2.51 1.63	2.81 1.41	X

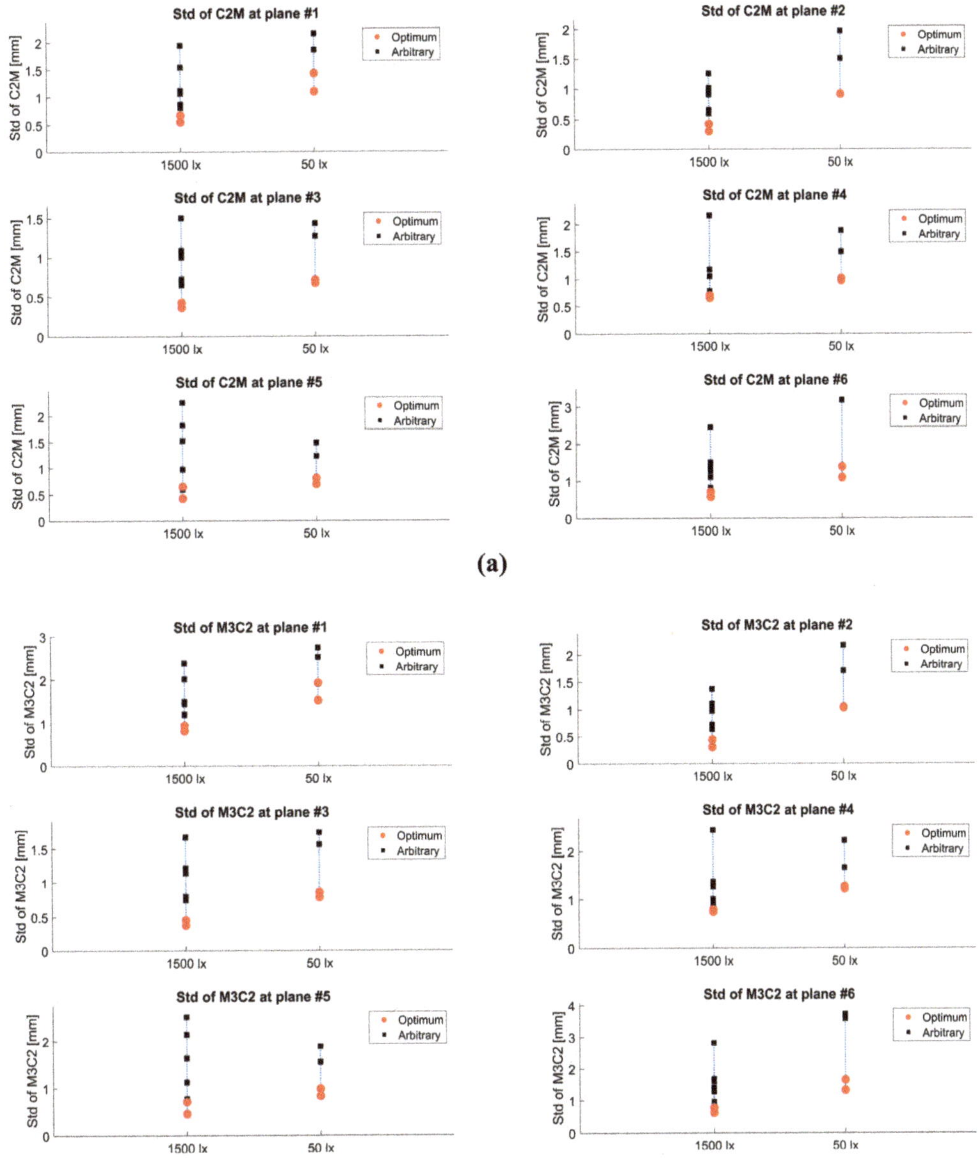

Figure 10. Comparison of std values for joint planes #1 to #6. Red dots represent the std values for the OPS, and black dots represent the std values for the APS: (**a**) StdC2M comparison; (**b**) stdM3C2 comparison.

4.2. The Potential of SfM–MVS Utilizing OPS for Enhanced 3D Modeling

In Section 3.3, we generated the point cloud of an underground tunnel rock face using both SfM–MVS with OPS and LiDAR. During the SfM process, the reprojection error was measured at 0.54 px, confirming the calibration accuracy. Although the OPS were determined with a time constraint of 4 min, allowing for drone approach and retreat as well as acceleration and deceleration phases, the actual shooting time was extended to 5 min, matching the duration required for LiDAR operations.

The LiDAR-generated point cloud is displayed in Figure 11a. The surveying team operated the LiDAR instrument to collect data over a 360-degree range, capturing a total of 12,100,000 points. However, the point resolution for the tunnel face was relatively low, at approximately 1,700,000 points, equivalent to 150 pts/m^2. This lower point density resulted from the chosen scanning approach, where the surveying team conducted a full 360-degree scan instead of restricting data collection to the tunnel face. Although the instrument is capable of achieving higher resolutions, the limited shooting time of 5 min constrained the resolution of the LiDAR-generated point cloud, as LiDAR collects discrete point samples rather than capturing continuous surface data [51]. As a result of the active nature of laser scanning [51], intensity data were captured, offering some insights into the surface characteristics. However, this method was ineffective in capturing the texture of the rock mass, even though texture is a crucial factor in rock characterization. According to its specifications (Table 5), the LiDAR method would have resulted in a point cloud error of 6 mm when measured from a distance of 10 m. While higher-end LiDAR units could achieve greater point density and accuracy within the same time frame, this would come at the cost of higher equipment expenses.

SfM–MVS generated a high-resolution point cloud of the tunnel face, containing approximately 35,100,000 points (670 pts/m), which increased to 39,700,000 points when including the surrounding area. This resolution is approximately 20 times higher than that achieved by LiDAR, as illustrated in Figure 11b. The significant contrast in point counts mainly stems from whether the instrument only targets the tunnel face. When considering the total number of points generated, the difference narrows to approximately 3.3 times. Although additional time is required to generate 3D point clouds, the duration varies based on computational power, resolution, and the number of images used. Since LiDAR data also require post-processing, our focus was on the potential to generate high-resolution 3D models within the allocated field shooting time. Since SfM–MVS is based on photographic images, it effectively captures the texture of the rock mass, though it lacks the ability to obtain intensity data like LiDAR. The passive nature of photographic images allows SfM–MVS to achieve high-resolution data over a broad area within a short shooting period [51]. When utilizing OPS, the SfM–MVS method would likely result in a point cloud error of 2 mm under the constraints of E = 100 lx and A = 0.3 m^2/s (refer to Section 2.1).

In addition to quantitative indices, we qualitatively compared the point clouds from both methods. Given that the target area was a tunnel construction site, we focused on the capability of each method to express joint planes and joint traces, which are critical for rock mass stability analysis. Figure 12a presents the comparison of joint plane expression capabilities. Both methods successfully depicted the normal vectors of joint planes within a five-degree deviation, which is better than typical human bias. However, both methods exhibited point resolutions lower than the commonly referenced laboratory values of 1000 pts/m [52] to 10,000 pts/m [53] used for measuring surface roughness. However, due to its lower resolution, LiDAR had more significant issues with underestimating the joint roughness compared to SfM–MVS [54].

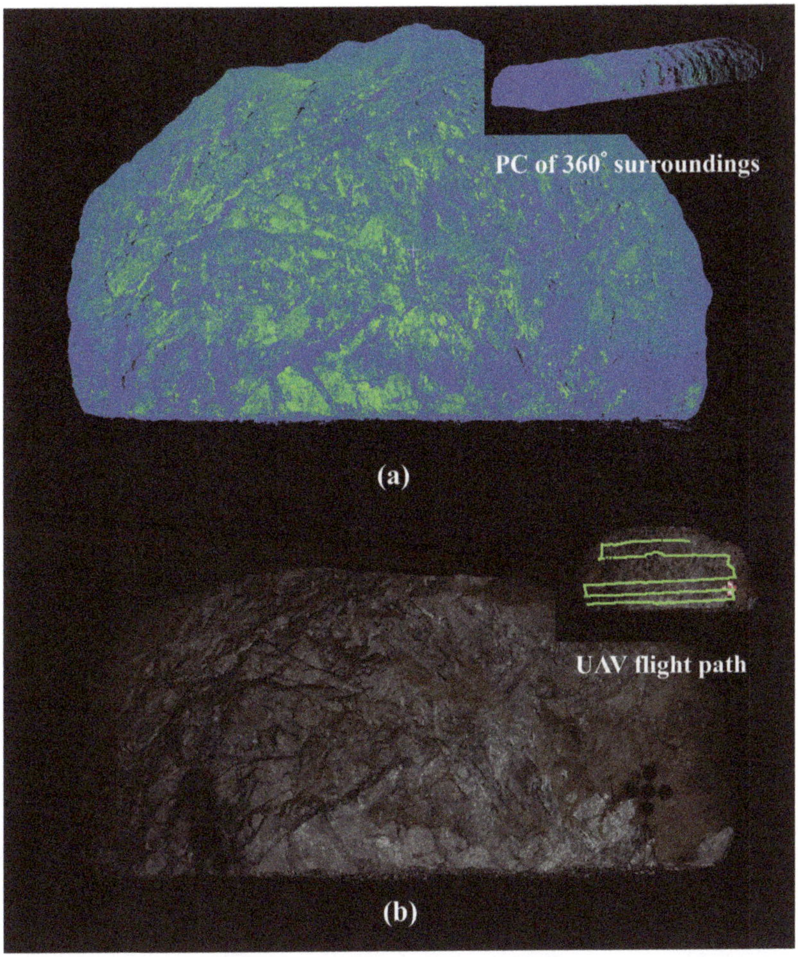

Figure 11. Point clouds generated by LiDAR and SfM–MVS with OPS: (**a**) point cloud generated by LiDAR; (**b**) point cloud generated by SfM–MVS utilizing OPS. The image acquisition trajectories (UAV flight path) are illustrated in the upper right section.

Shadow zones (yellow ellipse in Figure 12a), areas where no points are recorded, appeared in the LiDAR data due to the use of a terrestrial laser scanner (TLS). Since TLS emits light from a fixed position, the light often fails to reach areas obscured from its direct line of sight, resulting in shadow zones. These absences of points can complicate the recognition of joint planes and their orientations during data analysis. In contrast, dark regions appeared in the SfM–MVS point cloud; however, unlike in LiDAR data, dark-colored points were recorded in these areas. This difference arises because SfM–MVS captures images from multiple viewpoints. While these dark-colored points may be subject to potential systematic errors due to low saturation, they still provide useful information for detecting plane existence and determining orientation. It is important to note that the dark regions in the SfM–MVS data are not equivalent to the shadow zones found in LiDAR data. While LiDAR combined with UAV technology (ULS) may mitigate shadow zone issues, it introduces other problems, including reduced accuracy and resolution, particularly in GNSS-denied environments [1].

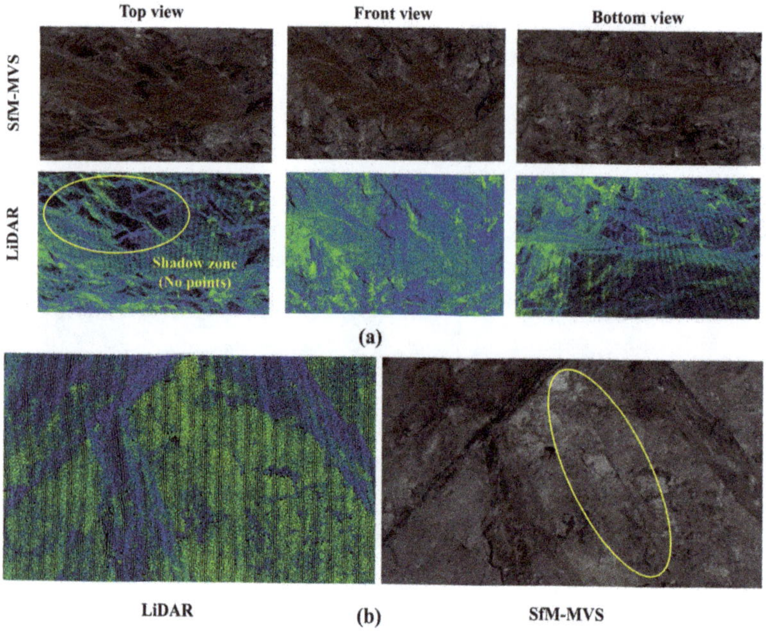

Figure 12. Comparison of rock mass characterization capabilities between SfM–MVS and LiDAR: (**a**) joint plane expression capabilities; (**b**) joint trace expression capabilities.

Figure 12b presents a comparison of the joint trace expression capabilities. Joint traces vary in aperture from sub-millimeter to several millimeters. While LiDAR produced a point cloud that was overly sparse to effectively depict the joint traces, SfM–MVS was able to represent them up to 2 mm. Beyond resolution, the texture data captured by SfM–MVS proved advantageous in depicting joint traces, which appeared as dark-colored points (yellow ellipse in Figure 12b). The overall performance comparison between the two methods is summarized in Table 8.

Table 8. Performance comparison between SfM-MVS utilizing the OPS and LiDAR at the Yeoju-si tunnel site.

	SfM–MVS	LiDAR
Instrument	DJI Mavic 2 Pro	Leica BLK 360
Capital cost	USD 1770	USD 15545
Survey time	5 min	5 min
Resolution	35,100,000 pts (670 pts/m)	Rock face: 1,700,000 pts (150 pts/m) Whole tunnel: 12,100,000 pts (400 pts/m)
Accuracy	2 mm (@ OPS under E = 25 lx and A = 0.3 m^2/s)	6 mm (@ D = 10 m)
Features	Texture data Joints well depicted Need for drone control	Intensity data Shadow zones Simple scanning (∵ TLS)

5. Discussion

Although the quality of input images in the SfM–MVS procedure significantly influences the resulting 3D model accuracy, limited research has been conducted on optimizing both camera settings and UAV flight methods in an integrated manner. In contrast, this study introduced a simultaneous multi-parameter optimization framework that jointly considers the effects of camera settings, flight methods, and environmental conditions (i.e., illumination and time) on reconstruction accuracy. By systematically analyzing the influence of each contributing factor on reconstruction error, the proposed method identifies an optimal combination of photography settings that minimizes overall error, resulting in a globally optimal configuration.

5.1. Robustness of the OPS Derivation Procedure

In this study, we introduced a procedure to determine the OPS under illuminance and time constraints and validated its significance through a comparison with APS and LiDAR. Nonetheless, practical limitations may sometimes prevent the application of the derived OPS due to the limited availability of several factors such as UAV speed, distance, shutter speed, or resolution. For example, UAV speed and distance may be constrained by speed regulations, physical performance limitations, or altitude restrictions. These factors then act as upper or lower bounds for UAV speed and distance. Despite these constraints, a sub-optimal D–v combination can still be determined within the restricted D–v zone, as long as this feasible D–v combination meets the required A. For instance, if the minimum v is restricted to 0.6 m/s, one cannot use the optimal UAV flight method of v = 0.5 m/s and D = 2.5 m when A = 1 m^2/s is required, as shown in Figure 3. In such cases, an alternative approach could be to adopt the second-best UAV flight method, which might involve approximately v = 0.6 m/s and D = 2 m, though this may result in a slightly higher error.

However, this represents an ideal scenario, and in practice, it is necessary to introduce the concept of a safety factor. For instance, drone battery performance can significantly influence UAV flight methods. Battery efficiency is closely linked to UAV speed, and in the case of large-scale sites, battery replacement must also be considered when designing an optimal flight strategy. Quantifying battery efficiency is challenging due to its strong dependency on hardware specifications and external environmental factors such as temperature and wind. Since the primary objective of this study was to propose a practical optimization strategy for UAV-based 3D reconstruction, battery performance considerations were not included in the optimization framework. Nevertheless, to ensure more stable flights and reliable image acquisition, particularly for large-scale surveys, future studies should explore the integration of a safety factor into the optimization of camera settings and UAV flight paths.

Occasionally, images captured under certain conditions may exhibit periodic noise, particularly under artificial lighting such as the LED strobe lights commonly used in underground mines or tunnels. This phenomenon often occurs when the camera's rolling shutter fails to synchronize with oscillating light sources. Although the negative effects of periodic noise were not specifically evaluated in this study, they can be mitigated by adjusting the photography settings. If images captured using the OPS display periodic noise, users can attempt alternative settings with varying shutter speeds to eliminate the noise, albeit with a potentially higher error margin. For example, if the optimal camera settings shown in Figure 1 result in periodic noise, alternative settings like N = 3.2, t = 1/120 s, and S = 3200 or N = 2.8, t = 1/80 s, and S = 1600 can be tested. If these adjustments do not display periodic noise, they can be adopted despite the slightly increased error. If not, further adjustments should be explored until the issue is resolved.

The resolution of the resultant point cloud is crucial for analyses such as surface roughness measurements. For instance, a resolution ranging from 1000 pts/m [52] to 10,000 pts/m [53] is typically required to accurately measure the surface roughness in laboratory settings. The maximum resolution of a point cloud is achieved at the ideal GSD, which involves deriving a point from each image pixel. Thus, the sampling distance of the resultant point cloud, in meter units, is approximately $(D \times d)/(f \times p)$, where d and f are generally fixed. Therefore, the resolution is primarily influenced by D and p.

Figure 13 illustrates the change in OPS when an additional GSD constraint is considered. Under E = 100 lx, D = 5 m, and v = 0.2 m/s, the optimal camera settings without the GSD constraint are t = 1/80 s, N = 2.8, S = 1600, and p = 1920, which achieve an ideal GSD of 3.3 mm and a minimum error of 2.3 mm (Figure 13a). When GSD is set to be less than 2 mm, settings that do not meet this requirement—including the previously optimal settings—are excluded (Figure 13c). The new alternative optimal camera settings in this case are t = 1/40 s, N = 2.8, S = 800, and p = 3840, resulting in a slightly higher minimum error of 2.5 mm. Considering that the resolution constraint increases the minimum achievable error, the revised settings still minimize the error while satisfying the resolution requirement. With these new E–D–v charts (note Figure 3a), it is possible to complete the new D–v chart for UAV flight method optimization (Figure 13d). It shows that regions with relatively far distances ($D \geq 4$ m) are strongly affected by the GSD constraint (compare Figure 13b,d), which significantly impacts the selection of optimal UAV flight methods, particularly in higher A values.

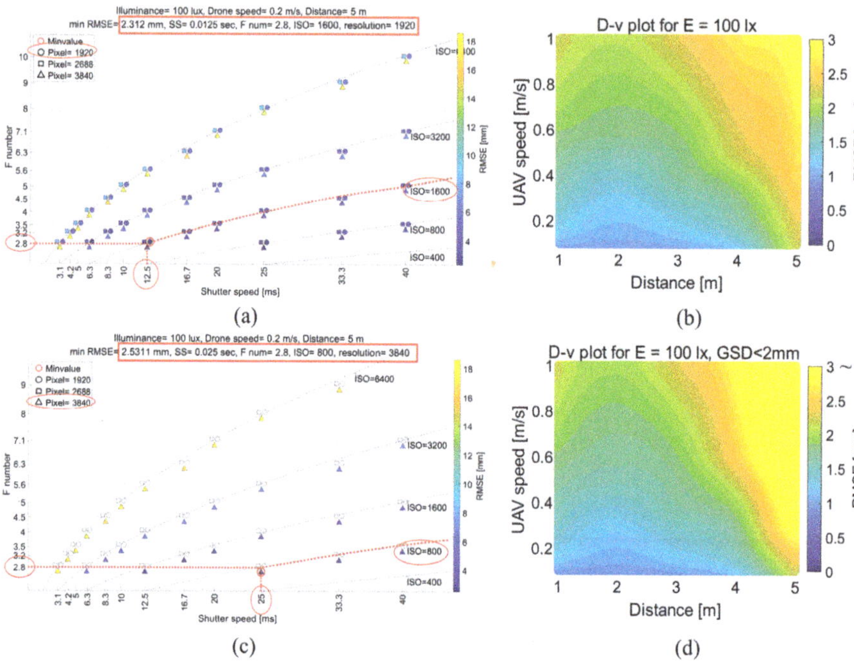

Figure 13. Comparison of the OPS considering the GSD constraint: (**a**) the E–D–v chart under E = 100 lx, v = 0.2 m/s, and D = 5 m; (**b**) the D–v chart for E = 100 lx, identical to Figure 3b; (**c**) the new E–D–v chart with a GSD constraint of 2 mm; (**d**) the new D–v chart for E = 100 lx, reflecting the GSD constraint.

We introduced a procedure for determining the OPS for UAV-based SfM–MVS with parallel image acquisition. However, establishing the OPS for still imaging is beyond the scope of this study, as convergent acquisition schemes may offer better accuracy for still images [7]. Additionally, when considering time constraints, it is important to account for the acceleration or deceleration between still photography locations, which can vary significantly based on the physical performance of the camera platform. Nonetheless, since motion blur is not a concern in still images, a slower shutter speed can be used. This allows for a higher SNR in the image, making the prevention of out-of-focus blur the top priority. Therefore, we recommend setting the highest possible F-number, then the lowest ISO, and finally the slowest shutter speed that still maintains appropriate image brightness (Equation (2)).

This study assumed uniform photography settings across the entire image set. Consequently, the performance of the OPS might be compromised when using mixed image sets to generate a point cloud. For instance, if the target area includes both brightly and darkly illuminated areas, such as shadows, a separate OPS would be required for each lighting condition. This would result in a transition zone where images used for 3D reconstruction are captured with different settings. The performance in this transition zone could be diminished due to the inconsistent distortions across image sets. A study on orthophotos by Aasen et al. [3] noted that details in overlapping images tend to blend, rather than selecting the best details from each. Similarly, we anticipate that errors in the transition zone may average between the errors observed in the brightly and darkly lit regions.

Another challenge is when the surveyed area exhibits significant terrain variation, making it difficult to maintain a uniform distance from the target. In such cases, it is challenging to clearly define the distance to the target, or multiple distances may coexist. Similar to the transition zone observed under mixed lighting conditions, it becomes impossible to optimize the photography settings using a single OPS. A practical solution to overcome this is to increase the minimum distance for the OPS analysis area, setting it significantly larger than the terrain variation (sub-optimal OPS selection), where the impact of height differences can be neglected. However, identifying a minimum distance that sufficiently mitigates the effects of terrain variation requires further research.

UAV-based image acquisition inherently involves minor vibrations during flight, and although gimbals are employed to stabilize the camera, complete elimination of vibrational motion is not achievable. At higher flight speeds, motion blur typically dominates image degradation, rendering vibration-induced blur largely negligible. However, when UAVs operate at sufficiently low speeds, vibration blur becomes significant. This introduces a trade-off in flight speed selection: reducing speed to improve image sharpness inadvertently increases susceptibility to vibration-induced distortions. The magnitude of the vibration blur is influenced by various factors, including drone hardware specifications, environmental conditions (e.g., wind), and the effectiveness of internal damping mechanisms. Due to the stochastic nature of these factors, quantifying vibration blur is challenging. Since the objective of this study was to propose a practical optimization scheme for photographic settings, vibration blur was not explicitly considered in this study. Nevertheless, for more robust image acquisition strategies—particularly in low-speed UAV operations where vibration effects are more pronounced—future research should explore integrating vibration effects into the optimization of photography settings.

5.2. Sensitivity of the OPS

In Section 4, we demonstrated the effectiveness and potential of the OPS. When the SfM–MVS utilized the OPS, it achieved significant improvements in point cloud quality compared to the APS, without requiring additional lighting or extra time. However, it is crucial to use the optimal camera settings in conjunction with the optimal UAV flight methods, as the lowest error is achieved only when both are applied together. If only partial settings are implemented, such as using only the optimal UAV flight method while selecting camera settings arbitrarily, the combination is no longer optimal, and alternative settings may produce better results. Since camera settings are less likely to be adjusted during image acquisition, it is essential to meticulously maintain the UAV flight method through precise UAV path programming to preserve the effectiveness of the OPS.

In Figure 14, we demonstrate the variation in optimal camera settings based on UAV speeds (0.1, 0.5, and 0.9 m/s) and distances (1–9 m) under specific illuminance levels (600,

770, 970, and 1550 lx). The figure illustrates trends in camera settings adjustments according to UAV flight methods (D, v). Along the lines representing constant UAV speeds (colored red, green, and blue in Figure 14), one trend observed is the decrease in the F-number at larger distances. As the distance increases, motion blur in the image decreases, shifting the priority to reducing out-of-focus blur by lowering the F-number. Another trend is the need for faster shutter speeds at higher UAV speeds for the same distance, moving from red to green and then from green to blue in Figure 14. This adjustment is necessary to reduce motion blur, thus taking precedence over reducing the F-number at higher UAV speeds. Once the maximum aperture (or minimum F-number) is reached, using a slower shutter speed becomes the only option. Importantly, the selection of optimal camera settings is more critical at close distances and slow UAV speeds and becomes less sensitive at greater distances and faster UAV speeds. Therefore, precise maintenance of the UAV flight method is crucial when operating under short time constraints.

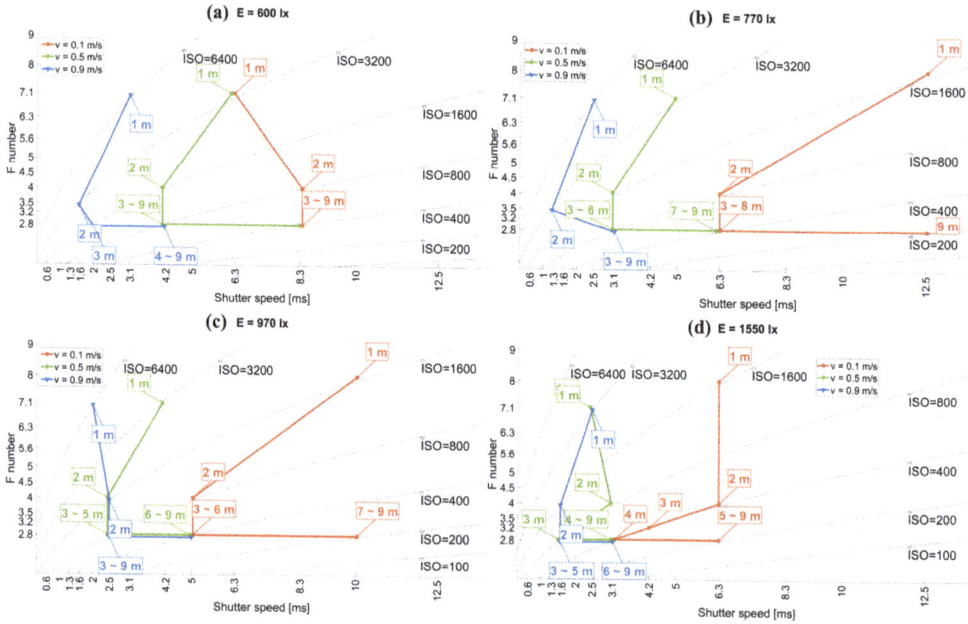

Figure 14. The sensitivity of selecting optimal camera settings for UAVs varies with changes in speed (0.1, 0.5, and 0.9 m/s), distance (1–9 m), and illuminance (600, 770, 970, and 1550 lx). Subfigures (**a**–**d**) correspond to illuminance levels of 600, 770, 970, and 1550 lx, respectively. Each point on the chart represents the optimal camera settings under specific E–D–v conditions.

The variation in optimal camera settings also correlates with illumination, though no clear relationship between reducing the F-number and using faster shutter speeds is evident when illumination changes. However, a consistent trend is observed in the camera settings moving toward the lower left of the charts from Figure 14a–d, despite varying illuminance intervals (ΔE = 170, 200, and 480 lx). For instance, at D = 1 m and v = 0.5 m/s (top points on the green lines), the F-number (N) decreased from 6.3 to 5 (a reduction to 0.8 times the original) when the illumination increased from 600 to 770 lx (ΔE = 170 lx; from Figure 14a,b). Meanwhile, N decreased from 4.2 to 2.5 (a reduction of 0.6 times the original) when the illumination jumped from 970 to 1550 lx (ΔE = 480 lx; from Figure 14c,d). Despite the near tripling in illuminance change, the alteration in camera settings was relatively minor. This indicates that the selection of optimal camera settings is particularly sensitive under low illumination conditions, similar to the sensitivity observed under limited time constraints. The balance between error minimization and sensitivity under varying conditions warrants further investigation. Consequently, extra caution is advised when employing the OPS under low illuminance and short time constraints.

Concerns have been raised about the sufficiency of SfM–MVS-generated point clouds in underground tunnels, where limited lighting and constrained surveying time might distort the captured images [29]. Traditionally, LiDAR has been used for 3D rock mass modeling in such underground spaces, despite its high capital costs, issues with shadow zones, lack of texture detail, and lengthy surveying times. However, when SfM–MVS is utilized with OPS, it shows potential for more effectively characterizing underground rock masses by generating high-quality 3D point clouds at a lower cost (refer to Table 8). Previous studies [1,6,7] have noted the superior performance of SfM–MVS compared to LiDAR, and our findings reinforce this advantage in challenging environments characterized by low lighting and limited time. The purpose of this study was not to claim the outright superiority of SfM–MVS over LiDAR but to demonstrate that SfM–MVS can achieve sufficient performance even in challenging environments. As discussed, each method has its own advantages and limitations, and they should ideally be used in a complementary manner.

Certain challenges may arise when applying UAV-derived SfM–MVS in underground spaces. For instance, a single UAV may not satisfy the time constraints due to ongoing construction activities. Operating multiple UAVs simultaneously could be a viable solution, as the cost of moderate UAVs is relatively low compared to LiDAR and other laser scanning methods. Similarly, the accuracy of the resultant point cloud could be improved with the use of multiple UAVs. For example, as shown in Figure 3, using a single UAV under a time constraint of $A = 1\ m^2/s$ might achieve a minimum error of approximately 1.7 mm. However, by deploying two UAVs simultaneously, where each UAV covers half of the target area and operates under a time constraint of $A = 0.5\ m^2/s$, the minimum error could be reduced to approximately 1.2 mm.

Excessively low illuminance conditions in underground tunnels pose a significant challenge. In such cases, no camera settings may achieve adequate image brightness (Equation (2)), and the impact of read or dark current noise becomes more pronounced. Addressing this lack of light with field lighting alone may not be feasible, as it is difficult to increase the illumination across an entire rock face in an underground tunnel. A more practical solution might involve equipping UAVs with their own lighting sources to illuminate the specific areas being photographed. However, to avoid systematic errors, the lighting must be homogeneous across the scene and capable of covering a larger area than the camera's FOV. Furthermore, the distance from the target affects the illumination intensity, influencing the selection of OPS. Future research should explore this relationship further and its impact on determining OPS.

6. Conclusions

Based on the error prediction model developed in our previous work [30], we identified the optimal photography settings that minimize error under specific illuminance and time constraints. The efficacy of the OPS was validated by comparing the point cloud quality obtained from both an indoor reference target and an outdoor rock mass, using optimal versus arbitrary photography settings. The use of OPS enabled the generation of higher-quality point clouds without requiring increased illumination or additional surveying time.

Furthermore, we demonstrated the potential of SfM–MVS, utilizing the OPS, at an underground tunnel construction site by comparing its performance with the LiDAR method. Given the low illuminance and limited surveying time typical of such environments, which are generally challenging for photography, SfM–MVS with the OPS showed the potential to be more effective than LiDAR, offering higher accuracy and resolution at a lower cost.

Author Contributions: Conceptualization, J.L.; methodology, J.L.; validation, J.L., S.M. and I.-S.K.; investigation, J.L. and D.-H.Y.; resources, I.-S.K. and D.-H.Y.; writing—original draft preparation, J.L., S.M., I.-S.K., Y.S. and J.J.; visualization, J.L. and S.M.; supervision, J.-J.S. and J.J.; project administration, J.-J.S.; funding acquisition, J.-J.S.; software, J.L. and S.M. All authors have read and agreed to the published version of the manuscript.

Funding: This research was supported by the Energy & Mineral Resources Development Association of Korea (EMRD) funded by the Korea government (MOTIE) (grant no. 2021060003, Training Program for Specialists in Smart Mining), and by the Korea Institute of Energy Technology Evaluation and Planning (KETEP) grant funded by the Korea government (MOTIE) (grant no. 20214000000500, Training program of CCUS for the green growth).

Data Availability Statement: The data that support the findings of this study are available from the author, J. L., upon reasonable request. A derivation of OPS has been developed as an app and can be accessed at the following repository: https://github.com/sooyalim/Optimal-Photographing-Settings-OPS-, accessed on 2 March 2025.

Conflicts of Interest: The authors declare no conflicts of interest.

Abbreviations and Symbols

The following abbreviations and symbols are used in this manuscript:

Abbreviations	
SfM	Structure from motion
MVS	Multiview stereo
LiDAR	Light Detection and Ranging
DEM	Digital elevation model
UAV	Unmanned aerial vehicle
PP	Point-to-point
PR	Point-to-raster
RR	Raster-to-raster
TS	Total station
dGPS	Differential global positioning system
RMSE	Root mean squared error
ME	Mean error
MAE	Mean absolute error
OPS	Optimal photography settings
APS	Arbitrary photography settings
ISO	International Organization for Standardization
SNR	Signal-to-noise ratio
BA	Bundle adjustment
GCP	Ground control point
FOV	Field of view
GSD	Ground sampling distance
C2M	Cloud-to-mesh
M3C2	Multiscale model-to-model cloud comparison
stdC2M	Standard deviation of the C2M distances
stdM3C2	Standard deviation of the M3C2 distances
ICP	Iterative closest point
GNSS	Global Navigation Satellite System

Symbols

t	Camera shutter speed
N	Camera F-number
S	Camera ISO value
p	Camera pixel resolution
D	Distance from the target
v	UAV or camera platform speed
d	Camera sensor size
f	Camera focal length
ε_{3D}	Errors in SfM–MVS-generated point clouds
ε_{2D}	Image errors
ε_{plane}	Point-to-plane-type RMSE
ε_{point}	Point-to-point-type RMSE, equivalent to ε_{3D}
s	Scaling factor
E	Illumination
M	Maximum template size
Q	Nosie sensitivity coefficient
A	Area photographed per unit of time

Appendix A

This section provides a brief summary of the theoretical background for the error predetermination for each photography setting [30]. Equation (1) suggests that the error present in the image is propagated to the resulting point cloud, where it is amplified by the scaling factor s. The error within the pixel domain of the image can be quantified using Equation (A1) [30].

$$\epsilon_{2D}^2 = \int_{-0.5}^{0.5}\int_{-0.5}^{0.5} \sum_i \sum_j \left\{ \mathrm{P}((i,j)|(x,y)) \times \left\{ (x-i)^2 + (y-j)^2 \right\} \right\} dx\, dy \left[\mathrm{px}^2 \right], \quad \text{(A1)}$$

where the indices i and j represent the pixel coordinates within an image and $\mathrm{P}((i,j)|(x,y))$ denotes the probability that a point (x,y), where x and y are within the range $[-0.5,0.5]$, is expected to be observed at the pixel location (0,0) but is instead observed at a different pixel location (i,j) due to distortion effects. This probability is determined by the extent of distortion within the pixel domain, represented by Δx and Δy, which arise from various sources such as motion blur, out-of-focus blur, and noise.

The magnitude of motion blur, denoted as B, is determined using Equation (A2), while the distortion effects caused by motion blur, represented by Δx_{mb} and Δy_{mb}, are calculated using Equation (A3) [30].

$$B = \frac{vtfp}{Dd}[\mathrm{px}], \quad \text{(A2)}$$

$$\begin{cases} \Delta x_{\mathrm{mb}} \sim \mathrm{U}\left(-\frac{B}{2}, \frac{B}{2}\right) \cdots B \leq M \\ \Delta x_{\mathrm{mb}} \sim \mathrm{Trap}\left(-\frac{B}{2}, \min\left(\frac{2M-B}{2}, \frac{B-2M}{2}\right), \max\left(\frac{2M-B}{2}, \frac{B-2M}{2}\right), \frac{B}{2}\right) \cdots B > M \\ \Delta y_{\mathrm{mb}} = 0, \end{cases} \quad \text{(A3)}$$

where d stands for the sensor size, U denotes a uniform probability density distribution, and Trap refers to a trapezoidal-shaped probability density distribution. The influence of the maximum template size M utilized in the MVS algorithm was analyzed by Leem et al. [30], highlighting its role in mitigating the motion blur distortion when B exceeds M.

The level of out-of-focus blur, represented by σ_{ob}, is determined using Equation (A4), while the distortion effects resulting from out-of-focus blur, denoted as Δx_{ob} and Δy_{ob}, are calculated using Equation (A5) [30].

$$\sigma_{\text{ob}} = \frac{|D-H|}{2D}\frac{f^2}{N(H-f)}\frac{p}{d}[\text{px}], \tag{A4}$$

$$\Delta x_{\text{ob}} \sim \text{N}(0, \sigma_{\text{ob}}), \Delta y_{\text{ob}} \sim \text{N}(0, \sigma_{\text{ob}}), \tag{A5}$$

where the notation N stands for a normal probability density distribution, and H denotes the focus distance, which, for this work, was set to the hyperfocal distance. This setting ensures that the camera maintains sharp focus from the hyperfocal distance to infinity, optimizing the depth of field for the broadest range of clear vision.

The noise-to-signal ratio or the reciprocal of SNR (signal-to-noise ratio), represented by σ_{n}, is derived from Equation (A6), while the distortion effects caused by noise, denoted as Δx_{n} and Δy_{n}, are calculated using Equation (A7) [30].

$$\sigma_{\text{n}} = \frac{\text{Noise}}{\text{Signal}} = Q\frac{Np}{d\sqrt{Et}}, \tag{A6}$$

$$\Delta x_{\text{n}} \sim \text{N}\left(0.8 \times 10^{-5} \times \left(Q\frac{Np}{d\sqrt{Et}}\right)^{3.24}\right), \Delta y_{\text{n}} = 0, \tag{A7}$$

where E stands for the level of illumination and Q denotes the noise sensitivity coefficient, which varies with the type of camera used. The probability $\text{P}((i,j)|(x,y))$ can then be determined through Monte Carlo simulations, incorporating the distortion effects caused by motion blur (Equation (A3)), out-of-focus blur (Equation (A5)), and noise (Equation (A7)).

References

1. Smith, M.W.; Carrivick, J.L.; Quincey, D.J. Structure from motion photogrammetry in physical geography. *Prog. Phys. Geogr.* **2016**, *40*, 247–275. [CrossRef]
2. Kovanič, Ľ.; Peťovský, P.; Topitzer, B.; Blišťan, P. Spatial Analysis of Point Clouds Obtained by SfM Photogrammetry and the TLS Method—Study in Quarry Environment. *Land* **2024**, *13*, 614. [CrossRef]
3. Aasen, H.; Burkart, A.; Bolten, A.; Bareth, G. Generating 3D hyperspectral information with lightweight UAV snapshot cameras for vegetation monitoring: From camera calibration to quality assurance. *ISPRS J. Photogramm. Remote Sens.* **2015**, *108*, 245–259. [CrossRef]
4. Reinprecht, V.; Kieffer, D.S. Application of UAV Photogrammetry and Multispectral Image Analysis for Identifying Land Use and Vegetation Cover Succession in Former Mining Areas. *Remote Sen.* **2025**, *17*, 405. [CrossRef]
5. Micheletti, N.; Chandler, J.H.; Lane, S.N. Investigating the geomorphological potential of freely available and accessible structure-from-motion photogrammetry using a smartphone. *Earth Surf. Process. Landf.* **2015**, *40*, 473–486. [CrossRef]
6. Smith, M.W.; Vericat, D. From experimental plots to experimental landscapes: Topography, erosion and deposition in sub-humid badlands from structure-from-motion photogrammetry. *Earth Surf. Process. Landf.* **2015**, *40*, 1656–1671. [CrossRef]
7. Eltner, A.; Kaiser, A.; Castillo, C.; Rock, G.; Neugirg, F.; Abellán, A. Image-based surface reconstruction in geomorphometry—Merits, limits and developments. *Earth Surf. Dyn.* **2016**, *4*, 359–389. [CrossRef]
8. Tonkin, T.N.; Midgley, N.G. Ground-control networks for image based surface reconstruction: An investigation of optimum survey designs using UAV derived imagery and structure-from-motion photogrammetry. *Remote Sens.* **2016**, *8*, 786. [CrossRef]
9. James, M.R.; Chandler, J.H.; Eltner, A.; Fraser, C.; Miller, P.E.; Mills, J.P.; Noble, T.; Robson, S.; Lane, S.N. Guidelines on the use of structure-from-motion photogrammetry in geomorphic research. *Earth Surf. Process. Landf.* **2019**, *44*, 2081–2084. [CrossRef]
10. Wheaton, J.M.; Brasington, J.; Darby, S.E.; Sear, D.A. Accounting for uncertainty in DEMs from repeat topographic surveys: Improved sediment budgets. *Earth Surf. Process. Landf.* **2010**, *35*, 136–156. [CrossRef]
11. Bakker, M.; Lane, S.N. Archival photogrammetric analysis of river–floodplain systems using structure from motion (SfM) methods. *Earth Surf. Process. Landf.* **2017**, *42*, 1274–1286. [CrossRef]
12. Vollgger, S.A.; Cruden, A.R. Mapping folds and fractures in basement and cover rocks using UAV photogrammetry, Cape Liptrap and Cape Paterson, Victoria, Australia. *J. Struct. Geol.* **2016**, *85*, 168–187. [CrossRef]

13. Becker, R.E.; Galayda, L.J.; MacLaughlin, M.M. Digital photogrammetry software comparison for rock mass characterization. In Proceedings of the 52nd US Rock Mechanics/Geomechanics Symposium, Seattle, WA, USA, 17–20 June 2018.
14. Zhang, Y.; Yue, P.; Zhang, G.; Guan, T.; Lv, M.; Zhong, D. Augmented reality mapping of rock mass discontinuities and rockfall susceptibility based on unmanned aerial vehicle photogrammetry. *Remote Sens.* **2019**, *11*, 1311. [CrossRef]
15. Javadnejad, F.; Slocum, R.K.; Gillins, D.T.; Olsen, M.J.; Parrish, C.E. Dense point cloud quality factor as proxy for accuracy assessment of image-based 3D reconstruction. *J. Surv. Eng.* **2021**, *147*, 04020021. [CrossRef]
16. Wenzel, K.; Rothermel, M.; Fritsch, D.; Haala, N. IMAGE ACQUISTION AND MODEL SELECTION FOR MULTI-VIEW. *Int. Arch. Photogramm. Remote Sens. Spatial Inf. Sci.* **2013**, *XL-5/W1*, 251–258. [CrossRef]
17. Rangelov, D.; Waanders, S.; Waanders, K.; Keulen, M.V.; Miltchev, R. Impact of camera settings on 3D Reconstruction quality: Insights from NeRF and Gaussian Splatting. *Remote Sens.* **2024**, *24*, 7594. [CrossRef]
18. Dunford, R.; Michel, K.; Gagnage, M.; Piégay, H.; Trémelo, M.-L. Potential and constraints of unmanned aerial vehicle technology for the characterization of Mediterranean riparian forest. *Int. J. Remote Sens.* **2009**, *30*, 4915–4935. [CrossRef]
19. Čerňava, J.; Chudý, F.; Tunák, D.; Saloň, Š.; Vyhnáliková, Z. The vertical accuracy of digital terrain models derived from the close-range photogrammetry point cloud using different methods of interpolation and resolutions. *Cent. Eur. For. J.* **2019**, *65*, 198–205. [CrossRef]
20. O'Connor, J.; Smith, M.J.; James, M.R. Cameras and settings for aerial surveys in the geosciences: Optimising image data. *Prog. Phys. Geogr.* **2017**, *41*, 325–344. [CrossRef]
21. Roncella, R.; Bruno, N.; Diotri, F.; Thoeni, K.; Giacomini, A. Photogrammetric Digital Surface Model Reconstruction in Extreme Low-Light Environments. *Remote Sens.* **2021**, *13*, 1261. [CrossRef]
22. Burdziakowski, P.; Bobkowska, K. UAV Photogrammetry under Poor Lighting Conditions-Accuracy Considerations. *Sensors* **2021**, *21*, 3531. [CrossRef] [PubMed]
23. Slaker, B.A.; Mohamed, K.M. A practical application of photogrammetry to performing rib characterization measurements in an underground coal mine using a DSLR camera. *Int. J. Min. Sci. Technol.* **2017**, *27*, 83–90. [CrossRef] [PubMed]
24. García-Luna, R.; Senent, S.; Jurado-Piña, R.; Jimenez, R. Structure from motion photogrammetry to characterize underground rock masses: Experiences from two real tunnels. *Tunn. Undergr. Space Technol.* **2019**, *83*, 262–273. [CrossRef]
25. Li, J.; Guan, J.; Chen, X.; Xi, J. Adaptive optimal exposure selection based on time cost function for 3D reconstruction of high dynamic range surfaces. *Meas. Sci. Technol.* **2023**, *34*, 125018. [CrossRef]
26. Kim, J.-H.; Sung, S.-M. Quality Analysis of Unmanned Aerial Vehicle Images Using a Resolution Target. *Appl. Sci.* **2024**, *14*, 2154. [CrossRef]
27. Abed, F.M.; Noori, A.M.; Al-Saedi, A.S.J. Optimizing Application of UAV-Based SfM Photogrammetric 3D Mapping in Urban Areas. *Iraqi J. Sci.* **2024**, *65*, 2958–2975. [CrossRef]
28. Young, D.J.N.; Koontz, M.J.; Weeks, J. Optimizing aerial imagery collection and processing parameters for drone-based individual tree mapping in structurally complex conifer forests. *Methods Ecol. Evol.* **2022**, *13*, 1447–1463. [CrossRef]
29. Korea Institute of Construction Technology (KICT). *Final Report on the Development of Online Rock Mass Classification Technology and Operation Model Through Digital Mapping in Tunnel Construction*; National Library of Korea: Seoul, Republic of Korea, 2017. [CrossRef]
30. Leem, J.; Kim, J.; Kang, I.S.; Choi, J.; Song, J.J.; Mehrishal, S.; Shao, Y. Practical error prediction in UAV imagery-based 3D reconstruction: Assessing the impact of image quality factors. *Int. J. Remote Sens.* **2025**, *46*, 1000–1030. [CrossRef]
31. International Organization for Standardization. Photography—General Purpose Photographic Exposure Meters (Photoelectric Type)—Guide to Product Specification. 1974. Available online: https://www.iso.org/standard/7690.html (accessed on 19 March 2025).
32. Zhou, Z.; Pain, B.; Fossum, E.R. Frame-transfer CMOS active pixel sensor with pixel binning. *IEEE Trans. Electron Devices* **1997**, *44*, 1764–1768. [CrossRef]
33. Lowe, D.G. Object recognition from local scale-invariant features. In Proceedings of the 7th IEEE International Conference on Computer Vision, Kerkyra, Greece, 20–27 September 1999. [CrossRef]
34. Bay, H.; Ess, A.; Tuytelaars, T.; Van Gool, L. Speeded-up robust features (SURF). *Comput. Vis. Image Underst.* **2008**, *110*, 346–359. [CrossRef]
35. Fischler, M.A.; Bolles, R.C. Random sample consensus: A paradigm for model fitting with applications to image analysis and automated cartography. *Commun. ACM* **1981**, *24*, 381–395. [CrossRef]
36. Torr, P.H.S.; Zisserman, A. MLESAC: A new robust estimator with application to estimating image geometry. *Comput. Vis. Image Underst.* **2000**, *78*, 138–156. [CrossRef]
37. Ballard, D.H.; Brown, C.M. *Computer Vision*; Prentice-Hall, Inc.: Englewood Cliffs, NJ, USA, 1982.
38. Dai, F.; Feng, Y.; Hough, R. Photogrammetric error sources and impacts on modeling and surveying in construction engineering applications. *Vis. Eng.* **2014**, *2*, 2. [CrossRef]

39. Piermattei, L.; Carturan, L.; Guarnieri, A. Use of terrestrial photogrammetry based on structure-from-motion for mass balance estimation of a small glacier in the Italian alps. *Earth Surf. Process. Landf.* **2015**, *40*, 1791–1802. [CrossRef]
40. Ren, C.; Jiao, Y.; Liu, Y.; Shang, H. Optimal camera focal length detection method for GPS-supported bundle adjustment in UAV photogrammetry. *Measurement* **2024**, *228*, 114329. [CrossRef]
41. Tan, C.; Chen, Z.; Chen, Z.; Liao, A.; Zeng, X.; Cao, J. Accuracy analysis of UAV aerial photogrammetry based on RTK mode, flight altitude, and number of GCPs. *Meas. Sci. Technol.* **2024**, *35*, 106310. [CrossRef]
42. Luo, W.; Shao, M.; Che, X.; Hesp, P.A.; Bryant, R.G.; Yan, C.; Xing, Z. Optimization of UAVs-SfM data collection in aeolian landform morphodynamics: A case study from the Gonghe Basin, China. *Earth Surf. Process. Landf.* **2020**, *45*, 3293–3312. [CrossRef]
43. Muhammad, M.; Tahar, K.N. Comprehensive Analysis of UAV Flight Parameters for High Resolution Topographic Mapping. *IOP Conf. Ser. Earth Environ. Sci.* **2021**, *767*, 012001. [CrossRef]
44. De Lima, R.S.; Lang, M.; Burnside, N.G.; Peciña, M.V.; Arumäe, T.; Laarmann, D.; Ward, R.D.; Vain, A.; Sepp, K. An Evaluation of the Effects of UAS Flight Parameters on Digital Aerial Photogrammetry Processing and Dense-Cloud Production Quality in a Scots Pine Forest. *Remote Sens.* **2021**, *13*, 1121. [CrossRef]
45. Paixão, A.; Muralha, J.; Resende, R.; Fortunato, E. Close-range photogrammetry for 3D rock joint roughness evaluation. *Rock Mech. Rock Eng.* **2022**, *55*, 3213–3233. [CrossRef]
46. Furukawa, Y.; Ponce, J. Accurate, dense, and robust multiview stereopsis. *IEEE Trans. Pattern Anal. Mach. Intell.* **2010**, *32*, 1362–1376. [CrossRef] [PubMed]
47. An, P.; Fang, K.; Jiang, Q.; Zhang, H.; Zhang, Y. Measurement of rock joint surfaces by using smartphone structure from motion (SfM) photogrammetry. *Sensors* **2021**, *21*, 922. [CrossRef] [PubMed]
48. Lee, S.; Jeon, S. Estimating Rock Mass Joint Roughness Using Terrestrial Laser Scanner and Artificial Neural Network. In Proceedings of the 56th U.S. Rock Mechanics/Geomechanics Symposium, Santa Fe, NM, USA, 26 June 2022. [CrossRef]
49. Huang, H.; Ye, Z.; Zhang, C.; Yue, Y.; Cui, C.; Hammad, A. Adaptive Cloud-to-Cloud (AC2C) Comparison Method for Photogrammetric Point Cloud Error Estimation Considering Theoretical Error Space. *Remote Sens.* **2022**, *14*, 4289. [CrossRef]
50. Lague, D.; Brodu, N.; Leroux, J. Accurate 3D comparison of complex topography with terrestrial laser scanner: Application to the Rangitikei canyon (NZ). *ISPRS J. Photogramm. Remote Sens.* **2013**, *82*, 10–26. [CrossRef]
51. Baltsavias, E.P. A comparison between photogrammetry and laser scanning. *ISPRS J. Photogramm. Remote Sens.* **1999**, *54*, 83–94. [CrossRef]
52. Ge, Y.; Kulatilake, P.H.; Tang, H.; Xiong, C. Investigation of natural rock joint roughness. *Comput. Geotech.* **2014**, *55*, 290–305. [CrossRef]
53. Park, J.W.; Lee, Y.K.; Song, J.J.; Choi, B.H. A constitutive model for shear behavior of rock joints based on three-dimensional quantification of joint roughness. *Rock Mech. Rock Eng.* **2013**, *46*, 1513–1537. [CrossRef]
54. Barton, N.; Wang, C.; Yong, R. Advances in joint roughness coefficient (JRC) and its engineering applications. *J. Rock Mech. Geotech. Eng.* **2023**, *15*, 3352–3379. [CrossRef]

remote sensing

MDPI

Article

Indeterminacy of Camera Intrinsic Parameters in Structure from Motion Using Images from Constant-Pitch Flight Design

Truc Thanh Ho *, Riku Sato, Ariyo Kanno, Tsuyoshi Imai , Koichi Yamamoto and Takaya Higuchi

Graduate School of Sciences and Technology for Innovation, Yamaguchi University, 2-16-1 Tokiwadai, Yamaguchi 755-8611, Japan; e021vev@yamaguchi-u.ac.jp (R.S.); kanno@yamaguchi-u.ac.jp (A.K.); imai@yamaguchi-u.ac.jp (T.I.); k_yama@yamaguchi-u.ac.jp (K.Y.); takaya@yamaguchi-u.ac.jp (T.H.)

* Correspondence: hothanhtruc@yucivil.onmicrosoft.com or hothanhtruc86@gmail.com

Abstract: Intrinsic parameter estimation by self-calibration is commonly used in Unmanned aerial vehicle (UAV)-based photogrammetry with Structure from Motion (SfM). However, obtaining stable estimates of these parameters from image-based SfM—which relies solely on images, without auxiliary data such as ground control points (GCPs)—remains challenging. Aerial imagery acquired with the constant-pitch (CP) flight pattern often exhibits non-linear deformations, highly unstable intrinsic parameters, and even alignment failures. We hypothesize that CP flights form a "critical configuration" that renders certain intrinsic parameters indeterminate. Through numerical experiments, we confirm that a CP flight configuration does not provide sufficient constraints to estimate focal length (f) and the principal point coordinate (cy) in image-based SfM. Real-world CP datasets further demonstrate the pronounced instability of these parameters. As a remedy, we show that by introducing intermediate strips into the CP flight plan—what we call a CP-Plus flight—can effectively mitigate the indeterminacy of f and cy in simulations and markedly improve their stability in all tested cases. This approach enables more effective image-only SfM workflows without auxiliary data, simplifies data acquisition, and improves three-dimensional reconstruction accuracy.

Keywords: UAV-based photogrammetry; structure from motion (SfM); indeterminacy; intrinsic parameters; critical configuration

1. Introduction

Unmanned aerial vehicle (UAV)-based photogrammetry has emerged recently as an effective technique to generate high-resolution three-dimensional (3D) models across multiple fields including infrastructure inspection [1–3], mining surveying [4–6], hazard modeling [7–9], and environmental condition monitoring [10,11]. Structure from Motion (SfM) underpins this technique, estimating intrinsic and extrinsic camera parameters. These parameters are then exploited in the Multi-View Stereo (MVS) to produce dense point clouds, enabling the generation of high-resolution deliverables, such as meshes, digital elevation models (DEMs), and orthophotos. Therefore, the precision of camera parameter estimation in SfM has a direct effect on the accuracy and precision of reconstructed models. While the accurate estimation of these parameters relies hardly on the number and distribution of auxiliary information such as Ground Control Points (GCPs), it is cost-intensive and time-consuming. Ensuring accurate parameter estimation without a heavy reliance on GCPs remains a primary concern for UAV-based photogrammetry.

Intrinsic parameters—such as focal length (f), principal point coordinates (cx, cy), radial distortion, and tangential distortion—can be estimated during the bundle adjustment

(BA) stage in a typical SfM pipeline (self-calibration), or alternatively, calculated using imagery captured specially for calibration purposes (pre-calibration). However, in UAV-based photogrammetry, where lightweight and compact drones with small cameras are commonly used, pre-calibrated intrinsic parameters are often considered unstable [12–15]. This instability can arise from mechanical effects during camera shooting, changes in image sensor temperature [16–18], or ambient temperature [19] during drone operations, potentially affecting the sensor geometry. Therefore, self-calibration has become the preferred approach for estimating these intrinsic parameters. By eliminating GCP surveys and dedicated calibration processes, self-calibration using image-based SfM significantly reduces fieldwork time and cost, enabling rapid, cost-effective surveys even in remote or hazardous areas.

A widely employed image acquisition method in UAV-based photogrammetry is the constant-pitch (CP) flight [18,20–24], where the optical axes of images within the same strip are parallel. In this flight design (Figure 1), the camera is tilted at a fixed pitch angle along the flight lines, and when drones fly in the opposing direction, the local convergence is achieved, thereby facilitating 3D reconstruction.

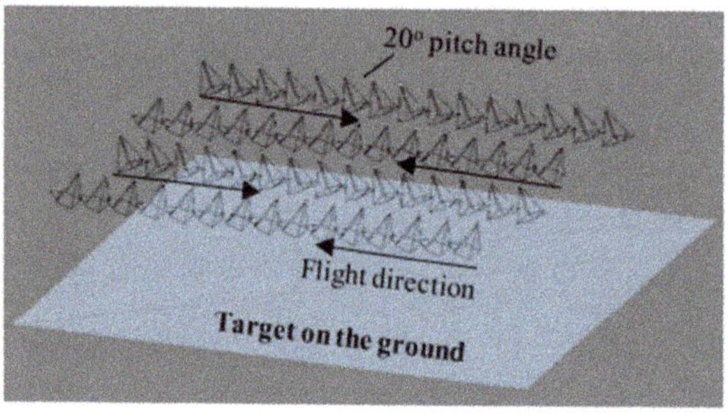

Figure 1. Constant-pitch flight at 20° pitch angle.

SfM using CP-based images has been reported to perform well when the camera positions measured by RTK-GNSS are incorporated in BA [21,23]. However, when camera position measurements are unavailable, referred to as image-based SfM, the performance of SfM becomes notably unstable. Sanz-Ablanedo et al. [18] reported poor image-based SfM performance using CP flight images captured at a 25° pitch angle, highlighting non-linear deformation (the dome effect). Their study also noted large mean vertical errors at validation points located at the four corners of the model, which can be caused by inaccuracy in the estimation of focal length (f). Similarly, Kon et al. [25] reported dome errors accompanied with large root mean square (RMS) errors of all validation points, and in some cases, alignment failures across various SfM analysis settings (1500 trials) for images captured by CP flights at a 20° pitch angle. For these 1500 SfM trials, the standard deviation of f and cy were notably large: 209 pixels and 84 pixels for the Phantom 4 RTK drone camera, and 53 pixels and 36 pixels for the SONY α7 III drone camera, respectively (unpublished). Observations of large mean vertical errors at validation points and alignment failures, coupled with the instability in estimations of f and cy, suggest that factors beyond the dome effect may contribute to the poor performance of image-based SfM. Despite these observations, the root causes of this instability have not been thoroughly investigated.

It has been mathematically shown that certain combinations of camera motion and scene geometry, known as critical configurations, can introduce indeterminacy into the bundle adjustment with self-calibration [26]. We suspected that the instability of camera parameters observed in image-based SfM under CP flights [21,23,27] stems from such critical configurations, leading to inaccurate or even incorrect model reconstructions.

Therefore, the objectives of this study are:

(a) To investigate the indeterminacy of two basic intrinsic parameters f and cy, which showed significant instabilities in the previous experiment as described above [25].

(b) To validate these numerical findings using real datasets acquired at three different ground sampling distances (GSDs).

(c) To discuss a practical mitigation strategy to stabilize intrinsic parameter estimation.

By bridging the theoretical–practical gap, we aim to provide a reliable UAV-SfM workflow without auxiliary information.

2. Materials and Methods

2.1. Fundamental of Self-Calibration in SfM

Suppose three 3D ground points P, T, and U that are visible in multiple images captured from different viewpoints (cameras A, B, and C), as shown in Figure 2. For the camera considered in this study, the geometric relationship between a 3D point and its 2D projection on each image is described by collinearity equations as follows:

$$\begin{cases} x = -f\frac{\alpha_{11}(X-X_0)+\alpha_{21}(Y-Y_0)+\alpha_{31}(Z-Z_0)}{\alpha_{13}(X-X_0)+\alpha_{23}(Y-Y_0)+\alpha_{33}(Z-Z_0)} \\ y - cy = -f\frac{\alpha_{12}(X-X_0)+\alpha_{22}(Y-Y_0)+\alpha_{32}(Z-Z_0)}{\alpha_{13}(X-X_0)+\alpha_{23}(Y-Y_0)+\alpha_{33}(Z-Z_0)} \end{cases} \quad (1)$$

where:

- (x, y): image coordinates of the projection of ground point.
- (X, Y, Z): object space coordinates of the ground point.
- (X_0, Y_0, Z_0): object space coordinates of the camera projection center.
- cy: y-coordinate of the principal point in the image.
- f: camera focal length.
- α_{1i}, α_{2i}, α_{3i} (i = 1, 2, 3): elements of the rotation matrix formed by three rotation angles.

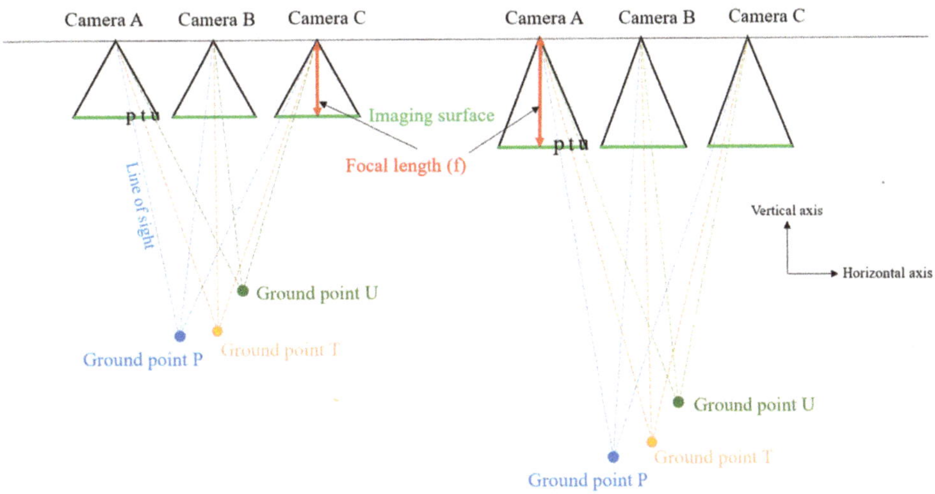

Figure 2. Example of a critical configuration where the focal length f cannot be determined from parallel shooting images. In this situation, multiple values of f can be estimated, making it impossible to uniquely determine the correct value.

The collinearity equations contain three categories of unknowns:

- Extrinsic parameters: parameters representing relative position and the orientation of each camera.
- Intrinsic parameters: parameters representing the geometric characteristics of the camera, such as focal length and principal points. Note that only two parameters f and cy are considered in this study and in Equation (1).
- Three-dimensional point coordinates: relative coordinates of scene points in object space.

These parameters are estimated simultaneously in the bundle adjustment (BA) of SfM by minimizing the overall reprojection error. The reprojection error of point i onto image j can be expressed as follows:

$$\begin{cases} F_{ij} := x_d - x_p = x_d + f\frac{\alpha_{11}(X-X_0)+\alpha_{21}(Y-Y_0)+\alpha_{31}(Z-Z_0)}{\alpha_{13}(X-X_0)+\alpha_{23}(Y-Y_0)+\alpha_{33}(Z-Z_0)} \\ G_{ij} := y_d - y_p = y_d + cy + f\frac{\alpha_{12}(X-X_0)+\alpha_{22}(Y-Y_0)+\alpha_{32}(Z-Z_0)}{\alpha_{13}(X-X_0)+\alpha_{23}(Y-Y_0)+\alpha_{33}(Z-Z_0)} \end{cases} \quad (2)$$

where (x_p, y_p) are the 2D projected coordinates computed via the collinearity equations and (x_d, y_d) are the 2D observed image coordinates of the point.

In SfM, for successful camera estimation, the objective function - root mean square (RMS) reprojection error - must have a unique global minimum. If multiple minima exist, the optimization becomes ambiguous, and SfM fails to produce reliable camera parameters. Parameters with multiple minimizing values are considered indeterminate. In such cases, a specific combination of camera motion and scene geometry that generates images impossible to unique estimation of certain intrinsic parameters is referred to as critical configuration (Figure 2).

2.2. Numerical Experiments

The indeterminacy of f and cy was investigated by analyzing RMS reprojection error, which is the objective function to be minimized at the BA stage in SfM. This was achieved by fixing either f or cy to incorrect values in BA and estimating the other parameters along with extrinsic parameters and tie points. If RMS reprojection error remains nearly unchanged regardless of the input values of f or cy, it indicates that these parameters do not influence the objective function, thereby demonstrating their indeterminacy in SfM.

(a) Synthetic Image Acquisition

We used Blender version 2.79b [28], an open-source computer graphic (CG) software program, to generate a flat plane with the dimensions of 400 m × 400 m and randomly place multiple rectangular parallelepiped objects of various sizes on it. Various random textures with a resolution of 5940 × 8410 pixels were applied to the plane and objects. These textures included local features suitable for matching in SfM, as shown in Figure 3. The details of the associated textures and the Blender project file are provided in the Supplementary Materials.

Figure 3. Two views of the target used for image acquisition in the numerical experiments: (**a**) side view; (**b**) front view.

The synthetic image acquisition of the target was performed using the image and camera specifications described in Table 1. In CG space, six image-sets were generated using six different flight designs, divided into two groups: (1) flight designs without an intermediate strip, including CP and CP random, and (2) flight designs with an intermediate strip such as CP flight plus one image in the intermediate strip, the CP flight plus three images randomly placed in the intermediate strip, CP-Plus flight, and CP random plus one image in the intermediate strip. The intermediate strip refers to a short flight path positioned between two primary strips, illustrated as green-blue arrows in Table 2. The pitch angle was set to 20° for all six flight designs, and the flight altitude was fixed at 73 m to achieve a Ground Sample Distance (GSD) of 40 mm. The details of flight designs are provided in Table 2. In this study, a 20° pitch angle was selected based on its proven effectiveness in enhancing UAV-SfM calibration from previous studies [18,24,29–31] and its confirmed robustness in SfM accuracy and precision across various analysis settings [21,25]. The indeterminacy of f and cy was examined with this best performing pitch angle, which serves as a representative case for indeterminacy under other pitch angles.

Table 1. Camera specifications used in CG space.

Parameters	Value
Pin hole camera	Linear camera with no distortion
Focal length (pixels)	1824
Image size (pixels)	2736 × 1824

Table 2. Descriptions of six image-sets taken by six flight designs in numerical experiments. Black-red arrows indicate image positions along the main flight strips, while green-blue arrows indicate those in the intermediate strips. The arrow size corresponds to the camera pitch angle. In this study, all CP flight designs used a fixed pitch angle of 20 degrees.

Flight Design	Description	Strip Interval/Images Interval in One Strip (Meters)/Number of Images	Illustration
CP flight	The camera is tilted at a fixed pitch angle along the flight lines, and when drones fly in the opposing direction, the local convergence is achieved.	20/20/121	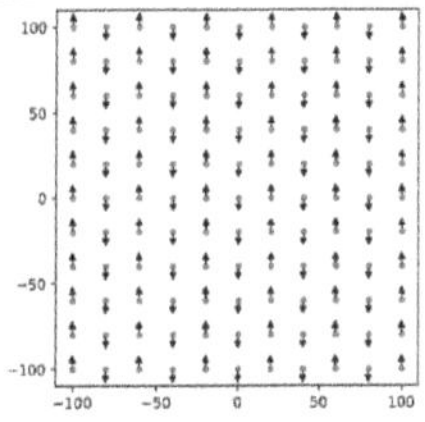
CP flight plus one image in the intermediate strip	Only one image with the same pitch angle was added to the middle of the intermediate strip (strip between two flight lines) of the CP flight.	20/20/122	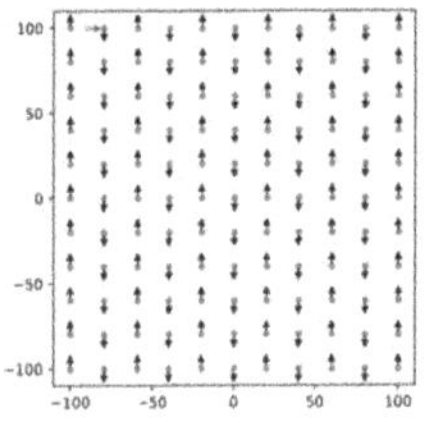

Table 2. *Cont.*

Flight Design	Description	Strip Interval/Images Interval in One Strip (Meters)/Number of Images	Illustration
CP flight plus three images randomly placed in the intermediate strip	Three images with the same pitch angle were randomly added to the middle of the intermediate strips of the CP flight.	20/20/124	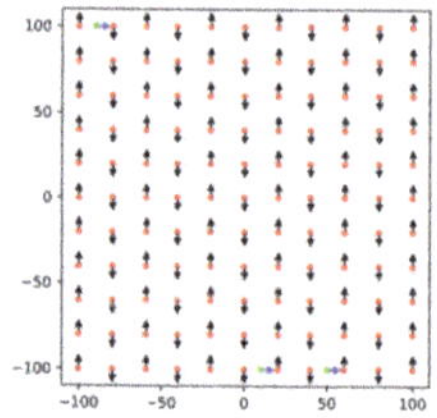
CP-Plus flight	One image with the same pitch angle was added to the middle of each intermediate strip of the CP flight.	20/20/131	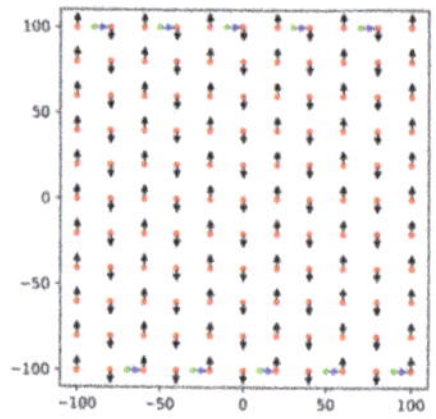
CP random flight	Cameras were randomly placed on the horizontal plane 73 m above the target plane. Each camera was oriented in one of the two orientations appearing in the CP flight.	Random/random/200	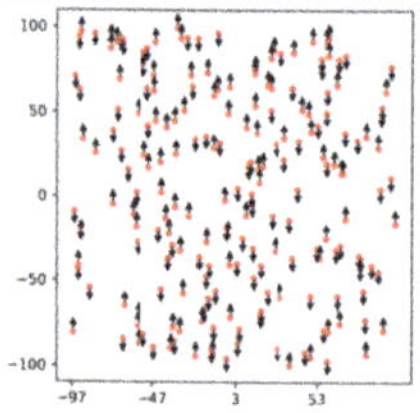
CP random plus one image in the intermediate strip	Only one image with the same pitch angle was added to the intermediate strip of the CP random flight.	Random/random/201	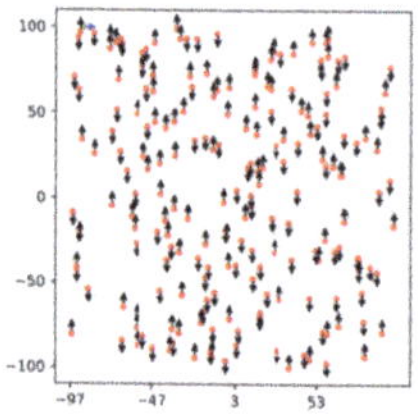

(b) SfM Processing

For each simulated image-set created in CG space, SfM was performed using the commercial software Agisoft Metashape version 1.5.5. The SfM workflow (Figure 4) began with "align photos", followed by bundle adjustment (BA) with the camera parameters effectively fixed to their true values to generate an accurate set of tie points. The settings used are detailed in Table 3.

Table 3. Settings for "Alignment" used to generate accurate tie points.

Parameters	Values
Input image size	High (original size) (2736 × 1824 pixels)
Key point limit	50,000
Intrinsic parameters estimation	Fixed to true value during BA

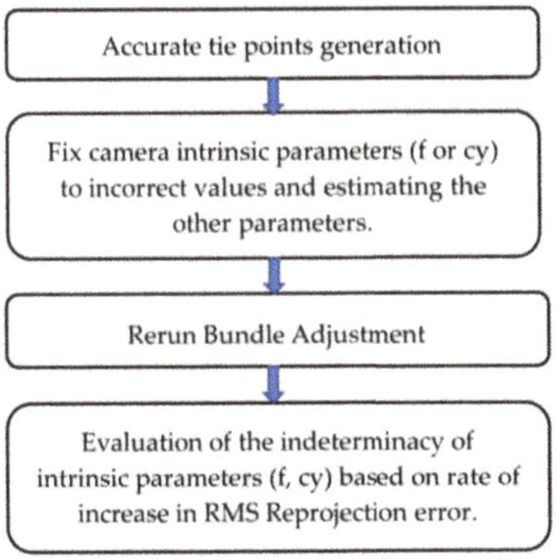

Figure 4. SfM flow in numerical experiments to evaluate the indeterminacy of each intrinsic parameter (f or cy).

Based on these tie points, an intrinsic camera parameter (either f or cy) was fixed to values higher than true value (v_+) and values lower than true values (v_-), while the other parameter (cy or f) was not fixed. Then, BA was rerun. In this process, BA was performed using the previously generated accurate tie points. This approach allowed us to observe the behavior of RMS reprojection error in relation to the fixed intrinsic parameters (f or cy).

(c) Evaluation of The Indeterminacy of Intrinsic Camera Parameters in SfM

To evaluate the indeterminacy of camera intrinsic parameters in SfM, we analyzed the variation in RMS reprojection error between two BA conditions: (1) BA with the intrinsic parameters (f and cy) fixed to their true values and (2) BA with f or cy fixed to incorrect values. The degree of variation was quantified by the rate of increase in RMS reprojection error, which serves as an indicator of how inaccurate intrinsic parameters influence the SfM and helps identify occurrences of their indeterminacy. In principle, the objective function minimized during BA - RMS reprojection error - must have a unique global minimum. The presence of multiple local minima, dependent on different fixed intrinsic parameter values, indicates that the SfM process is unable to uniquely determine these parameters. In such cases, the corresponding intrinsic parameters are considered as indeterminate.

Additionally, fixing either f or cy to incorrect values while estimating the remaining parameter (cy or f, respectively) provides insight into the correlation between f and cy.

The rate of increase in RMS reprojection error is calculated as follows:

$$R_{increase} = \frac{E_{incorrect} - E_{true}}{E_{true}} \quad (3)$$

where: $R_{increase}$ is the rate of increase in RMS reprojection error.

$E_{incorrect}$ (pixels) is RMS reprojection error when fixing either f or cy to the incorrect values and estimating the remaining parameter in BA.

E_{true} (pixels) is RMS reprojection error when fixing both f and cy to true values (1824 and 0 pixels, respectively) in BA.

2.3. Real-Data Analysis

(a) Study Site

The in situ data in this study utilized the data from Truc et al. [23]. These images captured an area of around 100 × 100 m^2, located in Tsukubamirai city, Ibaraki prefecture, Japan. This area was constructed for housing purposes. Figure 5 shows the ground condition of the target terrain, mostly bare land covered with short vegetation.

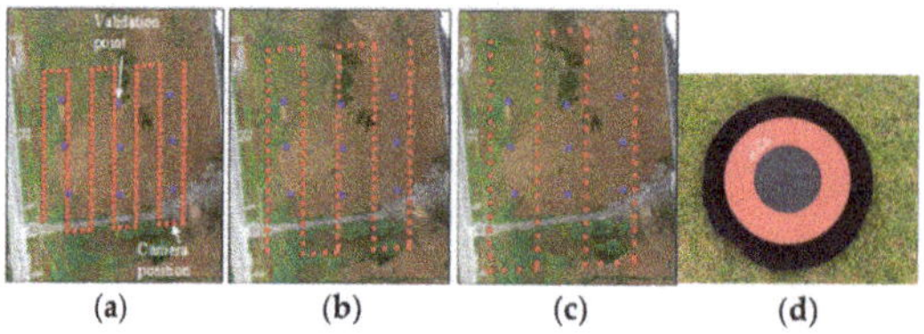

Figure 5. Distribution of camera positions and validation points at the study site overlaid with orthophotos generated by GSD10 (**a**), GSD15 (**b**), GSD20 (**c**). At each validation point, world coordinates were measured by Aerobo markers (**d**) with a built-in GNSS receiver.

(b) Image Acquisition

Aerial photos were collected using a high-precision 1-inch 20 MP CMOS camera sensor attached to a lightweight quad-copter UAV (DJI Phantom 4 RTK, hereafter named P4RTK) [32]. The Phantom 4 RTK used for image acquisition was manufactured by DJI, Shenzhen, Guangdong, China. The assembled camera lens has an approximate focal length of 8.8 mm and a mechanical shutter feature that avoids typical rolling shutter problems. This UAV can measure the camera position at the time of exposure by RTK-GNSS (real-time kinematic global navigation satellite system) positioning. A set of nine validation points evenly distributed across the study area (Figure 5) was employed to evaluate the accuracy of the SfM performance using the Aerobo marker [33]. The Aerobo markers used for ground validation were manufactured by Aerobo, Inc., Tokyo, Japan. Each Aerobo marker is equipped with an integrated GNSS receiver, enabling it to record the coordinates at the center of a 24 cm diameter circular target with sub-centimeter accuracy. The red-and-black reflective pattern of the marker is designed to enhance automatic detection by SfM processing software, such as Agisoft Metashape version 1.5.5, which was used in this study.

A flight design similar to the CP-Plus flight design was adopted for data acquisition. This flight pattern was generated using the 3D photogrammetry preset feature in the P4RTK system to capture images at two-second intervals and record camera coordinates via RTK-GNSS during image capture. In these CP-Plus flights, the camera was tilted at a fixed pitch angle of 20° along the flight lines, and the consecutive images with yaw angle deviation over 10° from the main flight line, taken during the drone rotation, were referred to as the intermediate strips. They were removed to simulate three CP flights (Figure 6).

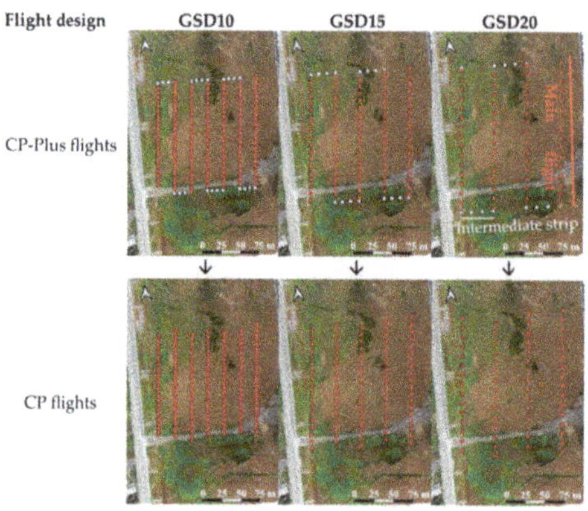

Figure 6. Generation of CP flights from CP-Plus flights used in real-data analysis (Red dots indicate image positions in the main flight strips, while white dots represent image positions in the intermediate strips.).

In this study, P4RTK was flown at three altitudes of 73, 55, and 36 m above ground level (at the drone's home point), which yielded three GSDs of 20, 15, and 10 mm, respectively. All photographs were captured along parallel flight lines with an overlap ratio of 80% forward and 60% side lap when the camera's optical axis was oriented toward nadir. The details of the image-sets used in the real-data analysis are described in Table 4.

Table 4. Summary of three CP-Plus flights acquired by P4RTK and three generated CP flights in real-data analysis.

Shooting Attitude [m]	Shooting Attitude [m]		Overlap Ratio [%]
	CP-Plus Flights	CP Flights	
73 (GSD20)	74	61	80% forward, 60% side lap.
55 (GSD15)	107	91	
36 (GSD10)	187	163	

(c) SfM Processing

For each image-set, SfM was performed following the "align photos" in Agisoft Metashape version 1.5.5 software. In this process, camera parameters were estimated by using images only, then the point clouds were georeferenced using projection centers recorded by an RTK-GNSS receiver embedded in the images taken by P4RTK. The absolute coordinates of the validation points evaluated by multi-view triangulation using camera parameters estimated in SfM were used to validate the accuracy of the SfM-derived model, in accordance with the nine validation points. It is worth noting that in this study, self-calibration was performed using images only, without additional information.

SfM was tested through a combination of SfM analysis settings summarized in Table 5, resulting in 30 settings with 50 trials for each image-set listed in Table 4. This analysis was conducted to observe the variation in the estimation of intrinsic parameters under different SfM settings.

Table 5. Summary of the analysis settings tested in this study.

Setting Items	Meaning	Setting Values
Aligned photo accuracy	Set the size of input images (shrinkage ratio) for feature extraction. • High: used original size. • Medium: image downscaled by a factor of two at each image side.	High, medium
Key point limit	Set the maximum number of feature points detected in each image.	1/2, 1/3, 1/4 of the average number of detectable key points for the input images, corresponding to 35,643; 23,762; 17,821 when the size of input images is set as high. 8229; 5486; 4115 when the size of input images is set as medium.
Tie point limit	Set the maximum number of tie points to be detected in each image.	0 It will attempt to detect as many tie points as possible in each image.

Table 5. *Cont.*

Setting Items	Meaning	Setting Values
Optimize camera alignment	Set the intrinsic parameters considered in Brown models.	① f, cx, cy, k1–k4, p1–p4, b1, b2 ② f, cx, cy, k1–k4, p1–p4 ③ f, cx, cy, k1–k4, p1–p2 ④ f, cx, cy, k1–k3, p1–p4 ⑤ f, cx, cy, k1–k3, p1–p2
2 × 3 × 5 = 30 analysis settings with 50 trials/setting, equivalent to 1500 trials for each image-set.		

In order to evaluate the variation of intrinsic parameters estimated in one SfM setting, we used a standard SfM setting repeated 50 times (trials), with the following parameters:

- Size of input images for feature detection: High
- Key point limit: a half (½) of the average number of detectable key points for the input images (35,643).
- Intrinsic parameters considered in Brown models: f, cx, cy, k1–k3, p1–p2.

(d) Intrinsic Parameters Estimation and Error Evaluation.

In this study, we quantified the variation of intrinsic parameter estimation (f and cy), in accordance with the accuracy and precision of SfM for each image-set. The accuracy and precision of SfM were evaluated using the RMS error of all nine validation points for each analysis setting, as recommended by James et al. [34]. The validation point error represents the difference between the estimated 3D coordinates of the validation points (calculated through triangulation using camera parameters estimated during the SfM process) and those obtained through static GNSS positioning by Aerobo markers. Additionally, statistical analyses were conducted on the total RMS error (Val_RMSE_T) and the vertical RMS error (Val_RMSE_Z) computed from all validation points for both a standard setting (with 50 trials) and 30 settings such as the maximum value, 75th percentile (Q3), 50th percentile (Q2), and interquartile range (IQR). For each of the 30 settings, the mean values of the estimated f, cy, Val_RMSE_T, and Val_RMSE_Z of all the validation points were first calculated over 50 repeated trials. These mean values were subsequently used to derive the statistical indicators (maximum, Q3, Q2, and IQR), thereby enabling a comparative assessment of SfM accuracy and precision across different settings. Moreover, we evaluated the mean vertical error of all nine validation points (MEZ) for each analysis setting. The mean vertical error is defined as the average difference between the estimated elevation of the validation points (derived using camera parameters estimated by SfM) and the measured elevation of corresponding validation points obtained by Aerobo markers.

3. Results

The results are divided into two main sections: numerical experiments and real-data analysis. In the numerical experiments, we present the verification of the indeterminacy of intrinsic parameters in image-based SfM for six flight designs. The real data analysis focuses on three points: the estimates of camera intrinsic parameters (f, cy) across multiple repetitions of a single SfM setting and in various SfM settings, and the correlation between the estimates of f and the mean vertical error (MEZ) for all validation points.

3.1. Numerical Experiments

This section verifies the indeterminacy of intrinsic parameters in image-based SfM by analyzing six different flight designs. The RMS reprojection error and its rate of increase were analyzed to evaluate the sensitivity of SfM when the intrinsic parameters were fixed to inaccurate values relative to their true values (Figures 7 and 8).

Fixed value setting (pixels)			CP flight		CP flight plus one image in intermediate strip		CP flight plus three images randomly placed in intermediate strip		CP-Plus flight		CP random flight		CP random plus one image in intermediate strip	
			$E_{incorrect}$ (pixels)	Estimated cy (pixels)	$E_{incorrect}$ (pixels)	Estimated cy (pixels)	$E_{incorrect}$ (pixels)	Estimated cy (pixels)	$E_{incorrect}$ (pixels)	Estimated cy (pixels)	$E_{incorrect}$ (pixels)	Estimated cy (pixels)	$E_{incorrect}$ (pixels)	Estimated cy (pixels)
f	v-	1641.6	0.06	141.20	0.21	135.81	0.33	126.91	0.47	111.38	0.06	141.20	0.17	137.51
	v+	2006.4	0.06	-162.12	0.25	-156.29	0.38	-139.35	0.51	-120.56	0.06	-162.12	0.20	-159.63
				Estimated f		Estimated f		Estimated f		Estimated f		Estimated f		Estimated f
cy	v-	-100	0.06	1,936.95	0.16	1,939.06	0.26	1,944.43	0.40	1,956.52	0.06	1,936.95	0.13	1,938.35
	v+	100	0.06	1,697.72	0.16	1,695.96	0.26	1,691.68	0.40	1,680.83	0.06	1,697.72	0.13	1,696.56
			E_{true} (pixels)		E_{true} (pixels)		E_{true} (pixels)		E_{true} (pixels)		E_{true} (pixels)		E_{true} (pixels)	
f	True value	1824	0.06		0.06		0.06		0.06		0.06		0.06	
cy	True value	0												

Figure 7. RMS Reprojection Error and Estimated Parameter Values When Fixing Parameter f or cy to Incorrect Values for Six Flight Designs. $E_{incorrect}$: RMS reprojection error when fixing either f or cy to the incorrect values and estimating the remaining parameter in BA. E_{true}: RMS reprojection error when fixing both f and cy to true values (1824 and 0 pixels, respectively) in BA. Red and blue indicate the estimated intrinsic parameters (f and cy), while green represents the RMS reprojection error (in pixels). The bar length is scaled relative to the maximum value within each group.

Fixed value setting (pixels)			Rate of increase in RMS reprojection error ($R_{increase}$)					
			CP flight	CP flight plus one image in intermediate strip	CP flight plus three images randomly placed in intermediate strip	CP-Plus flight	CP random flight	CP random plus one image in intermediate strip
f	v-	1641.6	0.00	2.66	4.71	6.98	0.00	1.81
	v+	2006.4	0.00	3.30	5.51	7.59	0.00	2.28
cy	v-	-100	0.00	1.79	3.48	5.77	0.00	1.17
	v+	100	0.00	1.76	3.42	5.76	0.00	1.15

Figure 8. Rate of Increase in RMS Reprojection Error When Fixing Parameter f or cy to Incorrect Values for Six Flight Designs. Green represents the rate of increase in RMS reprojection error ($R_{increase}$). The bar length is scaled relative to the maximum value in the table.

As shown in Figure 8, the rate of increase in RMS reprojection error was zero for images captured by flight designs without image in the intermediate strip including the CP flight and CP random flight. In contrast, for flight designs with images in the intermediate trip, including the CP flight plus one image in intermediate strip, CP flight plus three images randomly placed in intermediate strip, CP-Plus flight, and CP random plus one image in intermediate strip, the rate of increase in RMS reprojection error ranged from 1.1 to 7.6.

In CP and CP random flight, the RMS reprojection error remained unchanged regardless of whether the intrinsic parameters f or cy were fixed to true or incorrect values in BA. However, in flight designs with images in the intermediate strip, the addition of one image between two flight lines resulted in a noticeable increase in RMS reprojection error. Moreover, when f was fixed to incorrect values, the estimates of cy derived from BA were strongly inaccurate compared to the true value (0), and vice versa (Figure 7).

These results demonstrate that both f and cy are indeterminate in image-based SfM. In CP random flights, where camera positions were randomly distributed compared to CP flights, and the camera orientation remained fixed at 20° forward and backward, the RMS reprojection error remained unchanged regardless of whether the intrinsic parameters were fixed to incorrect or true values. This indicates that altering the shooting positions alone does not resolve the indeterminacy.

In contrast, for flight patterns with at least one image in the intermediate trip, the values of f or cy had a substantial impact on RMS reprojection error, with rates of increase in RMS reprojection error exceeding 7.6 for f and 5.8 for cy (Figure 8). This demonstrates that f and cy become determinable in image-based SfM when at least one image is added to the intermediate strip.

In summary, the numerical experiments showed that the intrinsic parameters f and cy are indeterminate in image-based SfM when using aerial images captured by flight designs without an intermediate strip such as the CP flight design. Adding images to the intermediate short strips effectively resolves the indeterminacy of f and cy.

In the next section, we will evaluate the estimation of intrinsic parameters in image-based SfM using in situ data captured by CP and CP-Plus flights.

3.2. *Real-Data Analysis*

3.2.1. Estimates of Intrinsic Camera Parameters (f, cy) Across 50 Trials of a Single SfM Setting

Figure 9 presents the results of a standard SfM setting for six image-sets captured at three different GSDs using two flight designs: the CP and CP-Plus. Across 50 repeated trials of a standard SfM setting, the standard deviation in the estimates of intrinsic parameters f and cy was zero for the image-sets taken by the CP-Plus flight design, indicating a highly stable estimation. In contrast, image-sets from the CP flight showed substantial variability with a standard deviation of 0.32 pixels for f and 0.08 pixels for cy. This instability also resulted in a 0.008 m and 0.009 m deviation in the interquartile range (IQR) of total RMS error (Val_RMSE_T) and vertical RMS error (Val_RMSE_Z), respectively.

	Image-sets	Standard deviation of f [pix]	Standard deviation of cy [pix]	50 percentile		75 percentile		Maximum		IQR	
				Val_RMSE_Z	Val_RMSE_T	Val_RMSE_Z	Val_RMSE_T	Val_RMSE_Z	Val_RMSE_T	Val_RMSE_Z	Val_RMSE_T
GSD20	CP flight	0.29	0.08	0.14	0.17	0.14	0.17	0.17	0.19	0.009	0.008
	CP-Plus flight	0.00	0.00	0.12	0.13	0.12	0.13	0.12	0.13	0.000	0.000
GSD15	CP flight	0.32	0.07	0.08	0.12	0.09	0.12	0.10	0.13	0.007	0.007
	CP-Plus flight	0.00	0.00	0.09	0.10	0.09	0.10	0.09	0.10	0.000	0.000
GSD10	CP flight	0.20	0.07	0.08	0.12	0.08	0.13	0.08	0.13	0.004	0.006
	CP-Plus flight	0.07	0.05	0.05	0.07	0.06	0.07	0.06	0.08	0.002	0.003

Figure 9. Intrinsic Parameter Estimation and Statistics of Total RMS Error (Val_RMSE_T) and Vertical RMS Error (Val_RMSE_Z) of All Validation Points for Each Image-Set (One SfM Analysis Setting with 50 Trials). Dark blue represents the standard deviation of the estimated intrinsic parameters (f and cy) in pixels, calculated from 50 trials of one SfM analysis settings. Green and bright blue represent Val_RMSE_Z and Val_RMSE_T, respectively, at the 50th, 75th, and maximum values, calculated from 50 trials of one SfM analysis settings. Orange indicates the interquartile range (IQR) of Val_RMSE_Z and Val_RMSE_T. In all cases, the bar length is scaled relative to the maximum value within each group.

These results demonstrate that the estimation of f and cy is unstable in image-based SfM when using images from the CP flight. However, in the CP-Plus flight, where images were added to the intermediate strips of the CP flight, the estimates of the intrinsic parameters f and cy became more stable, even within a single SfM setting.

3.2.2. Estimates of Intrinsic Camera Parameters (f and cy) Across 30 SfM Settings

The estimates of f and cy in image-based SfM obtained from 30 different SfM settings, each repeated 50 times, exhibited instability (Figure 10). The standard deviation of f ranged from over 1 pixel to 7.3 pixels, while the standard deviation of cy ranged from 0.6 to 3.1 pixels. This instability was particularly pronounced when using images from the CP flight, where the standard deviation was about six times higher for f and five times higher for cy (GSD20 image-sets) compared to those taken by the CP-Plus flight.

As shown in Figure 10, the RMS error of all the validation points was substantially higher for image-sets from the CP flight compared to those from the CP-Plus flight. For instance, in GSD15 image-set, the median RMS error of all the validation points (Val_RMSE_T) was 0.151 m, nearly four times greater than the 0.039 m observed for the CP-Plus flight. Similarly, the median vertical RMS error (Val_RMSE_Z) was approximately five times larger for the CP flight (0.139 m) compared to that for the CP-Plus flight (0.027 m). Furthermore, the accuracy and precision of SfM varied considerably across the 30 SfM settings for the CP flight image-sets. The interquartile range (IQR) of the RMS error of all validation points was

0.105 m for CP flights, compared to 0.047 m for CP-Plus flights (GSD15). Similar patterns were observed for image-sets from GSD10 and GSD20.

Image-sets		Standard deviation of f [pix]	Standard deviation of cy [pix]	50 percentile		75 percentile		Maximum		IQR	
				Val_RMSE_Z	Val_RMSE_T	Val_RMSE_Z	Val_RMSE_T	Val_RMSE_Z	Val_RMSE_T	Val_RMSE_Z	Val_RMSE_T
GSD20	CP flight	7.290	3.108	0.078	0.089	0.167	0.186	0.739	0.757	0.132	0.131
	CP-Plus flight	1.252	0.627	0.050	0.059	0.102	0.109	0.138	0.144	0.078	0.058
GSD15	CP flight	6.244	2.383	0.139	0.151	0.183	0.205	0.426	0.453	0.110	0.105
	CP-Plus flight	1.267	0.689	0.027	0.039	0.066	0.076	0.102	0.117	0.054	0.047
GSD10	CP flight	3.708	2.357	0.055	0.085	0.099	0.157	0.275	0.330	0.069	0.102
	CP-Plus flight	1.022	0.860	0.026	0.046	0.043	0.066	0.055	0.073	0.029	0.031

Figure 10. Intrinsic Parameter Estimation and Statistics of Total RMS Error (Val_RMSE_T) and Vertical RMS Error (Val_RMSE_Z) of All Validation Points for Each Image-Set (30 Analysis Settings). Dark blue represents the standard deviation of the estimated intrinsic parameters (f and cy) in pixels, calculated from 30 SfM analysis settings. Green and bright blue represent Val_RMSE_Z and Val_RMSE_T, respectively, at the 50th, 75th, and maximum values, calculated from 30 SfM analysis settings. Orange indicates the interquartile range (IQR) of Val_RMSE_Z and Val_RMSE_T. In all cases, the bar length is scaled relative to the maximum value within each group.

Overall, the estimates of f and cy were significantly less stable for images captured by CP flights than CP-Plus flights, both across multiple trials of a single setting and across different settings.

3.2.3. Correlation Between the Estimates of f and the Mean Vertical Error for All Validation Points

Theoretically, incorrect estimates of f lead to mean vertical error (MEZ) for all validation points in model reconstruction. As illustrated in Figure 11, a negative correlation was observed between the estimates of f and MEZ across 30 different SfM settings. This relationship can be explained by a simple pin hole camera model: as the estimated f increases, the triangulated ground point moves farther from the camera, causing its estimated vertical coordinate to decrease.

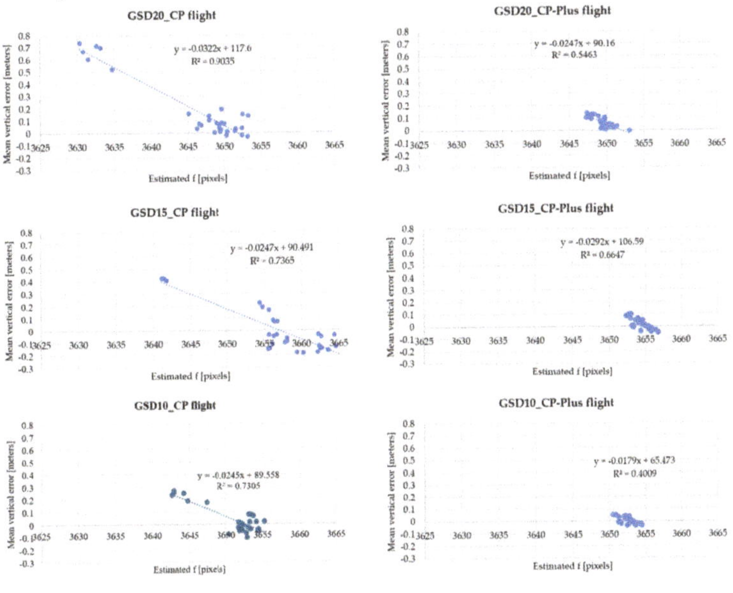

Figure 11. The correlation between estimated f and mean vertical error (MEZ).

To assess the contribution of vertical error due to incorrect f estimation to the total vertical error, the vertical error ratio was calculated as the proportion of the absolute mean vertical error (MEZ) to the total vertical error of all validation points (Val_RMSE_Z). As shown in Table 6, the average vertical error ratio for each image-set exceeded 80%, indicating that vertical error resulting from incorrect f estimation served as the primary contributor to the total vertical error, regardless of flight design.

Table 6. Average of Vertical Error Ratio for Each Image-Set (30 SfM Analysis Settings).

Image Sets	Average of Vertical Error Ratio (%)	
	CP–Plus Flight	**CP Flight**
GSD10	80.65	84.09
GSD15	80.71	97.84
GSD20	88.60	88.37

In summary, the indeterminacy of intrinsic parameters, especially f and cy, in image-based SfM has been demonstrated using six flight designs in numerical experiments, and its effect has been demonstrated in a real-data analysis using six image-sets. In the next section, we will discuss the implications of these findings on existing studies and suggestions for practical application.

4. Discussion

4.1. Indeterminacy of Intrinsic Camera Parameters in Image-Based SfM

To our knowledge, this study provides the first combined numerical and in situ analysis of the indeterminacy of intrinsic parameters in image-based SfM for CP flights. The numerical experiments show that when images are captured under a CP flight design, f and cy are indeterminable in image-based SfM.

Critical configurations - specific combinations of camera motion and scene geometry - are known to induce intrinsic parameter indeterminacy. Previous studies have extensively documented critical configurations related to linear images (images captured by a pin hole camera without lens distortion) [26,35]. For example, planar motion over general terrain, in which the camera moves on a plane while rotating around an axis normal to that plane, prevents the estimation of a single intrinsic parameter [26].

CP flight represents a restrictive form of planar motion in which the camera's movement is constrained into a plane, and rotation is limited to only two orientations (pitch and yaw) rather than allowing free rotation around the roll axis. Therefore, it is possible that these additional constraints cause more than one intrinsic parameter to become indeterminate, and in fact, our study confirmed that two parameters (f and cy) became indeterminate for CP flights.

In real CP flight datasets acquired at three different GSDs, we observed significant variations in f and cy across 30 distinct SfM analysis settings and across 50 repeat trials of a single setting. These results demonstrated that the theoretical indeterminacy of intrinsic parameters observed in numerical experiments under ideal conditions manifested in real-world scenarios as pronounced instability.

Our findings suggest that the previously reported poor performance of image-based SfM using CP flight images [18,25] - commonly attributed to the dome effect - is also influenced by the inaccurate estimation of f, which causes large mean vertical errors. Images acquired under CP flight, as confirmed in our study, significantly compromise the accuracy and precision of image-based SfM and can even lead to an incorrect model reconstruction in MVS. Therefore, practical surveys should avoid employing CP flight design or apply appropriate countermeasures such as augmenting camera coordinates measured by RTK-GNSS or adding many GCPs to ensure stable SfM results when using CP flights.

4.2. Remedy for Indeterminacy of Intrinsic Parameters in Image-Based SfM

Our numerical experiments demonstrate that adjustments to the shooting positions of CP flights, such as a CP random flight, do not resolve the indeterminacy of intrinsic parameters (f and cy). However, introducing intermediate short strips between primary CP lines, as in the CP-Plus flight, effectively eliminated this indeterminacy in numerical experiments and substantially reduced f and cy instability in the real-data analysis. This result aligns with the known fact in aerial triangulation that introducing "cross strips" to a block can improve the accuracy of block adjustment [36,37].

Our finding suggests that practitioners can largely mitigate concerns regarding the significant instability of intrinsic parameters estimated by image-based SfM by incorporating intermediate short strips into CP flights. Some built-in flight missions in flight planning software packages automatically include intermediate strips. Such strips can be achieved using timed-interval or distance-interval shot modes, as seen in the 3D photogrammetry function of the DJI P4RTK drone or the "area route" option of the DJI Mavic 3E drone, for example.

In contrast, there are other flight missions that do not include intermediate strips, such as "waypoint" flight, where the camera shutter is triggered at specific predefined points. Additionally, some flight planning software packages, such as Pix4D Capture [18,24,29,38] or Pix4D Capture Pro, do not generate intermediate strips by default. These choices preserve the CP critical configuration and risk that the intrinsic parameter estimation will be unstable. Overall, embedding short cross-strips within CP flight plans prevents theoretical indeterminacy of intrinsic parameters, delivering more stable and accurate SfM reconstructions in practice without a reliance on camera coordinates measured by RTK-GNSS or extensive GCP networks.

4.3. Implications for Common Flight Configurations

The commonly used nadir-viewing flight represents a special case of critical configuration, where the pitch angle is zero. Therefore, our findings suggest that nadir-viewing flights are also unable to determine f and cy.

Although not explicitly analyzed in this study, the results of the CP-Plus flight imply that images captured by a "double grid" flight—commonly generated by many flight planning software packages and built-in drone missions—are expected to support the determination of f and cy. In this "double grid" flight, the images captured in the reverse direction function as intermediate strips, thereby mitigating the indeterminacy of f and cy.

5. Conclusions

In this study, we investigated how certain intrinsic camera parameters in image-based SfM become indeterminate using aerial photos from CP flights. Our numerical experiments (Section 3.1) confirmed that the CP flight is a critical configuration that causes the indeterminacy of intrinsic parameters, particularly focal length (f) and principal point coordinate (cy), under idealized conditions in CG simulations. Further analyses with real data (Section 3.2), involving three image-sets over 30 different SfM settings, demonstrated that estimates of f and cy are highly unstable when using CP flight imagery.

As a countermeasure, we found that integrating intermediate short strips into the CP flight design eliminates the indeterminacy of intrinsic parameters (f and cy) in numerical experiments and markedly improves the stability of these parameters in real-data analyses.

Our study reveals the potential to improve the accuracy of self-calibration using only aerial images in SfM, without the need for GCPs or RTK-GNSS data. This advancement not only simplifies the data acquisition process but also makes it more accessible and cost-effective for practitioners. By stabilizing SfM calibration, all the subsequent stages—such as MVS and dense point cloud generation—can be effectively enhanced, leading to more accurate 3D reconstructions.

This study initiated the investigation of the indeterminacy from two basic intrinsic parameters, f and cy. Investigations of other intrinsic parameters, including the principal point in the x-direction (cx) and nonlinear distortion coefficients, need to be considered in future research. Additionally, beyond CP flight designs, various other combinations of camera motion and scene geometry may form critical configurations and highlight the need for further exploration.

Supplementary Materials: The following supporting information can be downloaded at: https://www.mdpi.com/article/10.3390/rs17122030/s1.

Author Contributions: Conceptualization: A.K.; methodology and software: T.T.H. and R.S.; formal analysis: T.T.H. and R.S.; writing—original draft preparation: T.T.H.; writing—review and editing: A.K.; T.I.; K.Y. and T.H.; visualization: T.T.H.; supervision: A.K. All authors have read and agreed to the published version of the manuscript.

Funding: This research was partially funded by JSPS KAKENHI Grant Number 22K04333.

Data Availability Statement: The original contributions presented in this study are included in the article and Supplementary Materials. Further inquiries can be directed to the corresponding author.

Acknowledgments: The in situ data (photos and ground surveying data) was obtained by Fujita Corporation, Japan.

Conflicts of Interest: The authors declare no conflicts of interest.

Abbreviations

The following abbreviations are used in this manuscript:

BA	Bundle Adjustment
CP	Constant Pitch
DEM	Digital Elevation Model
GCP	Ground Control Point
GNSS	Global Navigation Satellite System
MEZ	Mean vertical error of all validation points
MVS	Multi-View Stereo
RMS Error	Root Mean Square Error
RTK-GNSS	Real-Time Kinematic Global Navigation Satellite System
SfM	Structure from Motion
UAV	Unmanned Aerial Vehicle
Val_RMSE_T	Total RMS error of all validation points
Val_RMSE_Z	Vertical RMS error of all validation points

References

1. Xu, Y.; Zhang, J. UAV-Based Bridge Geometric Shape Measurement Using Automatic Bridge Component Detection and Distributed Multi-View Reconstruction. *Autom. Constr.* **2022**, 104376. [CrossRef]
2. Wang, F.; Zou, Y.; del Rey Castillo, E.; Ding, Y.; Xu, Z.; Zhao, H.; Lim, J.B.P. Automated UAV Path-Planning for High-Quality Photogrammetric 3D Bridge Reconstruction. *Struct. Infrastruct. Eng.* **2024**, *20*, 1595–1614. [CrossRef]
3. Khaloo, A.; Lattanzi, D.; Cunningham, K.; Dell'Andrea, R.; Riley, M. Unmanned Aerial Vehicle Inspection of the Placer River Trail Bridge through Image-Based 3D Modelling. *Struct. Infrastruct. Eng.* **2018**, *14*, 124–136. [CrossRef]

4. Siqueira, H.L.; Marcato, J.; Matsubara, E.T.; Eltner, A.; Colares, R.A.; Santos, F.M. The Impact of Ground Control Point Quantity on Area and Volume Measurements with UAV SFM Photogrammetry Applied in Open Pit Mines. In Proceedings of the IGARSS 2019 IEEE International Geoscience and Remote Sensing Symposium, Yokohama, Japan, 28 July 2019; pp. 9093–9096.
5. Hao, J.; Zhang, X.; Wang, C.; Wang, H.; Wang, H. Application of UAV Digital Photogrammetry in Geological Investigation and Stability Evaluation of High-Steep Mine Rock Slope. *Drones* **2023**, *7*, 198. [CrossRef]
6. Dawei, Z.; Lizhuang, Q.; Demin, Z.; Baohui, Z.; Lianglin, G. Unmanned Aerial Vehicle (UAV) Photogrammetry Technology for Dynamic Mining Subsidence Monitoring and Parameter Inversion: A Case Study in China. *IEEE Access* **2020**, *8*, 16372–16386. [CrossRef]
7. Bonali, F.L.; Corti, N.; Russo, E.; Marchese, F.; Fallati, L.; Pasquaré Mariotto, F.; Tibaldi, A. Commercial-UAV-Based Structure from Motion for Geological and Geohazard Studies. In *Building Knowledge for Geohazard Assessment and Management in the Caucasus and Other Orogenic Regions*; Springer: Dordrecht, The Netherlands, 2021; pp. 389–427.
8. Yakar, M.; Ulvi, A.; Yiğit, A.Y.; Alptekin, A. Discontinuity Set Extraction from 3D Point Clouds Obtained by UAV Photogrammetry in a Rockfall Site. *Surv. Rev.* **2023**, *55*, 416–428. [CrossRef]
9. Butcher, B.; Walton, G.; Kromer, R.; Gonzales, E.; Ticona, J.; Minaya, A. High-Temporal-Resolution Rock Slope Monitoring Using Terrestrial Structure-from-Motion Photogrammetry in an Application with Spatial Resolution Limitations. *Remote Sens.* **2023**, *16*, 66. [CrossRef]
10. Saito, T.; Kawai, T.; Takahara, Y.; Nagamatsu, D.; Inosako, K. Sand Movement Analysis of the Tottori Sand Dunes Using Aerial Photogrammetry by UAVs. *J. Arid. Land Stud.* **2024**, *34*, 87–91. [CrossRef]
11. Laporte-Fauret, Q.; Marieu, V.; Castelle, B.; Michalet, R.; Bujan, S.; Rosebery, D. Low-Cost UAV for High-Resolution and Large-Scale Coastal Dune Change Monitoring Using Photogrammetry. *J. Mar. Sci. Eng.* **2019**, *7*, 63. [CrossRef]
12. Fraser, C.S.; Al-Ajlouni, S. Zoom-Dependent Camera Calibration in Digital Close-Range Photogrammetry. *Photogramm. Eng. Remote Sens.* **2006**, *72*, 1017–1026. [CrossRef]
13. Fraser, C.S. Photogrammetry and Remote Sensing Digital Camera Self-Calibration. *ISPRS J. Photogramm. Remote Sens.* **1997**, *52*, 149–159. [CrossRef]
14. Abeho, D.R.; Shoko, M.; Odera, P.A. Effects of Camera Calibration on the Accuracy of Unmanned Aerial Vehicle Sensor Products. *Int. J. Eng. Geosci.* **2024**, *9*, 314–323. [CrossRef]
15. Senn, J.A.; Mills, J.; Walsh, C.L.; Addy, S.; Peppa, M. On-site Geometric Calibration of RPAS Mounted Sensors for SfM Photogrammetric Geomorphological Surveys. *Earth Surf. Process Landf.* **2022**, *47*, 1615–1634. [CrossRef]
16. Daakir, M.; Zhou, Y.; Pierrot Deseilligny, M.; Thom, C.; Martin, O.; Rupnik, E. Improvement of Photogrammetric Accuracy by Modeling and Correcting the Thermal Effect on Camera Calibration. *ISPRS J. Photogramm. Remote Sens.* **2019**, *148*, 142–155. [CrossRef]
17. Elias, M.; Eltner, A.; Liebold, F.; Maas, H.-G. Assessing the Influence of Temperature Changes on the Geometric Stability of Smartphone- and Raspberry Pi Cameras. *Sensors* **2020**, *20*, 643. [CrossRef]
18. Sanz-Ablanedo, E.; Chandler, J.H.; Ballesteros-Pérez, P.; Rodríguez-Pérez, J.R. Reducing Systematic Dome Errors in Digital Elevation Models through Better UAV Flight Design. *Earth Surf. Process Landf.* **2020**, *45*, 2134–2147. [CrossRef]
19. Smith, M.J.; Cope, E. The Effects of Temperature Variation on Single-Lens-Reflex Digital Camera Calibration Parameters. In Proceedings of the International Archives of Photogrammetry, Remote Sensing and Spatial Information Sciences, Commission V Symposium, Part 5, Newcastle upon Tyne, UK, 21–24 June 2010; Volume 38, pp. 554–559.
20. Luo, W.; Shao, M.; Che, X.; Hesp, P.A.; Bryant, R.G.; Yan, C.; Xing, Z. Optimization of UAVs-SfM Data Collection in Aeolian Landform Morphodynamics: A Case Study from the Gonghe Basin, China. *Earth Surf. Process Landf.* **2020**, *45*, 3293–3312. [CrossRef]
21. Kanno, A.; Yuji, M.; Otsu, Y.; Sugimoto, K.; Takada, M. Effect of Constant Pitch Meandering Flight Method and Analysis Settings in SfM without GCPs. In Proceedings of the Annual Conference of the Japan Society of Photogrammetry and Remote Sensing, 20–21 May 2021; pp. 49–52. Available online: https://ds.cc.yamaguchi-u.ac.jp/~kanno/pdf/JSPRS_2021_1/paper_JSPRS_2021_1.pdf (accessed on 7 June 2025). (In Japanese)
22. Ariyo, K.; Junnosuke, K.; Truc, T.H.; Masaki, H.; Tatsuji, F.; Kyouhei, K. Accuracy Verification Experiment of UAV Photogrammetry Using Optical Axis Optimization Shooting with Airpeak S1 + A7III. In Proceedings of the Fall Conference of the Japan Society of Photogrammetry and Remote Sensing, 10–11 November 2022; pp. 41–42. Available online: https://ds.cc.yamaguchi-u.ac.jp/~kanno/pdf/JSPRS_2022_2/proceedings_JSPRS22_C-5.pdf (accessed on 7 June 2025). (In Japanese).
23. Truc, T.H.; Ariyo, K.; Yuji, M.; Masahiko, S.; Tsuyoshi, I.; Koichi, Y.; Takaya, H. Robustness of Structure from Motion Accuracy/Precision against the Non-Optimality in Analysis Settings: Case Study in Constant-Pitch Flight Design. *Int. J. Autom. Technol.* **2024**, *18*, 621–631.
24. Nesbit, P.R.; Hugenholtz, C.H. Enhancing UAV-SfM 3D Model Accuracy in High-Relief Landscapes by Incorporating Oblique Images. *Remote Sens.* **2019**, *11*, 239. [CrossRef]

25. Kon, J.; Kanno, A.; Haruna, M.; Fujii, T.; Kurita, K. The Robustness of SfM Settings and Vulnerability to Dome-Shaped Deformation in Four Types of UAV Oblique Photography Methods. In Proceedings of the Annual Conference of the Japan Society of Photogrammetry and Remote Sensing, 18–19 May 2023; pp. 37–38. Available online: https://jglobal.jst.go.jp/detail?JGLOBAL_ID=202302237161111141 (accessed on 7 June 2025). (In Japanese)
26. Hartley, R.; Zisserman, A. *Multiple View Geometry in Computer Vision*, 2nd ed.; Cambridge University Press: Cambridge, UK, 2004; ISBN 9780521540513.
27. Štroner, M.; Urban, R.; Seidl, J.; Reindl, T.; Brouček, J. Photogrammetry Using UAV-Mounted GNSS RTK: Georeferencing Strategies without GCPs. *Remote Sens.* **2021**, *13*, 1336. [CrossRef]
28. Blender Foundation Blender 2.79 Release Notes. Available online: https://developer.blender.org/docs/release_notes/2.79/ (accessed on 7 June 2025).
29. Ahmed, S.; El-Shazly, A.; Abed, F.; Ahmed, W. The Influence of Flight Direction and Camera Orientation on the Quality Products of UAV-Based SfM-Photogrammetry. *Appl. Sci.* **2022**, *12*, 10492. [CrossRef]
30. James, M.R.; Robson, S. Mitigating Systematic Error in Topographic Models Derived from UAV and Ground-Based Image Networks. *Earth Surf. Process. Landf.* **2014**, *39*, 1413–1420. [CrossRef]
31. Dai, W.; Zheng, G.; Antoniazza, G.; Zhao, F.; Chen, K.; Lu, W.; Lane, S.N. Improving UAV-SfM Photogrammetry for Modelling High-relief Terrain: Image Collection Strategies and Ground Control Quantity. *Earth Surf. Process. Landf.* **2023**, *48*, 2884–2899. [CrossRef]
32. DJI Enterprise Phantom 4 RTK. Available online: https://www.dji.com/jp/support/product/phantom-4-rtk (accessed on 19 November 2024).
33. Aerosense Inc. Aerobo Marker. Available online: https://aerosense.co.jp/en/products/equipment_systems/ (accessed on 19 November 2024).
34. James, M.R.; Chandler, J.H.; Eltner, A.; Fraser, C.; Miller, P.E.; Mills, J.P.; Noble, T.; Robson, S.; Lane, S.N. Guidelines on the Use of Structure-from-Motion Photogrammetry in Geomorphic Research. *Earth Surf. Process. Landf.* **2019**, *44*, 2081–2084. [CrossRef]
35. Kahl, F.; Triggs, B.; Åström, K. Critical Motions for Auto-Calibration When Some Intrinsic Parameters Can Vary. *J. Math. Imaging. Vis.* **2000**, *13*, 131–146. [CrossRef]
36. Ackermann, F.; Schade, H. Application of GPS for Aerial Triangulation. *Photogramm. Eng. Remote Sens.* **1993**, *59*, 1625–1632.
37. Forstner, W. The Reliability of Block Triangulation. *Photogramm. Eng. Remote Sens.* **1985**, *51*, 1137–1149.
38. Iheaturu, C.; Okolie, C.; Ayodele, E.; Stanley, E.A.; Musa, S.; Ifejika Speranza, C.I. A Simplified Structure-from-Motion Photogrammetry Approach for Urban Development Analysis. *Remote Sens. Appl.* **2022**, *28*, 100850. [CrossRef]

remote sensing

MDPI

Article

Improving UAV Remote Sensing Photogrammetry Accuracy Under Navigation Interference Using Anomaly Detection and Data Fusion

Chen Meng [1], Haoyang Yang [2], Cuicui Jiang [3], Qinglei Hu [1,4] and Dongyu Li [1,*]

1 School of Cyber Science and Technology, Beihang University, Beijing 100191, China; xingchen_irc@buaa.edu.cn (C.M.); huql_buaa@buaa.edu.cn (Q.H.)
2 School of Engineering, University of Warwick, Coventry CV4 7AL, UK; haoyang_ac@outlook.com
3 Department of Electronic and Computer Engineering, Hong Kong University of Science and Technology, Hong Kong 999077, China; eeccjiang@ust.hk
4 School of Automation Science and Electrical Engineering, Beihang University, Beijing 100191, China
* Correspondence: dongyuli@buaa.edu.cn

Abstract

Accurate and robust navigation is fundamental to Unmanned Aerial Vehicle (UAV) remote sensing operations. However, the susceptibility of UAV navigation sensors to diverse interference and malicious attacks can severely degrade positioning accuracy and compromise mission integrity. Addressing these vulnerabilities, this paper presents an integrated framework combining sensor anomaly detection with a Dynamic Adaptive Extended Kalman Filter (DAEKF) and federated filtering algorithms to bolster navigation resilience and accuracy for UAV remote sensing. Specifically, mathematical models for prevalent UAV sensor attacks were established. The proposed framework employs adaptive thresholding and residual consistency checks for the real-time identification and isolation of anomalous sensor measurements. Based on these detection outcomes, the DAEKF dynamically adjusts its sensor fusion strategies and noise covariance matrices. To further enhance the fault tolerance, a federated filtering architecture was implemented, utilizing adaptively weighted sub-filters based on assessed trustworthiness to effectively isolate faults. The efficacy of this framework was validated through rigorous experiments that involved real-world flight data and software-defined radio (SDR)-based Global Positioning System (GPS) spoofing, alongside simulated attacks. The results demonstrate exceptional performance, where the average anomaly detection accuracy exceeded 99% and the precision surpassed 98%. Notably, when benchmarked against traditional methods, the proposed system reduced navigation errors by a factor of approximately 2-3 under attack scenarios, which substantially enhanced the operational stability of the UAVs in challenging environments.

Keywords: UAV photogrammetry; multi-source fusion; sensor attacks; interference detection

1. Introduction

Leveraging their efficiency and flexibility, Unmanned Aerial Vehicles (UAVs) have achieved widespread deployment in the field of remote sensing, with applications spanning critical domains such as environmental monitoring, precision agriculture, and engineering surveying [1–6]. Particularly in applications utilizing UAV photogrammetry—the process of generating high-accuracy three-dimensional (3D) digital models and orthomosaics from aerial imagery—extremely stringent requirements are placed on the accuracy of the flight platform's navigation and positioning system since navigation errors directly impact the geometric fidelity and georeferencing precision of the final remote sensing product [7,8]. Consequently, achieving high-precision navigation has emerged as a critical challenge that warrants further investigation in the field of UAV photogrammetry [9–11].

In the UAV navigation system, different sensors provide specific measurements. For example, the Global Positioning System (GPS) utilizes the time of arrival data from multiple satellites to calculate the distance between the satellites and the receivers, thereby estimating the UAV's position. Moreover, the Inertial Navigation System (INS), based on the Inertial Measurement Unit (IMU), processes the sensor inputs from the gyroscopes and accelerometers to determine the current navigation attitude, speed, and position. It provides high short-term accuracy and strong resistance to external interference. Furthermore, the magnetometer utilizes the Earth's magnetic field or the artificial magnetic fields to estimate the UAV's yaw angle. However, a single sensor cannot consistently guarantee accurate and seamless navigation for the UAV [12,13]. For example, the accumulating INS navigation errors over time reduce the long-term accuracy [14]. Magnetometers are susceptible to interference from nearby magnetic fields, significantly affecting their attitude estimation, while the GPS is vulnerable to signal blockage and multipath distortions in constrained environments [15].

In addition to static environmental factors, the GPS is particularly vulnerable to human-induced malicious interference and attacks. With the rapid expansion of GPS and magnetometer use, combined with advances in software-defined radio (SDR) technology, UAV sensors are increasingly exposed to security threats [16,17]. For example, the availability of affordable radio hardware enables SDR devices capable of GPS signal spoofing, thereby potentially compromising UAV positioning [18,19]. Moreover, the reliance on wireless communication between UAVs, ground control stations, and satellites renders these systems susceptible to a variety of attacks, including GPS spoofing, replay attacks, and magnetic jamming [20–22]. Although recent real-time detectors—such as the isolation-forest-based approach in [23]—can identify GPS spoofing and correct positioning errors using secure infrastructure data, they necessitate extensive trusted coverage. Notably, inertial sensors (e.g., accelerometers and gyroscopes) can perform consistency checks with GPS data at the acceleration and angular velocity levels to detect spoofing attacks; these inertial sensors are inherently immune to radio frequency (RF)-based spoofing or jamming attacks [24].

Due to the limitations inherent in relying on a single sensor for attack detection, numerous multi-source fusion (MSF) methods have been proposed to enhance state estimation accuracy in UAV navigation [25]. Dynamic black-box analysis has shown that MSF schemes can significantly reinforce security against GPS spoofing [26]. For example, Tao et al. [27] improved the positioning accuracy and stability by integrating low-cost sensors. Wang Y et al. [28] categorized vehicular cybersecurity challenges into two main areas: information-oriented and control-oriented. The control-oriented approaches were further divided into data-driven methods [29,30] and model-based methods [31]. However, while data-driven methods are effective only for attacks represented in the training data, model-based techniques detect deviations in sensor measurements from the expected behavior, as predicted by the UAV's mathematical model—and have been validated over long-term operations [32]. In addition, Michieletto et al. [33] applied information-theoretic tools for threshold testing by assessing the consistency of multimodal measurements, yet this approach requires collaboration between multiple UAVs, thus limiting its applicability to single-UAV operations.

Motivated by these findings, we introduce an integrated navigation framework fusing the GPS, an IMU, and magnetic sensor data. This framework leverages mathematical model-based techniques within an adaptive MSF architecture to detect and isolate abnormal measurements effectively. A key feature is the dynamic selection of reliable sensor combinations based on real-time reliability assessments, ensuring that robust, high-precision navigation can be achieved using cost-effective sensors. The proposed framework for anomaly detection and multi-source fusion is illustrated in Figure 1. Accordingly, this paper presents the following three key contributions:

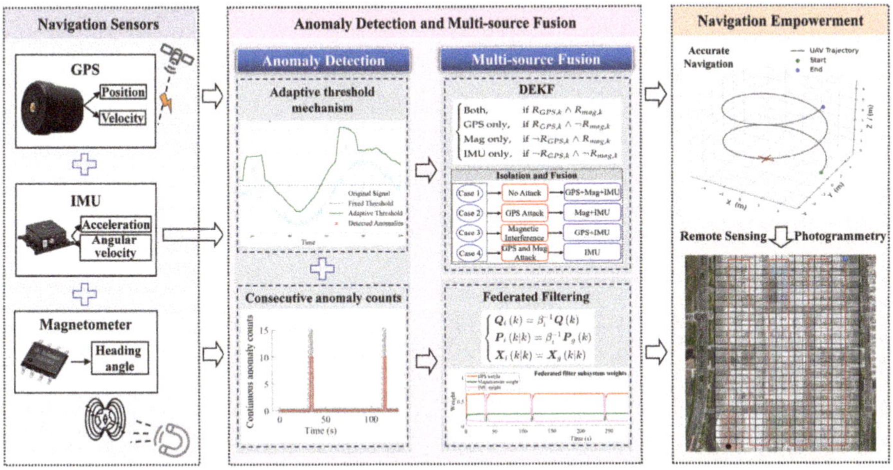

Figure 1. An illustration of the proposed anomaly detection and multi-source fusion algorithm framework.

1. Problem: How can we achieve precise UAV navigation in environments subject to complex interference and deliberate attacks to ensure the success of photogrammetry missions?

2. Solution: To address this challenge, we propose a trusted navigation framework integrating sensor anomaly detection with adaptive sensor fusion.

3. Platform experiments: We conducted spoofing experiments using the SDR and rigorously evaluated the proposed framework's performance with UAV flight data. The evaluation included comprehensive assessments of navigation error metrics and anomaly detection accuracy.

The remainder of this paper is organized as follows: Section 2 details the kinematic and measurement models for the UAV. In Section 3, we introduce the sensor attack detection method and two adaptive fusion algorithms. Section 4 presents experimental validations based on real flight data. Finally, Section 5 concludes the paper.

2. Preliminary Problem

This section presents the kinematic and measurement models for the UAV.

2.1. Kinematic Model of UAV

Kinematic and measurement models are critical for the accurate state estimation and control of UAVs. The UAV was assumed to satisfy the following conditions:

(1) The UAV was modeled as a bilaterally symmetrical rigid body;

(2) It was subject only to gravitational and aerodynamic forces;

(3) The effects of Earth's rotation and curvature on the navigation system were neglected.

The kinematic model was established using two coordinate systems: an inertial (Earth-fixed) frame $\sum_n$ and a body-fixed frame $\sum_b$ attached to the UAV. Euler angles (ϕ, θ, ψ) were used to represent the orientation between $\sum_b$ and $\sum_n$, with a rotation order of $z-y-x$. The rotation matrix $\boldsymbol{R}$, which transforms coordinates from the inertial system to the body-fixed frame, is given as follows:

$$\boldsymbol{R} = \boldsymbol{R}_x(\phi)\boldsymbol{R}_y(\theta)\boldsymbol{R}_z(\psi) \tag{1}$$

Subsequently, the rates of change of the Euler angles are expressed as

$$\begin{bmatrix} \dot{\phi} \\ \dot{\theta} \\ \dot{\psi} \end{bmatrix} = \begin{bmatrix} 1 & \sin\phi\tan\theta & \cos\phi\tan\theta \\ 0 & \cos\theta & -\sin\phi \\ 0 & \sin\phi\sec\theta & \cos\phi\sec\theta \end{bmatrix} \omega \tag{2}$$

where $\omega = \left[\omega_x, \omega_y, \omega_z\right]^{\mathrm{T}}$ denotes the UAV's angular velocity, reflecting its attitude changes. The UAV's position is represented as $\boldsymbol{P} = [P_n, P_e, P_d]^{\mathrm{T}}$, and its linear velocity as $\boldsymbol{V} = [V_n, V_e, V_d]^{\mathrm{T}}$. This framework links sensor measurements from onboard gyroscopes to the UAV's rotational motion.

2.2. *Measurement Model of UAV*

In this study, three sensors were utilized: the GPS, magnetometer, and an IMU. First, the GPS provides the UAV's position measurement, $\tilde{\boldsymbol{P}}$, and linear velocity measurement, $\tilde{\boldsymbol{V}}$, modeled as follows:

$$\begin{aligned} \tilde{\boldsymbol{P}} &= \boldsymbol{P} + \boldsymbol{e}_P \\ \tilde{\boldsymbol{V}} &= \boldsymbol{V} + \boldsymbol{e}_V \end{aligned} \tag{3}$$

where $\boldsymbol{e}_P$ and $\boldsymbol{e}_V$ are Gaussian white noise terms with covariances $\boldsymbol{Q}_P$ and $\boldsymbol{Q}_V$, respectively.

The magnetometer provides the UAV's yaw angle measurement, $\tilde{\psi}$, as follows:

$$\tilde{\psi} = \psi + e_\psi \tag{4}$$

where e_ψ represents Gaussian white noise with covariance Q_ψ.

The IMU, comprising a gyroscope and an accelerometer, measures the UAV's angular velocity and acceleration, respectively. The gyroscope measurement model is given by

$$\tilde{\omega} = \omega + \boldsymbol{b}_w + \boldsymbol{e}_\eta \tag{5}$$

where $\tilde{\omega}$ is the measured angular velocity, $\boldsymbol{b}_w = [b_{\omega_x}, b_{\omega_y}, b_{\omega_z}]^{\mathrm{T}}$ denotes the constant gyroscope bias, and $\boldsymbol{e}_\eta = [e_{\eta_x}, e_{\eta_y}, e_{\eta_z}]^{\mathrm{T}}$ is Gaussian zero-mean noise. Similarly, the accelerometer measurement model is given by

$$\tilde{\boldsymbol{a}} = \boldsymbol{R}(\boldsymbol{a} + \boldsymbol{b}_a) + \boldsymbol{e}_a \tag{6}$$

where $\tilde{\boldsymbol{a}}$ is the acceleration measurement, $\boldsymbol{b}_a$ represents the constant accelerometer bias, and $\boldsymbol{e}_a$ denotes Gaussian zero-mean noise.

2.3. *State Estimation Based on EKF*

Due to the nonlinear nature of the UAV's kinematic model (state equations), this paper employs an adaptive EKF and federated filtering to implement an IMU/GPS/magnetometer loosely coupled integration model. The EKF addresses nonlinear problems via local linear approximations and has been widely applied [34,35].

First, we define the system state vector as

$$\boldsymbol{x} = [\phi \quad \theta \quad \psi \quad \boldsymbol{P} \quad \boldsymbol{V} \quad \boldsymbol{b}_w \quad \boldsymbol{b}_a]^{\mathrm{T}} \tag{7}$$

The system motion model of the EKF is

$$\dot{x} = f(x) + w(t) \tag{8}$$

where $f(x)$ is a nonlinear function of the state vector that represents the product of the state vector and the system matrix, and $w(t)$ is the Gaussian white noise term.

Combining Equations (3) and (4), the observation model is expressed as

$$z = h(x) + v(t) = \begin{bmatrix} z^{\mathrm{GPS}} \\ z^{\mathrm{Mag}} \end{bmatrix} = \begin{bmatrix} \tilde{P} \\ \tilde{V} \\ \tilde{\psi} \end{bmatrix} + v(t) \tag{9}$$

where $h(x)$ is a linear function that maps the state to the observation space that is updated using actual measurements.

Consequently, the measurement matrices for the GPS and magnetometer are defined accordingly:

$$H(t) = \left[H^{\mathrm{GPS}}(t)^{\mathrm{T}}, H^{\mathrm{Mag}}(t)^{\mathrm{T}}\right]^{\mathrm{T}} \tag{10}$$

where

$H^{\mathrm{GPS}}(t) = [\mathbf{0}_{6\times 3}, I_{6\times 6}, \mathbf{0}_{6\times 6}], H^{\mathrm{Mag}}(t) = [0, 0, 1, \mathbf{0}_{1\times 12}]$

The EKF comprises two primary steps:

(1) Prediction step:

In this step, the current state is estimated from the previous state using the state transition equations, and the error covariance matrix is likewise propagated.

Discretizing Equations (8)–(10) using an improved Euler discretization method [36] introduces a subscript k to denote the discrete time index:

$$\begin{aligned} \hat{x}_{k/k-1} &= F_{k/k-1}\hat{x}_{k-1/k-1} \\ P_{k/k-1} &= F_{k/k-1}P_{k-1/k-1}F_{k/k-1}^{\mathrm{T}} + Q_{k-1} \end{aligned} \tag{11}$$

where Q_k denotes the discrete process noise covariance at step $k-1$.

(2) Update step:

In the update step, the state estimate and error covariance are corrected based on the innovation—the difference between the observed and predicted values. The Kalman gain is calculated as follows:

$$K_k = P_{k/k-1}H_k^{\mathrm{T}}\left(H_kP_{k/k-1}H_k^{\mathrm{T}} + R_k\right)^{-1} \tag{12}$$

The state update process utilizes the innovation $r_k = [(r_k^{\mathrm{GPS}})^{\mathrm{T}}, r_k^{\mathrm{Mag}}]^{\mathrm{T}}$, comprising the GPS residual r_k^{GPS} and the magnetometer residual r_k^{Mag}. The residual r_k can be estimated and calculated as the difference between the actual measurement and the predicted value:

$$\begin{aligned} r_k^{\mathrm{GPS}} &= z^{\mathrm{GPS}} - H^{\mathrm{GPS}}\hat{x}_{k/k-1} \\ r_k^{\mathrm{Mag}} &= z^{\mathrm{Mag}} - H^{\mathrm{Mag}}\hat{x}_{k/k-1} \end{aligned} \tag{13}$$

3. Trusted Navigation Framework for UAV in Different Sensor Interference Environments

UAVs are vulnerable to various sensor interferences during flight, which can significantly compromise the navigation accuracy. This study mainly investigated the sensor attacks on the GPS and magnetometer, and modeled GPS denial as a jamming attack. To mitigate these threats, we first outline the primary types of UAV sensor attacks. Subsequently, two robust navigation algorithms were developed for abnormal detection and for fusing trusted sensor information.

3.1. Types and Modeling of Sensor Attacks on UAV

GPS communications are frequently targeted by attackers, while magnetometers are particularly susceptible to environmental magnetic interference. Navigation sensor attacks are broadly categorized into jamming and spoofing attacks [37]. The characteristics of these two threats to UAV communications are outlined below:

(1) Jamming attacks

Jamming attacks employ specialized equipment to inject noise into sensor signals, degrading data quality and hindering the accurate collection of environmental information. For example, electromagnetic or physical interference can disrupt GPS signals. Mathematically, a jamming attack on the signal is modeled by reducing the signal to NaN:

$$\begin{aligned} \tilde{\boldsymbol{P}}_{\text{GPS},k}^{\text{interrupted}} &= NaN \\ \tilde{\boldsymbol{V}}_{\text{GPS},k}^{\text{interrupted}} &= NaN \end{aligned} \tag{14}$$

where $\tilde{\boldsymbol{P}}_{\text{GPS},k}^{\text{interrupted}}$ and $\tilde{\boldsymbol{V}}_{\text{GPS},k}^{\text{interrupted}}$ represent the jammed GPS position and velocity data, respectively.

Similarly, magnetic jamming attacks are modeled by injecting a random bias:

$$\tilde{\psi}^{\text{Mag}} = \tilde{\psi} + \delta\psi_{\text{interference}} \tag{15}$$

where $\tilde{\psi}^{\text{Mag}}$ denotes the jammed magnetic data and $\delta\psi_{\text{interference}}$ is a random bias introduced as the attack parameter.

(2) Spoofing attacks

Spoofing attacks involve the injection of forged signals during data collection, which disrupts environmental perception, impairs decision-making, and compromises UAV flight safety. For instance, attackers may transmit counterfeit GPS signals to spoof a UAV. Mathematically, a spoofing attack is modeled by adding a fixed deviation:

$$\begin{aligned} \tilde{\boldsymbol{P}}_{\text{GPS},k}^{\text{spoofed}} &= \tilde{\boldsymbol{P}} + \delta_P^{\text{spoofed}} \\ \tilde{\boldsymbol{V}}_{\text{GPS},k}^{\text{spoofed}} &= \tilde{\boldsymbol{V}} + \delta_V^{\text{spoofed}} \end{aligned} \tag{16}$$

where $\delta_P^{\text{spoofed}}$ and $\delta_V^{\text{spoofed}}$ are the position and velocity offsets, respectively, used in the spoofing attack model.

Replay attacks involve recording previously transmitted signals and resending them later. The replay attack is modeled by introducing two distinct times t and t':

$$\begin{aligned} \tilde{\boldsymbol{P}}_{\text{GPS},k}^{\text{replayed}} &= \tilde{\boldsymbol{P}}_{\text{GPS},k-\Delta} \\ \tilde{\boldsymbol{V}}_{\text{GPS},k}^{\text{replayed}} &= \tilde{\boldsymbol{V}}_{\text{GPS},k-\Delta} \end{aligned} \tag{17}$$

where $\tilde{\boldsymbol{P}}_{\text{GPS},k}^{\text{replayed}}$ and $\tilde{\boldsymbol{V}}_{\text{GPS},k}^{\text{replayed}}$ represent the received GPS data corresponding to a previous time $\Delta = t' - t$.

3.2. Anomaly Detection Mechanism for UAV in Sensor Attack Environment

Traditional methods for UAV sensor attack detection, such as model-driven mechanisms and statistical techniques (e.g., fixed-threshold and least-squares methods [38]), are widely applied. However, the complex operational environments and variable mission requirements of UAVs, coupled with their inherent strong nonlinearity and high dynamics, often limit the flexibility and adaptability of these methods when processing complex flight data, thereby hindering efficient and reliable detection in intricate scenarios.

To overcome the limitations of traditional methods, this paper proposes an anomaly detection mechanism for sensor attacks on UAVs. This mechanism employs a residual consistency check strategy that integrates continuous anomaly counting with dynamic adaptive thresholding. It is designed to monitor persistent inconsistencies between the UAV's predicted state and actual sensor measurements, thereby enabling the more reliable detection of sensor attacks. Specifically, the detection logic initially identifies potential sensor attacks by evaluating the quadratic forms of the GPS residuals d_k^{GPS} and magnetometer residuals d_k^{Mag}, along with their dynamic variations.

$$\begin{aligned} (d_k^{\text{GPS}})^2 &= (\boldsymbol{r}_k^{\text{GPS}})^{\text{T}} Cov^{-1}(\boldsymbol{r}_k^{\text{GPS}}) \boldsymbol{r}_k^{\text{GPS}} \\ (d_k^{\text{Mag}})^2 &= (r_k^{\text{Mag}})^{\text{T}} Cov^{-1}(r_k^{\text{Mag}}) r_k^{\text{Mag}} \end{aligned} \tag{18}$$

where $Cov(\boldsymbol{r}_k)$ represents the covariance matrix, and d_k^2 is expected to follow a chi-square distribution with degrees of freedom equal to $\dim(z_k)$.

If d_k^2 ($d_{\text{GPS},k}^2$, $d_{\text{Mag},k}^2$) exceeds the corresponding threshold τ_{GPS} (set to 20) or τ_{Mag} (set to 25), the respective sensor measurement is deemed abnormal:

$$\begin{cases} d_{\text{GPS},k}^2 > \tau_{\text{GPS}} & \Rightarrow GPS\ anomaly \\ d_{\text{Mag},k}^2 > \tau_{\text{Mag}} & \Rightarrow Mag\ anomaly \end{cases} \tag{19}$$

3.2.1. Adaptive Threshold and Heading Rate Compensation

In highly dynamic environments, using only a fixed threshold can result in numerous misjudgments. To address this, we propose an adaptive threshold mechanism that is particularly optimized for scenarios with rapid heading angle changes:

$$\tau_{\text{Mag}}(k) = \tau_{\text{Mag,base}} \cdot f(\dot{\psi}_k) \tag{20}$$

where $\tau_{\text{Mag,base}}$ is the initial threshold of the magnetic sensor mentioned in Equation (19); $f(\dot{\psi}_k)$ denotes the heading angle change rate function:

$$f(\dot{\psi}_k) = \begin{cases} 1, & \text{if } |\dot{\psi}_k| \le \dot{\psi}_{\text{threshold}} \\ 1 + \lambda(|\dot{\psi}_k| - \dot{\psi}_{\text{threshold}}), & \text{if } |\dot{\psi}_k| > \dot{\psi}_{\text{threshold}} \end{cases} \tag{21}$$

where $\dot{\psi}_k = \frac{\psi_k - \psi_{k-1}}{\Delta t}$ represents the rate of change of the heading angle; λ is the compensation coefficient (set to 3.0); and $\dot{\psi}_{\text{threshold}}$ is the threshold that triggers the compensation (set to 0.15 rad/s).

3.2.2. Consecutive Anomaly Counting and Residual Consistency Detection

To mitigate misjudgments caused by instantaneous fluctuations, a continuous anomaly counting mechanism is introduced:

$$C_{\text{sensor},k} = \begin{cases} \min(C_{\text{sensor},k-1} + 1, C_{\max}), & \text{if sensor abnormal at k} \\ \max(C_{\text{sensor},k-1} - 1, 0), & \text{otherwise} \end{cases} \tag{22}$$

where $C_{\text{sensor},k}$ is a non-negative integer, and the upper limit of the count is $C_{\max}$ (set to 6).

At this time, the conditions judged as abnormal are

$$A_{\text{sensor},k} = \begin{cases} \text{true}, & \text{if } C_{\text{sensor},k} \geq C_{\text{threshold}} \\ \text{false}, & \text{otherwise} \end{cases} \tag{23}$$

where $C_{\text{threshold}}$ (set to 6) is the threshold for the consistency of continuous residual signs. When the threshold is exceeded, it is considered that enough continuous anomalies are detected to reduce the interference of short-term peak noise.

Furthermore, to enhance the detection accuracy, a residual consistency check is also incorporated:

$$consistent_{\text{bias}} = \left(\left| \sum_{i=0}^{N-1} \text{sign}(r_{k-i}) \right| \geq \eta_1 N \right) \wedge \left(\left| \frac{1}{N} \sum_{i=0}^{N-1} r_{k-i} \right| \geq \eta_2 \right) \tag{24}$$

where r_k represents the measurement residual, N is the history window length (set to 10), η_1 is the consistency threshold (set to 0.6), and η_2 is the deviation amplitude threshold (set to 0.05).

By combining d_k^2 with the residual consistency measure, we define a comprehensive reliability criterion for each sensor:

$$Rel_{\text{sensor},k} = \neg\left(d^2_{\text{sensor},k} > \tau_{\text{sensor},k} \vee \text{consistent bias}_{\text{sensor},k}\right) \tag{25}$$

where $Rel_{\text{sensor},k}(Rel_{\text{GPS},k}, Rel_{\text{Mag},k})$ denotes the reliability of the sensor at time k, and a true value means that the sensor is reliable.

Based on this analysis, if the $Rel_{\text{sensor},k}$ value is true, the corresponding sensor measurement is used in the EKF update step. Conversely, if a sensor abnormality is detected, the measurement from that sensor is isolated (not used directly in the update), and the system relies on the fusion of the remaining reliable sensors or prediction only:

$$\begin{aligned} \hat{\boldsymbol{x}}_k &= \hat{\boldsymbol{x}}_{k/k-1} + \boldsymbol{K}_k \boldsymbol{r}_k \\ \boldsymbol{P}_k &= (\boldsymbol{I} - \boldsymbol{K}_k \boldsymbol{H}_k) \boldsymbol{P}_{k/k-1} \end{aligned} \tag{26}$$

3.3. Multi-source Fusion Based on DAEKF

Following the sensor data anomaly detection, additional data processing and multi-source fusion are conducted. The abnormal data isolation and fusion scheme proposed in this paper is illustrated in Figure 2. Specifically, the primary innovation of the DAEKF is its ability to adaptively select reliable measurements for the update step based on the anomaly detection results. Based on the sensor reliability flags Rel_{GPS} and Rel_{Mag}, four update modes are defined:

$$\text{Mode}_k = \begin{cases} \text{Both}, & \text{if } Rel_{\text{GPS},k} \wedge Rel_{\text{Mag},k} \\ \text{GPS only}, & \text{if } Rel_{\text{GPS},k} \wedge \neg Rel_{\text{Mag},k} \\ \text{Mag only}, & \text{if } \neg Rel_{\text{GPS},k} \wedge Rel_{\text{Mag},k} \\ \text{IMU only}, & \text{if } \neg Rel_{\text{GPS},k} \wedge \neg Rel_{\text{Mag},k} \end{cases} \tag{27}$$

The covariance is updated using the Joseph form to ensure numerical stability:

$$\mathbf{P}_{k|k} = (\mathbf{I} - \mathbf{K}_k \mathbf{H}_k) \mathbf{P}_{k|k-1} (\mathbf{I} - \mathbf{K}_k \mathbf{H}_k)^{\mathrm{T}} + \mathbf{K}_k \mathbf{R}_k \mathbf{K}_k^{\mathrm{T}} \tag{28}$$

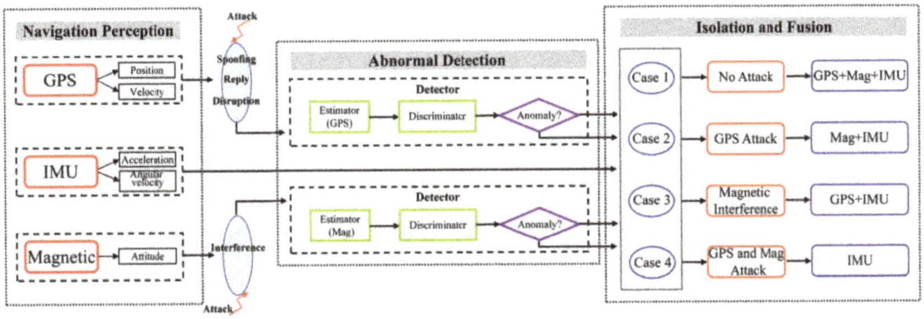

Figure 2. Diagram of the proposed anomaly detection and adaptive fusion framework.

To further improve the system adaptability, both the process noise covariance matrix Q_k and the measurement noise covariance matrix R_k are dynamically adjusted based on the anomaly status:

$$\mathbf{Q}_k = \begin{cases} \gamma_Q \cdot \mathbf{Q}_{\text{base}}, & \text{if in anomaly mode} \\ \mathbf{Q}_{\text{base}}, & \text{otherwise} \end{cases} \tag{29}$$

$$\mathbf{R}_k = \begin{cases} \gamma_R \cdot \mathbf{R}_{\text{base}}, & \text{if in effect period} \\ \mathbf{R}_{\text{base}}, & \text{otherwise} \end{cases} \tag{30}$$

where γ_Q and γ_R are the augmentation factors, which are set to 2 and 3, respectively. This adjustment allows the system to increase the process uncertainty during anomalies and to augment the measurement uncertainty during the recovery phase, thereby achieving a smooth transition.

The proposed anomaly isolation and adaptive fusion mechanism was validated through four representative attack scenarios, as summarized in Table 1, that encompassed the following: (1) no attacks or anomalies; (2) GPS-only attacks; (3) only magnetic interference attack exists; (4) simultaneous GPS and magnetic jamming attacks.

Table 1. Proposed fusion scheme.

Case	GPS Status	Magnetic Status	Result	Fusion Strategy
1	No Alarm	No Alarm	No Attack	GPS + Magnetic + IMU
2	Alarm	No Alarm	GPS Attack	Magnetic + IMU
3	No Alarm	Alarm	Magnetic Attack	GPS + IMU
4	Alarm	Alarm	GPS and Magnetic Attack	IMU

Case 1: No warning was generated, indicating an attack-free environment. In this scenario, the GPS, magnetometer, and IMU data were fused for navigation.

Case 2: Only a GPS anomaly was detected, which triggered a GPS attack warning. Consequently, the navigation system relied primarily on the magnetometer and IMU data.

Case 3: Only a magnetometer anomaly was detected, which led to a magnetic sensor warning; the GPS and IMU data were then used for navigation.

Case 4: Both GPS and magnetometer anomalies were detected. These sensors were isolated, and the system used solely IMU data to maintain short-term precise navigation.

We simulated these attack scenarios using GPS and magnetometer data to represent potential threats during UAV flights. When the anomaly detection mechanism did not flag any anomaly (as in Case 1), the sensors were considered reliable and their data was fused using the EKF. Conversely, when a sensor attack was identified, the compromised data was isolated, and information from the unaffected sensors was fused to ensure reliable navigation in complex environments.

Detector 1 is dedicated to detecting GPS attacks. When GPS sensor data is flagged as abnormal, a GPS attack warning is issued and the state estimate along with the error covariance are updated accordingly:

$$\begin{aligned} K_k &= P_{k/k-1} H_k^{\mathrm{T}} \left(H_k P_{k/k-1} H_k^{\mathrm{T}} + R_{\mathrm{GPS},k} \right)^{-1} \\ \hat{x}_k &= \hat{x}_{k/k-1} + K_k r_k^{\mathrm{GPS}}, P_k = (I - K_k H_k) P_{k/k-1} \end{aligned} \tag{31}$$

In scenarios where only a GPS attack is present, the compromised GPS data is isolated while the unaffected magnetometer and IMU data are fused to update the UAV's attitude and position (corresponding to Case 2 in Table 1).

Detector 2 is responsible for detecting magnetic interference attacks. When an anomaly is detected in the magnetometer data, a magnetic interference warning is triggered, and the state estimate along with the error covariance is updated accordingly:

$$\begin{aligned} K_k &= P_{k/k-1} H_k^{\mathrm{T}} \left(H_k P_{k/k-1} H_k^{\mathrm{T}} + R_{\mathrm{Mag},k} \right)^{-1} \\ \hat{x}_k &= \hat{x}_{k/k-1} + K_k r_k^{\mathrm{Mag}}, P_k = (I - K_k H_k) P_{k/k-1} \end{aligned} \tag{32}$$

In scenarios where only magnetic interference is present (Case 3 in Table 1), the system isolates the affected magnetometer data. If both the GPS and magnetic sensors are compromised, both data sources are isolated, and the UAV relies solely on the IMU for short-term precise navigation (Case 4 in Table 1).

3.4. Multi-source Fusion Based on Federated Filtering

When facing severe sensor attacks, a local single-point failure can result in a global deviation in navigation. To address this, we further investigated federated filtering for information isolation and global fusion. The federated filter stands as an advanced decentralized filtering architecture, leveraging the information sharing principle for robust state estimation and fault tolerance. It achieves this through meticulously designed update steps, information fusion, and information distribution mechanisms.

During the filtering process, each sub-filter performs a time update and measurement update independently, and the main (global) filter fuses the results of each sub-filter to obtain the global optimal estimate or conservative suboptimal estimate [39,40]. In our adaptive federated filtering algorithm, three sub-filters are implemented: (1) GPS sub-filter: mainly processes position and velocity information; (2) magnetometer sub-filter: mainly processes yaw angle information; (3) IMU sub-filter: mainly processes acceleration and angular velocity information.

The information allocation in the federated filter is governed by the following principle [41]:

$$\mathbf{P}_{0,i}^{-1} = \beta_i \mathbf{P}_0^{-1}, \quad \sum_i \beta_i = 1 \tag{33}$$

where β_i denotes the proportion of information allocated to the i-th sub-filter. The initial weights are configured as 0.6 for the GPS sub-filter, 0.3 for the IMU sub-filter, and 0.1 for the magnetometer sub-filter, respectively.

The core innovation of the adaptive federated fusion framework lies in its dynamic adjustment of the trust levels of sensor subsystems based on real-time anomaly detection results. The trust assigned to each subsystem is updated continuously over time:

$$trust_{i,k} = trust_{i,k-1} + \Delta trust_{i,k} \tag{34}$$

When an anomaly is detected:

$$\Delta trust_{i,k} = -\alpha_{\text{decay},i} \cdot trust_{i,k-1} \tag{35}$$

When the sensor is normal:

$$\Delta trust_{i,k} = \alpha_{\text{increase},i} \cdot (trust_{i,\max} - trust_{i,k-1}) \tag{36}$$

where $\alpha_{\text{decay},i}$ and $\alpha_{\text{increase},i}$ are the decay rate and growth rate, and $trust_{i,\max}$ is the maximum trust limit.

Based on the trustworthiness of the subsystems, normalized weights are calculated:

$$\omega_{i,k} = \frac{trust_{i,k}}{\sum_j trust_{j,k}} \tag{37}$$

Subsequently, fusion is performed in the information filter framework by combining the individual sub-filter information matrices.

$$\begin{aligned} \mathbf{P}^{-1}_{\text{global},k} &= \sum_i \omega_{i,k} \mathbf{P}^{-1}_{i,k} \\ \mathbf{P}^{-1}_{\text{global},k} \hat{\mathbf{x}}_{\text{global},k} &= \sum_i \omega_{i,k} \mathbf{P}^{-1}_{i,k} \hat{\mathbf{x}}_{i,k} \end{aligned} \tag{38}$$

Thus, the global optimal estimate $\hat{\mathbf{x}}_{\text{global},k}$ is obtained:

$$\hat{\mathbf{x}}_{\text{global},k} = \mathbf{P}_{\text{global},k} \sum_i \omega_{i,k} \mathbf{P}^{-1}_{i,k} \hat{\mathbf{x}}_{i,k} \tag{39}$$

To prevent individual sub-filters from diverging and to contain the impact of an attack locally—thus avoiding a global collapse due to a single-point failure—a feedback mechanism was implemented. For example, if an attacker spoofs or jams the data of a particular sensor, only that sub-filter's local estimation $\hat{\mathbf{x}}_{i,k}$ will be affected. The main filter then mitigates the impact on the overall estimation by appropriately adjusting the corresponding weights $\mathbf{P}_{i,k}$:

$$\begin{aligned} \hat{\mathbf{x}}_{i,k} &\leftarrow \alpha_{\text{feedback},i} \cdot \hat{\mathbf{x}}_{\text{global},k} + (1 - \alpha_{\text{feedback},i}) \cdot \hat{\mathbf{x}}_{i,k} \\ \mathbf{P}_{i,k} &\leftarrow \alpha_{\text{feedback},i} \cdot \mathbf{P}_{\text{global},k} + (1 - \alpha_{\text{feedback},i}) \cdot \mathbf{P}_{i,k} \end{aligned} \tag{40}$$

where $\alpha_{\text{feedback},i}$ is the feedback coefficient.

4. Experimental Analysis

This section presents the experimental validation of our proposed navigation framework, specifically assessing its robustness for high-precision applications, like UAV photogrammetry. The fidelity of 3D environmental reconstructions, such as those of a university campus, is critically dependent on accurate vehicle trajectory data. However, the navigation sensors that provide this data are vulnerable to both environmental interference and malicious attacks. Therefore, the experiments herein were designed to quantitatively evaluate the framework's ability to mitigate these threats. The evaluation focused on the efficacy of the dynamic adaptation and federated filtering algorithms in maintaining a high navigation accuracy under anomalous conditions, which ensured the data integrity required for successful 3D reconstruction in complex or adversarial environments.

The experimental hardware platform, depicted in Figure 3, comprised a DJI M350 RTK UAV manufactured by DJI (Shenzhen, China) and a ground station. The ground station included a GPS spoofing attack device (USRP B210) from Ettus Research (Austin, TX, USA), a portable power bank, and a laptop terminal for operation. The USRP B210 features a broad RF coverage range from 70 MHz to 6 GHz. During the experiments, the UAV's flight controller logged a comprehensive suite of telemetry data, including the geodetic position (longitude, latitude, height), attitude (roll, pitch, yaw), inertial measurements (linear acceleration, angular velocity), velocity vectors, and magnetometer readings. All flight experiments, including the GPS attack scenarios, were conducted in an open-field environment at the Beihang University Hangzhou International Innovation College in Hangzhou, Zhejiang Province, China (coordinates: 119°58′14″E, 30°21′34″N).

Figure 3. UAV flight and GPS attack experimental setup.

Initially, a GPS attack experiment was conducted. In this experiment, a USRP B210 was connected to a laptop and used to generate a forged ephemeris file that corresponded to the target location (Xining, coordinates: 101°48′25″E, 36°38′50″N). The generated spoofing signal, mimicking GPS satellite signals, was then transmitted via a connected antenna. As shown in Figure 4, the position information on both Google Maps and the test software "Satellite Companion" confirmed that the smartphone's location was successfully spoofed to the preset target. Furthermore, Figure 5 shows the map interface on the UAV controller, which presents map data corresponding to the spoofed location—thus demonstrating the feasibility of launching GPS attacks on the UAV.

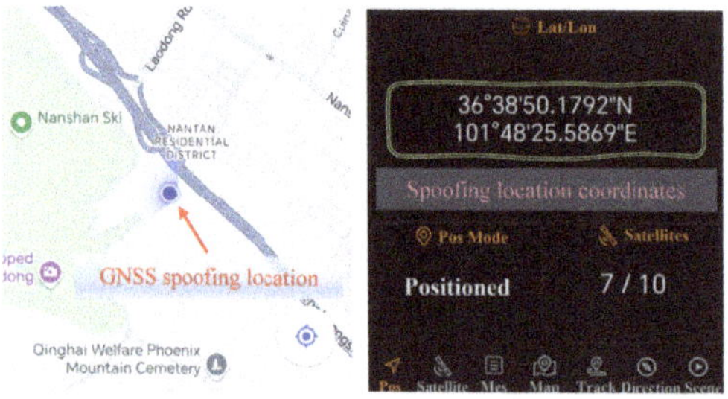

(a) (b)

Figure 4. Illustrative results of GPS spoofing attack on a map-based testing tool: (**a**) Google Maps; (**b**) satellite companion application.

Figure 6 demonstrates the comparison of the UAV photogrammetry under two different conditions. Figure 6a shows the normal flight state: the UAV flew along the planned route (red trajectory line) to cover the target mapping area. During this process, the UAV flew from the starting point to the end point and systematically collected a series of aerial images. The overlap rate between these images (with black rectangular boxes indicating their ground coverage)—primarily encompassed the forward overlap rate (the overlap between two consecutive photos as the UAV advanced along a single flight line) and the sidelap overlap rate (the overlap between adjacent photos from two parallel flight lines)—was precisely controlled. This ensured the integrity and spatial continuity of the data, which laid the foundation for subsequent high-precision three-dimensional scene reconstruction through the principles of photogrammetry. In contrast, Figure 6b demonstrates a scenario where the UAV encountered external interference or GPS spoofing attacks during flight. As a result, the actual flight trajectory of the UAV deviated significantly from the planned route, which presented an irregular and chaotic pattern. This directly led to disordered aerial image acquisition positions and uneven ground coverage. The overlap rate between the images became uncontrollable, which potentially resulted in rates that were too low, missing, or even large areas of coverage gaps. This unstructured and defective data collection mode greatly increased the difficulty of the 3D reconstruction and was likely to result in reconstruction failure or data loss in the final model, which made it impossible to fully and accurately reproduce the target scene.

Figure 5. Map interface of a UAV remote controller displaying the effects of a GPS spoofing attack.

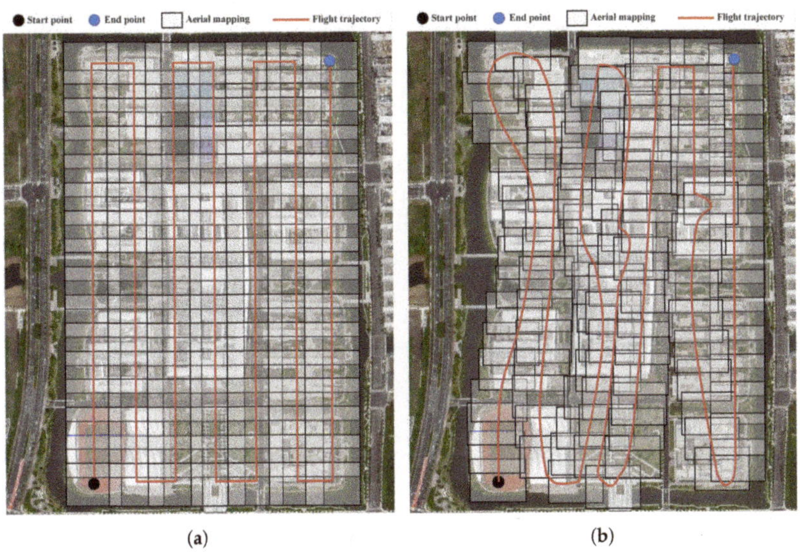

Figure 6. Comparison of UAV photogrammetry under different flight conditions: (**a**) normal flight; (**b**) flight under simulated navigation interference or spoofing attack.

To validate the proposed adaptive dynamic fusion and federated filtering algorithms, real flight data were collected from a UAV over a 380 s period. In the post-processing, various attack scenarios were simulated using the collected data: from 32 to 36 s, a spoofing attack was applied with a positional offset of [15, 15, 20] m; from 113 to 118 s, a replay attack was performed wherein the GPS received data from 4 s earlier; from 240 to 243 s, an interference/interruption attack was implemented, during which the GPS data became NaN; and from 320 to 323 s, a magnetic interference attack imposed a yaw angle offset of 0.25 rad (approximately 14.32°).

The performance of the navigation algorithms was evaluated using the RMSE [42,43], while the anomaly detection performance was assessed using the accuracy and precision metrics, which evaluated both the localization robustness and detection capability.

The position accuracy was quantified using the RMSE:

$$\mathrm{RMSE}_{\mathrm{pos}} = \sqrt{\frac{1}{N}\sum_{k=1}^{N} \|\hat{\mathbf{p}}_k - \mathbf{p}_k^{\mathrm{true}}\|^2} \tag{41}$$

where $\hat{\mathbf{p}}_k$ is the position estimate, and $\mathbf{p}_k^{\mathrm{true}}$ is the real position.

The attitude accuracy was evaluated by the Euler angle RMSE:

$$\mathrm{RMSE}_{\phi,\theta,\psi} = \sqrt{\frac{1}{N}\sum_{k=1}^{N} \|\hat{\mathbf{\Phi}}_k - \mathbf{\Phi}_k^{\mathrm{true}}\|^2} \tag{42}$$

where $\hat{\mathbf{\Phi}}_k$ is the pose angle estimate.

Figure 7 illustrates the UAV flight trajectories. It is evident that the Classic EKF suffered from significant position deviations during the GPS attack periods. In practice, GPS spoofing may cause the UAV to be redirected to unintended locations—posing collision risks—while outdated data and interference interruptions can lead to severe positioning inaccuracies. The isolation and fusion algorithms proposed in this paper can mitigate these risks. As illustrated in Figure 8 (middle and bottom panels), the anomaly detection algorithm proposed in this study accurately identified GPS spoofing and magnetic interference attacks. Simultaneously, the federated filter adjusted the weights of the sub-filters (Figure 8, top panel) based on the detected anomalies. Moreover, by dynamically selecting more reliable data for subsequent fusion, the approach significantly mitigated the risks and navigation errors induced by sensor attacks. Even under sensor attacks, these algorithms delivered robust navigation.

To evaluate the performance of the proposed MSF algorithm, a comparative analysis was conducted against a Classic EKF that utilized a fixed-threshold anomaly detection mechanism. The experimental protocol was designed to compare the RMSE in the navigation solutions and the computational latency associated with the anomaly detection and judgment across the entire operational duration. Ten distinct sets of comparative experiments were executed, incorporating specific attack benchmarks: signal interruption (3 s), replay attack (5 s), magnetic interference (3 s), spoofing attack (3 s), yaw offset (0.25 rad), position offset ([10, 10, 20] m), and replay time offset (4 s). In addition, a series of interference and spoofing attacks of progressively increasing intensity were implemented, with detailed outcomes presented in Table 2. The experimental results reveal that the DAEKF achieved the shortest latency for algorithmic anomaly detection and judgment, whereas the federated filtering algorithm exhibited marginally longer processing times. Regarding the navigation accuracy, the conventional EKF demonstrated comparatively elevated the RMSE values; its performance was most substantially degraded under replay attack scenarios. In contrast, both the proposed DAEKF and the federated filtering algorithm yielded lower average RMSE values, indicating superior robustness and enhanced navigation accuracy across diverse attack and interference conditions.

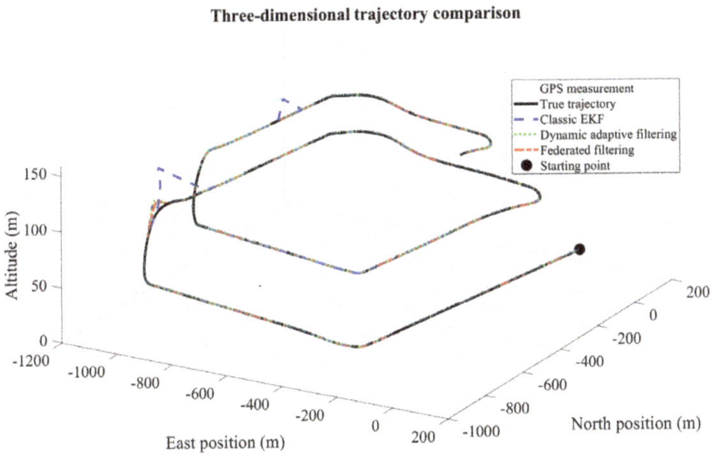

Figure 7. Comparative Three-dimensional trajectory analysis of three algorithms.

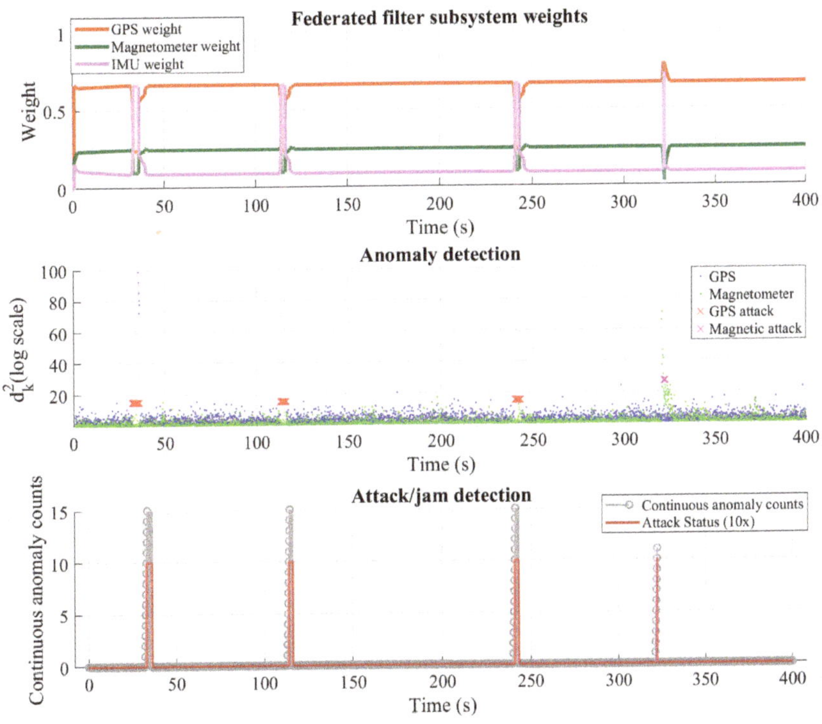

Figure 8. Federated filtering-based anomaly detection and global fusion process diagram: (**top**) real-time sub-filter weight adjustments; (**middle**) sensor attack anomaly detection statistics; (**bottom**) consecutive anomaly counts and attack status flags.

Figure 9 presents the performance evaluation metrics for the anomaly detection concerning sensor attacks. Each experiment covered four typical attack events, namely, GPS spoofing attacks, replay attacks, interference/interruption attacks, and magnetic interference attacks. The results clearly demonstrate that the proposed algorithm achieved an average accuracy that exceeded 99% and an average precision that surpassed 98% across these attack scenarios.

Table 2. Comparison of RMSE and anomaly detection times for three algorithms.

	Attack Events	Classic EKF (m ǀ s)	DAEKF (m ǀ s)	FF (m ǀ s)
1	Attack benchmarks	3.00 ǀ 0.264	1.54 ǀ 0.079	1.01 ǀ 0.481
2	Signal interruption 5 s	3.04 ǀ 0.191	1.69 ǀ 0.048	1.17 ǀ 0.324
3	Replay attack 7 s	17.21 ǀ 0.197	3.51 ǀ 0.050	2.55 ǀ 0.335
4	Mag interference 8 s	2.98 ǀ 0.186	1.59 ǀ 0.047	1.09 ǀ 0.317
5	Spoofing attack 6 s	3.67 ǀ 0.192	1.72 ǀ 0.048	1.18 ǀ 0.323
6	Spoofing offset [20, 20, 30] m	1.65 ǀ 0.189	1.64 ǀ 0.048	1.08 ǀ 0.324
7	Spoofing offset [40, 40, 60] m	1.70 ǀ 0.206	1.69 ǀ 0.052	1.08 ǀ 0.347
8	Spoofing yaw offset 0.35 rad	2.88 ǀ 0.193	1.60 ǀ 0.048	1.10 ǀ 0.325
9	Spoofing yaw offset 0.50 rad	3.07 ǀ 0.197	1.73 ǀ 0.049	1.15 ǀ 0.333
10	Replay time offset 8 s	2.95 ǀ 0.197	1.63 ǀ 0.049	1.03 ǀ 0.329

FF: federated filtering.

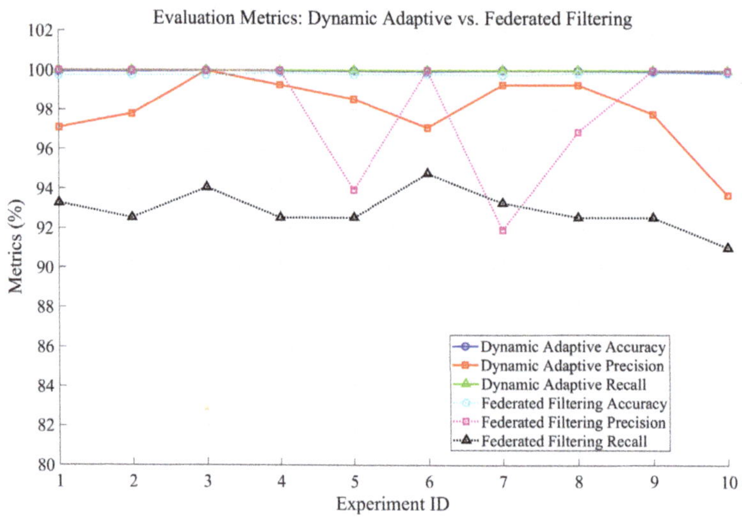

Figure 9. Anomaly detection performance of DAEKF and federated filtering.

5. Conclusions

This study addressed the critical challenge of maintaining navigation integrity for a UAV under sensor attacks or severe interference, particularly for precision-dependent applications, such as UAV-based remote sensing. We propose a robust navigation fusion framework that integrates an anomaly detection mechanism based on adaptive thresholding and residual consistency checks, with an adaptive sensor fusion strategy. By mathematically modeling sensor attacks, the system efficiently identifies and isolates anomalous data, thereby preventing error propagation. The proposed DAEKF adaptively reconfigures the fusion strategy upon attack detection, prioritizing trusted data sources, such as the IMU. To further bolster the resilience, a federated filtering architecture was implemented, which isolates sensor faults at the sub-filter level and dynamically adjusts information fusion weights based on assessed reliability, effectively mitigating fault propagation. Rigorous experimental evaluations that utilized real-world UAV flight data alongside simulated GPS spoofing, replay, interference, and magnetic interference attacks, empirically validated the framework's performance. The anomaly detection mechanism achieved a high accuracy

(>99%) and precision (>98%). Critically, compared with conventional methods, the proposed DAEKF and federated filtering algorithms significantly enhanced the navigation accuracy under attack conditions, where the DAEKF also substantially reduced the anomaly detection latency. These findings underscore the efficacy of the proposed framework in sustaining precise and reliable UAV navigation in complex or adversarial environments, thereby holding significant promise for improving UAV operational safety and mission success rates. Future research will focus on integrating reinforcement learning techniques to enhance system autonomy and robustness in unknown complex environments, and on optimizing the federated filter structure to further reduce the anomaly detection time.

Author Contributions: Conceptualization, C.M. and D.L.; methodology, C.M., Q.H., and H.Y.; validation, C.M. and H.Y.; formal analysis, C.M. and D.L.; investigation, C.M. and C.J.; writing—original draft preparation, C.M.; writing—review and editing, C.M. and C.J. All authors have read and agreed to the published version of the manuscript.

Funding: This research was funded by the National Natural Science Foundation of China under Grant 62473023, the International Innovation Institute of Beihang University under Grant 2024KQ018, and the Foundation of Science and Technology on Space Intelligent Control Laboratory under Grant HTKJ2024KL502030.

Data Availability Statement: This article presents the original contributions of the study, along with equations for the primary algorithms. Additional information or specific data inquiries can be directed to the author.

Conflicts of Interest: The authors declare no conflicts of interest.

References

1. Yang, B.; Zhu, W.; Rezaei, E.E.; Li, J.; Sun, Z.; Zhang, J. The optimal phenological phase of maize for yield prediction with high-frequency UAV remote sensing. *Remote Sens.* **2022**, *14*, 1559. [CrossRef]
2. Yin, Q.; Zhang, Y.; Li, W.; Wang, J.; Wang, W.; Ahmad, I.; Zhou, G.; Huo, Z Estimation of winter wheat SPAD values based on UAV multispectral remote sensing. *Remote Sens.* **2023**, *15*, 3595. [CrossRef]
3. Řeřicha, J.; Kohútek, M.; Vandírková, V.; Krofta, K.; Kumhála, F.; Kumhálová, J. Assessment of UAV Imageries for Estimating Growth Vitality, Yield and Quality of Hop (*Humulus lupulus* L.) Crops. *Remote Sens.* **2025**, *17*, 970. [CrossRef]
4. Lyu, X.; Li, X.; Dang, D.; Dou, H.; Wang, K.; Lou, A. Unmanned aerial vehicle (UAV) remote sensing in grassland ecosystem monitoring: A systematic review. *Remote Sens.* **2022**, *14*, 1096. [CrossRef]
5. Jiang, S.; Jiang, W.; Wang, L. Unmanned Aerial Vehicle-Based Photogrammetric 3D Mapping: A survey of techniques, applications, and challenges. *IEEE Geosci. Remote Sens. Mag.* **2021**, *10*, 135–171. [CrossRef]
6. Feroz, S.; Abu Dabous, S. Uav-based remote sensing applications for bridge condition assessment. *Remote Sens.* **2021**, *13*, 1809. [CrossRef]
7. Ahmed, S.; El-Shazly, A.; Abed, F.; Ahmed, W. The influence of flight direction and camera orientation on the quality products of UAV-based SfM-photogrammetry. *Appl. Sci.* **2022**, *12*, 10492. [CrossRef]
8. Lu, W.; Okayama, T.; Komatsuzaki, M. Rice height monitoring between different estimation models using UAV photogrammetry and multispectral technology. *Remote Sens.* **2021**, *14*, 78. [CrossRef]
9. Lu, H.; Shen, H.; Tian, B.; Zhang, X.; Yang, Z.; Zong, Q. Flight in GPS-denied environment: Autonomous navigation system for micro-aerial vehicle. *Aerosp. Sci. Technol.* **2022**, *124*, 107521. [CrossRef]
10. Yang, Z.; Ying, J.; Shen, J.; Feng, Y.; Chen, Q.A.; Mao, Z.M.; Liu, H.X. Anomaly detection against GPS spoofing attacks on connected and autonomous vehicles using learning from demonstration. *IEEE Trans. Intell. Transp. Syst.* **2023**, *24*, 9462–9475. [CrossRef]
11. Huang, X.; Yang, C.; Wen, M. A sliding window filter with GNSS-state constraint for RTK-visual-inertial navigation. *IEEE Trans. Robot.* **2024**, *40*, 1920–1937. [CrossRef]
12. Zhuang, Y.; Sun, X.; Li, Y.; Huai, J.; Hua, L.; Yang, X.; Cao, X.; Zhang, P.; Cao, Y.; Qi, L. Multi-sensor integrated navigation/positioning systems using data fusion: From analytics-based to learning-based approaches. *Inf. Fusion* **2023**, *95*, 62–90. [CrossRef]
13. Ji, T.; Sivakumar, A.N.; Chowdhary, G.; Driggs-Campbell, K. Proactive anomaly detection for robot navigation with multi-sensor fusion. *IEEE Robot. Autom. Lett.* **2022**, *7*, 4975–4982. [CrossRef]

14. Li, X.; Chen, W.; Chan, C.; Li, B.; Song, X. Multi-sensor fusion methodology for enhanced land vehicle positioning. *Inf. Fusion* **2019**, *46*, 51–62. [CrossRef]
15. Li, S.; Li, X.; Wang, H.; Zhou, Y.; Shen, Z. Multi-GNSS PPP/INS/Vision/LiDAR tightly integrated system for precise navigation in urban environments. *Inf. Fusion* **2023**, *90*, 218–232. [CrossRef]
16. Lee, J.; Taha, A.F.; Gatsis, N.; Akopian, D. Tuning-free, low memory robust estimator to mitigate GPS spoofing attacks. *IEEE Control Syst. Lett.* **2019**, *4*, 145–150. [CrossRef]
17. Narain, S.; Ranganathan, A.; Noubir, G. Security of GPS/INS based on-road location tracking systems. In Proceedings of the 2019 IEEE Symposium on Security and Privacy (SP), San Francisco, CA, USA, 19–23 May 2019; pp. 587–601.
18. Zeng, K.C.; Liu, S.; Shu, Y.; Wang, D.; Li, H.; Dou, Y.; Wang, G.; Yang, Y. All your {GPS} are belong to us: Towards stealthy manipulation of road navigation systems. In Proceedings of the 27th USENIX Security Symposium (USENIX Security 18), Baltimore, MD, USA, 15–17 August 2018; pp. 1527–1544.
19. Liu, S.; Cheng, X.; Yang, H.; Shu, Y.; Weng, X.; Guo, P.; Zeng, K.C.; Wang, G.; Yang, Y Stars can tell: A robust method to defend against {GPS} spoofing attacks using off-the-shelf chipset. In Proceedings of the 30th USENIX Security Symposium (USENIX Security 21), Virtual, 11–13 August 2021; pp. 3935–3952.
20. He, D.; Chan, S.; Guizani, M. Communication security of unmanned aerial vehicles. *IEEE Wirel. Commun.* **2016**, *24*, 134–139. [CrossRef]
21. Bianchin, G.; Liu, Y.C.; Pasqualetti, F. Secure navigation of robots in adversarial environments. *IEEE Control Syst. Lett.* **2019**, *4*, 1–6. [CrossRef]
22. Darvish, K.; Penco, L.; Cisneros, R.; Pratt, J.; Yoshida, E.; Ivaldi, S.; Pucci, D. Teleoperation of humanoid robots: A survey. *IEEE Trans. Robot.* **2023**, *39*, 1706–1727. [CrossRef]
23. Wang, F.; Hong, Y.; Ban, X. Infrastructure-enabled GPS spoofing detection and correction. *IEEE Trans. Intell. Transp. Syst.* **2023**, *24*, 13878–13892. [CrossRef]
24. Tan, H.; Li, Z.; Xie, N.; Lu, J.; Niyato, D. Detection of jamming attacks for the physical-layer authentication. *IEEE Trans. Wirel. Commun.* **2023**, *22*, 9579–9594. [CrossRef]
25. Xu, H.; Luo, H.; Wu, Z.; Wu, F.; Bao, L.; Zhao, F. Towards predicting the measurement noise covariance with a transformer and residual denoising autoencoder for GNSS/INS tightly-coupled integrated navigation. *Remote Sens.* **2022**, *14*, 1691. [CrossRef]
26. Shen, J.; Won, J.Y.; Chen, Z.; Chen, Q.A. Drift with devil: Security of {Multi-Sensor} fusion based localization in {High-Level} autonomous driving under {GPS} spoofing. In Proceedings of the 29th USENIX security symposium (USENIX Security 20), Boston, MA, USA, 12–14 August 2020; pp. 931–948.
27. Tao, X.; Zhu, B.; Xuan, S. A multi-sensor fusion positioning strategy for intelligent vehicles using global pose graph optimization. *IEEE Trans. Veh. Technol.* **2021**, *71*, 2614–2627. [CrossRef]
28. Wang, Y.; Liu, Q.; Mihankhah, E.; Lv, C.; Wang, D. Detection and isolation of sensor attacks for autonomous vehicles: Framework, algorithms, and validation. *IEEE Trans. Intell. Transp. Syst.* **2021**, *23*, 8247–8259. [CrossRef]
29. Wang, Y.; Masoud, N.; Khojandi, A. Real-time sensor anomaly detection and recovery in connected automated vehicle sensors. *IEEE Trans. Intell. Transp. Syst.* **2020**, *22*, 1411–1421. [CrossRef]
30. Van Wyk, F.; Wang, Y.; Khojandi, A.; Masoud, N. Real-time sensor anomaly detection and identification in automated vehicles. *IEEE Trans. Intell. Transp. Syst.* **2019**, *21*, 1264–1276. [CrossRef]
31. Mousavinejad, E.; Yang, F.; Han, Q.L.; Ge, X.; Vlacic, L. Distributed cyber attacks detection and recovery mechanism for vehicle platooning. *IEEE Trans. Intell. Transp. Syst.* **2019**, *21*, 3821–3834. [CrossRef]
32. Biron, Z.A.; Dey, S.; Pisu, P. Real-time detection and estimation of denial of service attack in connected vehicle systems. *IEEE Trans. Intell. Transp. Syst.* **2018**, *19*, 3893–3902. [CrossRef]
33. Michieletto, G.; Formaggio, F.; Cenedese, A.; Tomasin, S. Robust localization for secure navigation of UAV formations under GNSS spoofing attack. *IEEE Trans. Autom. Sci. Eng.* **2022**, *20*, 2383–2396. [CrossRef]
34. Wang, M.; Cui, J.; Huang, Y.; Wu, W.; Du, X. Schmidt ST-EKF for autonomous land vehicle SINS/ODO/LDV integrated navigation. *IEEE Trans. Instrum. Meas.* **2021**, *70*, 8504909. [CrossRef]
35. Chen, W.; Jing, Y.; Zhao, S.; Yan, L.; Liu, Q.; He, Z. A Distributed Collaborative Navigation Strategy Based on Adaptive Extended Kalman Filter Integrated Positioning and Model Predictive Control for Global Navigation Satellite System/Inertial Navigation System Dual-Robot. *Remote Sens.* **2025**, *17*, 721. [CrossRef]
36. Park, T.H.; D'Amico, S. Adaptive neural-network-based unscented kalman filter for robust pose tracking of noncooperative spacecraft. *J. Guid. Control Dyn.* **2023**, *46*, 1671–1688. [CrossRef]
37. Deng, Y.; Zhang, T.; Lou, G.; Zheng, X.; Jin, J.; Han, Q.L. Deep learning-based autonomous driving systems: A survey of attacks and defenses. *IEEE Trans. Ind. Inform.* **2021**, *17*, 7897–7912. [CrossRef]
38. Keipour, A.; Mousaei, M.; Scherer, S. Automatic real-time anomaly detection for autonomous aerial vehicles. In Proceedings of the 2019 International Conference on Robotics and Automation (ICRA), Montreal, QC, Canada, 20–24 May 2019; pp. 5679–5685.

39. Xu, X.; Pang, F.; Ran, Y.; Bai, Y.; Zhang, L.; Tan, Z.; Wei, C.; Luo, M. An indoor mobile robot positioning algorithm based on adaptive federated Kalman filter. *IEEE Sens. J.* **2021**, *21*, 23098–23107. [CrossRef]
40. Li, H.; Ao, L.; Guo, H.; Yan, X. Indoor multi-sensor fusion positioning based on federated filtering. *Measurement* **2020**, *154*, 107506. [CrossRef]
41. Wang, Y.; Zhao, B.; Zhang, W.; Li, K. Simulation experiment and analysis of GNSS/INS/LEO/5G integrated navigation based on federated filtering algorithm. *Sensors* **2022**, *22*, 550. [CrossRef]
42. Hodson, T.O. Root mean square error (RMSE) or mean absolute error (MAE): When to use them or not. *Geosci. Model Dev. Discuss.* **2022**, *15*, 5481–5487. [CrossRef]
43. Chivasa, W.; Mutanga, O.; Burgueño, J. UAV-based high-throughput phenotyping to increase prediction and selection accuracy in maize varieties under artificial MSV inoculation. *Comput. Electron. Agric.* **2021**, *184*, 106128. [CrossRef]

remote sensing

MDPI

Article

Optimization of Spatial Sampling in Satellite–UAV Integrated Remote Sensing: Rationale and Applications in Crop Monitoring

Zhen Zhao [1,2], Hang Xiong [1,3,*], Yawen Yu [1,3], Baodong Xu [1,4] and Jian Zhang [1,4]

1 Digital Agriculture Research Institute, Huazhong Agricultural University, Wuhan 430070, China
2 College of Environmental and Resource Sciences, Zhejiang University, Hangzhou 310058, China
3 College of Economics and Management, Huazhong Agricultural University, Wuhan 430070, China
4 College of Resources and Environment, Huazhong Agricultural University, Wuhan 430070, China
* Correspondence: hxiong@hzau.edu.cn

Highlights

What are the main findings?

- Layout configuration of satellite–UAV integrated remote sensing was transformed into a spatial sampling problem.
- An SSO (spatial sampling optimization) model was proposed.

What is the implication of the main finding?

- Sampling efficiency requires considering both cost and accuracy.
- The SSO-optimized plan improved efficiency by at least 38.7% over conventional plans.

Abstract

Satellite and UAV-based remote sensing have been widely used for agricultural systems monitoring jointly. How to quantitatively optimize the efficiency of integrating these two techniques remains largely understudied. To address this gap, we, for the first time, formulate the configuration of satellite–UAV integrated system as a spatial sampling optimization problem and propose an SSO (spatial sampling optimization) model that jointly optimizes the spatial locations and flight paths of UAV sampling within the satellite monitoring area. The SSO model enables maximizing the accuracy of monitoring under a given cost constraint. We obtained comprehensive data in rapeseed fields and conducted experiments based on the SSO model. We compared the sampling effectiveness of the SSO model with that of simple random sampling, systematic sampling, equal stratified sampling and Neyman stratified sampling. The results showed that the SSO-optimized plan had the highest sampling efficiency, which was at least 38.7% higher than that of the best-performing conventional method (Neyman stratified sampling). Under the same cost constraint, the SSO-optimized sampling scheme can have 11.1% more sampling points than the conventional sampling scheme. The Elite Genetic Algorithm (EGA) performed well in solving the SSO model. The error of the SSO-optimized scheme was reduced by 27.3% and the sampling distance was reduced by 7000 to 8000 m on average. In conclusion, the proposed SSO model helps to optimize the configuration of satellite–UAV integrated remote sensing, thereby improving the cost-effectiveness of agricultural monitoring systems. We call for considering cost constraints and increasing efficiency in agricultural system monitoring and government censuses in the future.

Keywords: spatial sampling; remote sensing; UAV; optimization; genetic algorithm (GA)

1. Introduction

Satellites and UAVs (unmanned aerial vehicles) are playing an increasing role in monitoring agricultural systems [1,2]. Satellite remote sensing and UAV-based remote sensing are highly complementary in terms of coverage and accuracy of acquired data [3]. Satellite remote sensing offers broad spatial coverage and low observation costs but lacks accuracy, while UAV-based remote sensing provides high observation accuracy and can potentially replace ground sampling in specific scenarios [4], albeit at a higher cost.

Integrating satellite and UAV data can provide higher-accuracy observation over a larger spatial area [5]. One of the approaches that previous studies focus on is using UAV data to calibrate satellite models and thus indirectly enhance the accuracy of large-scale monitoring [6]. Another approach is merging satellite and UAV image data to generate new products for spatial-temporal fusion [7]. Despite this, the challenges of directly and instantly integrating the extensive coverage of satellite remote sensing with the high accuracy of UAV-based remote sensing remain unresolved [8]. Since UAV-based remote sensing focuses on localized areas, whereas satellite remote sensing covers large regions, the relationship between UAV observations and satellite scenes can be regarded as analogous to sampling within a population. Thus, it is meaningful to consider UAV configuration as a spatial sampling problem within satellite-monitored regions. Spatial sampling optimization is crucial in agricultural remote sensing for achieving the desired observation accuracy at minimal cost [9].

Spatial sampling refers to the design and implementation of sampling methods for spatially correlated targets. A central goal is to use as few sampling points as possible to characterize the spatial pattern of the target variable, thus improving parameter estimation at unsampled locations [10]. In agricultural monitoring, spatial sampling optimization aims to balance accuracy and cost by either achieving the highest accuracy under a fixed cost or minimizing cost for a predefined accuracy level [11].

A broad range of spatial sampling methods has been proposed, including design-based approaches and model-based approaches [10]. Classic methods include simple random sampling [12], systematic sampling [13], stratified sampling [14], MSN (means of surfaces with nonhomogeneity) sampling [15], and Latin hypercube sampling [16]. These studies provide important foundations for spatial monitoring in environmental and agricultural systems.

However, existing spatial sampling approaches applied in remote sensing still share two critical limitations: (1) they primarily focus on single-platform systems (either satellite or UAV) [17,18], and (2) they neglect the cost-accuracy trade-offs in multi-source integration [19,20]. Such limitations are particularly restrictive for agricultural systems, where satellite coverage and UAV accuracy offer natural complementarities that could be exploited through optimized cross-platform sampling designs.

This study aims to develop an SSO (spatial sampling optimization) model that combines the characteristics of satellite and UAV-based remote sensing. By doing so, it seeks to balance accuracy and cost in spatial sampling surveys, addressing the configuration optimization problem of satellite–UAV integrated remote sensing systems for agricultural monitoring.

Specifically, our work focuses on the optimal allocation of observation resources for both satellite remote sensing and UAV-based remote sensing, and, to the best of our knowledge, is the first to transform the configuration optimization problem of a satellite–UAV integrated agricultural monitoring system into a formal spatial sampling problem. This approach enriches the theory of spatial sampling by incorporating satellite–UAV integrated monitoring techniques and improves the estimation of sampling errors in UAV observations within satellite observation areas. Furthermore, the optimization analysis method under constraints is utilized to introduce the cost dimension as an optimization objective, enabling the resolution of the resource allocation problem for multiple observation techniques.

Overall, the proposed SSO model contributes to improving the efficiency of integrating satellite and UAV-based remote sensing for agricultural monitoring. By jointly optimizing sampling accuracy and cost, the model enables fast, cost-effective, and accurate monitoring, offering a practical and efficient solution for large-scale agricultural sampling surveys.

2. Spatial Sampling Optimization Model

2.1. Problem Formulation

In this study, agricultural condition monitoring was employed as the application context for developing the SSO model for a satellite–UAV integrated remote sensing system. Farmland areas, characterized by large spatial extents and few obstacles, are well-suited for large-scale monitoring through the combined use of satellites and low-altitude UAV platforms. The overall structure and spatial sampling layout of the satellite–UAV integrated system for agricultural monitoring are illustrated in Figure 1.

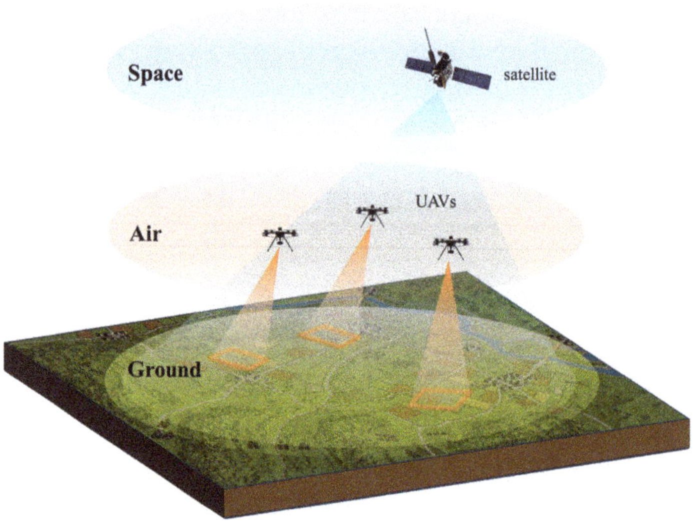

Figure 1. Satellite–UAV integrated remote sensing system for agricultural monitoring.

When observing a single area at the same time each day, the satellite requires only one pass to acquire an image, whereas UAV monitoring must rely on multiple sampling points, repeated flights, or even simultaneous operations of multiple UAVs. Therefore, optimizing the configuration of the satellite–UAV system essentially involves multi-objective optimization of UAV sampling locations and flight paths to achieve both accuracy and efficiency.

In the field of geostatistics, model-based sampling techniques often incorporate prior knowledge to address multi-objective optimization problems. Given the distinct data acquisition characteristics and sensing mechanisms of satellite and UAV platforms, a spatial sampling model was designed to enable UAV-based low-altitude sampling and monitoring within satellite observation areas.

Sampling size and spatial distribution are key factors that influence both the accuracy and implementation cost of spatial surveys [21]. Hence, the proposed spatial sampling model focuses on optimizing the number and spatial arrangement of samples while balancing sampling accuracy and cost.

During the generation of the sampling plan, accuracy is the primary consideration. Generally, increasing the number of sampling points improves accuracy. Moreover, properly selected sampling locations can ensure that the sample data adequately represent the overall spatial pattern; that is, minimizing the error between sampled and population values. Thus, the initial optimization objective is to minimize sampling error, which characterizes the accuracy of sampling results.

In the process of generating the sampling plan, the primary consideration is the accuracy of the sampling survey. In general, the greater the number of sampling points is, the better the accuracy of sampling. Proper locations enable the sampling survey data to fully represent the overall situation in the region; that is, the error between the sampling data and the population data is minimized. Therefore, the initial optimization goal is to minimize the sampling error, which is used to characterize the sampling accuracy; the smaller the error is, the better the accuracy.

While ensuring sufficient accuracy, the sampling cost must also be constrained. The total flight time required for UAV sampling can be used as a quantitative indicator of cost; longer flight times correspond to higher operational costs of the integrated system.

The normalized difference vegetation index (NDVI), a widely used indicator of crop growth and nutritional status, was selected as the target variable. By optimizing the spatial configuration of the satellite–UAV integrated system, this study aims to obtain accurate large-scale data at limited cost, thereby improving the efficiency of agricultural condition monitoring.

2.2. Assumptions

Due to the flexibility in the design of the satellite–UAV integrated remote sensing system, the parameters of the UAV are set based on the research objectives and experimental plans. For conciseness and simplicity, the following assumptions and simplifications are made for the UAV sampling parameters:

- The UAV sampling frame was fixed at 10 × 10 m. Each sampling point required 5 s to complete, and the UAV could operate continuously for 25 min (1500 s) on a single battery charge without replacement during the entire sampling process. These parameters were determined based on actual flight tests and the study area characteristics.
- The UAV flight speed was fixed at 6 m/s, which meets the minimum cruising speed requirement for multirotor UAVs specified in the Chinese national standard GB/T 39612-2020 [22].
- Instrument reuse precluded exact cost calculation; thus, sampling cost was proxied by total time, with accuracy evaluated via sampling error per conventional practice.

2.3. Model Establishment

The number and spatial distribution of UAV sampling points directly affect both the accuracy and cost of data acquisition. Too few points may fail to meet monitoring requirements, whereas too many lead to unnecessary resource consumption. Selecting sampling locations rationally can reduce the number of points while maintaining sufficient representativeness. An unreasonable spatial layout, however, may cause deviations in overall estimates. Therefore, determining the optimal number and distribution of sampling points is essential for layout optimization in satellite–UAV integrated agricultural monitoring systems.

Spatial sampling design is a typical NP-hard (non-deterministic polynomial-time hard) optimization problem [23] involving multiple conflicting objectives such as accuracy, cost, and representativeness. In this study, sampling accuracy, jointly affected by the number and spatial configuration of sampling points, was used as the objective function, while sampling cost served as the constraint to search for a satisfactory solution.

Given the known number and locations of UAV sampling points and the UAV's flying speed, the flight path planning task was formulated as a classic traveling salesman problem (TSP) [24], in which the UAV starts from a fixed point, visits each sampling location exactly once, and returns to the origin. In this study, the starting point was set as the center of the research area. Since the TSP is an NP-hard problem whose computational complexity grows exponentially with problem size, traditional algorithms are often inefficient. Therefore, intelligent optimization algorithms such as the genetic algorithm (GA), ant colony optimization, and simulated annealing are commonly employed to obtain near-optimal solutions [25].

Based on the above analysis, the SSO model was developed. The overall workflow of the model is shown in Figure 2. First, for a given number of sampling points N, the spatial locations of these points are optimized to achieve the maximum possible accuracy. This maximum accuracy value is then compared with the empirical threshold α. If the value is lower than α, the number of sampling points is increased, and the optimization process is repeated. When the maximum accuracy exceeds α, the UAV flight sequence among the sampling points is optimized to minimize the total sampling cost. The minimum cost obtained from this step is compared with the empirical cost threshold β. If the minimum cost is below β, the corresponding sampling configuration is retained as a feasible alternative. Otherwise, the number of sampling points is considered to have reached its upper limit.

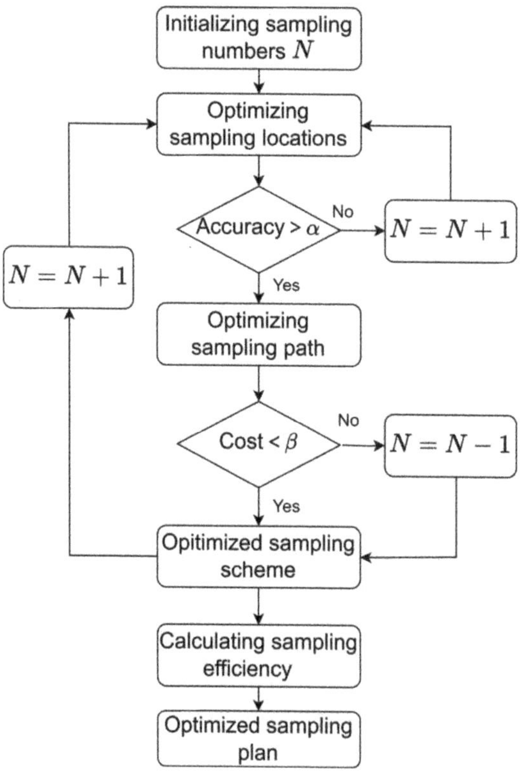

Figure 2. Workflow of the SSO model.

All feasible alternatives are then ranked according to their sampling accuracy and cost to produce an optimized set of sampling schemes. Finally, the sampling efficiency of these schemes is evaluated, and the one with the highest efficiency is selected as the optimal plan. The main parameters of the SSO model used in this study are summarized in Table 1.

Table 1. Parameters of the SSO model and EGA.

Parameter	Value	Description
α	3.33×10^2	Threshold for acceptable sampling accuracy
β	1500	Threshold for acceptable time cost
Population size	60	Number of individuals per generation
Max generations	50	Termination condition of evolution
Mutation probability	0.50	Probability of mutation
Crossover rate	0.70	Probability of crossover
Convergence threshold	1.00×10^{-6}	Criterion for stopping evolution
Selection method	Tournament	Strategy for selecting parents
Mutation operator	Inversion	Reverses a random gene segment
Crossover operator	PMX	Partially matched crossover (permutation encoding)
Encoding type	Permutation	Suitable for route/layout optimization

2.3.1. Objective Function

The core task of spatial sampling with UAVs in satellite observation areas is to use the NDVI sample data acquired by UAVs to obtain an accurate representation of the overall distribution of the NDVI in the study area. It is assumed that there are m satellite image pixels in the observation area, where each image pixel includes k UAV observation pixels; thus, the study area includes a total of mk UAV pixels. For the parameters to be estimated, the NDVI obtained from UAV imagery can be used as prior knowledge to update the spatial distribution, that is, by adding relative true values [4]. The representative satellite error of UAV sampling can be expressed by the following formula [26]:

$$\sigma^2 = \frac{1}{k} MSE \left[1 + \frac{1}{n} + \frac{S_X^2 + (\bar{X} - \bar{x})^2}{n s_x^2} \right] \tag{1}$$

$$MSE = \frac{1}{n} \sum_{i=1}^{n} (y_i - \hat{y}_i)^2 \tag{2}$$

where MSE represents the mean squared error between the satellite observations and the true values in the UAV sampling area. $\bar{x}$ and s_x^2 represent the mean and variance of the sampled UAV pixels, respectively. $\bar{X}$ and S_X^2 represent the mean and variance of the population distribution in the study area, respectively, and n represents the number of UAV pixels sampled.

Formula (1) shows that the errors that affect sampling include the number n of sampling points, the degree to which UAV sampling represents the population distribution $(\bar{X} - \bar{x})^2$, and the standard deviation of the UAV sampling pixels s_x. In addition, to minimize the UAV sampling error, the following optimization conditions need to be satisfied: the distribution of UAV sampling points needs to be as close to the population distribution as possible, that is, $\bar{X} - \bar{x} \to 0$; the variance between the UAV sampling points should be as large as possible, that is, $s_x \to max$; and the UAV sampling points should be distributed as widely as possible in geographic space.

In Formula (1), the number of UAV observation pixels in each satellite image is a constant value k. In the process of optimizing the sampling error, because the true value of the data in the area cannot be obtained in advance, only the satellite remote sensing data are used as prior knowledge for sampling optimization, with all data weighted equally. Therefore, the MSE is also regarded as a constant. In the process of using the algorithm for error optimization, the following formula, which is used to assess the optimization result, is used as the objective function for error optimization: $Obj = \frac{1}{n} + \frac{S_X^2+(\bar{X}-\bar{x})^2}{ns_x^2}$.

2.3.2. Constraints

This study explores how to minimize survey error under a given cost constraint. Compared with the cost function used in conventional sampling, the cost function applied in spatial surveys is also affected by the distribution of sampling points, as well as the sample size. The impact of the sampling point distribution on cost is mainly reflected in the density of the sample distribution. With other factors unchanged, the looser the sample distribution in space is, the greater the traffic cost incurred when transferring between sampling points. We used sampling time to characterize the cost of the sampling survey, and the sampling time could be calculated with the following formula:

$$T = N \cdot \bar{t}_N + \frac{min\left(\sum_{i=1}^{N} d_i\right)}{v} \tag{3}$$

The total time required to complete the UAV sampling task T is composed of two factors: the total time spent on data sampling at the sampling point and the flight time between the sampling points. $\bar{t}_N$ is the average time spent on UAV data sampling at a single sampling point, and N is the number of sampling points for the UAV. d_i is the flight speed of the UAV between the sampling points, v is the flight distance of the UAV from sampling point i to sampling point $i+1$, and $min\left(\sum_{i=1}^{N} d_i\right)$ is the shortest path among all the UAV sampling points.

When conducting spatial sampling surveys, surveyors usually move to the nearest sampling point to continue data collection after completing data collection at each sampling point. The transportation cost can be estimated by measuring the distance between each sampling point. It is simple to implement and works well when there are few sampling points. However, as the number of sampling points increases, the distance between subsequent sampling points may increase sharply. The final total distance may be very long due to the lack of global route planning. In this study, the UAV needs to traverse all the sampling points, so the sum of the distances for all the sampling tasks must be minimized. Based on the previous analysis, the problem of calculating the distance between sampling points can be abstracted as a TSP, which is solved using an intelligent algorithm.

2.3.3. Solution Assessment

We used sampling efficiency to assess each sampling plan, that is, the sampling accuracy that could be obtained per unit cost of input. The formula for calculating the sampling efficiency is as follows:

$$E = \frac{\frac{1}{\sigma^2}}{T} = \frac{1}{T\sigma^2} \tag{4}$$

where σ^2 is the sampling error, $\frac{1}{\sigma^2}$ characterizes the sampling accuracy, and T is the time required for sampling.

2.4. Solution Algorithm

GA is a stochastic optimization method that simulates natural evolution [27]. The evolution of biological populations depends on the crossover and mutation of chromosomes.

Similarly, in the GA, genetic operators act on the population $P(t)$ to perform selection, crossover, and mutation operations, thereby generating a new population $P(t+1)$ and gradually improving the fitness of solutions.

The key components of the GA are the following three operators:

Selection: Sampling locations with smaller sampling errors are preferentially selected from the previous generation and retained in the next generation.

Crossover: Pairs of sampling locations are randomly matched, and portions of their coordinate values are exchanged with a certain probability (the crossover rate).

Mutation: Some coordinate values within the sampling locations are replaced by those of other locations with a given mutation probability, enhancing population diversity.

To avoid premature convergence and preserve optimal solutions, the elitist retention strategy [28] is adopted. In this approach, the best individuals (elite solutions) found during evolution are directly carried over to the next generation without crossover or mutation. This modification significantly enhances the global convergence capability of the standard GA.

Therefore, the elitist genetic algorithm (EGA), incorporating the elitist retention mechanism, was applied in this study to solve the SSO model. During the optimization process, the sampling configuration with the minimum sampling error is preserved for the next iteration to ensure convergence toward the global optimum. The main parameters of the EGA used in this study are summarized in Table 1.

3. Materials and Methods

3.1. Study Area

The experiment was conducted in a rapeseed field located in Baimasi Town, Jiangling County, Hubei Province, China (Figure 3). The study area covers approximately 2.5 ha of farmland characterized by flat terrain with gentle slopes and an elevation of 26.1–31.2 m. The soil type is predominantly paddy soil with relatively uniform physical properties. At the time of sampling, rapeseed was planted across all fields, exhibiting uniform growth conditions. The mean NDVI for the entire farmland was 0.40, with a standard deviation of 0.22.

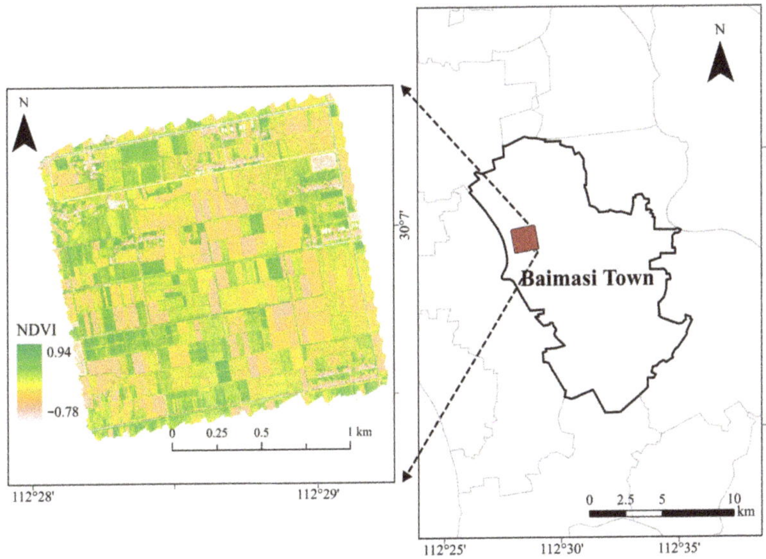

Figure 3. Overview map of the study area.

Baimasi Town has an average annual temperature of 16.6 °C and a mean annual precipitation of approximately 1045 mm. The main crops cultivated in this region include rice, wheat, rapeseed, and cotton. Benefiting from sufficient water and thermal resources, the area represents a typical farmland ecosystem of the Jianghan Plain. The site was intentionally selected for its relatively homogeneous topography and soil conditions, which minimize external variability and provide a suitable environment for evaluating the effectiveness of the proposed SSO model in optimizing spatial sampling design.

3.2. Satellite Image Data

The Sentinel-2 satellite, composed of two high-resolution satellites (2A and 2B), has a revisit period of 5 days. In this study, the Sentinel-2 satellite image data provided was downloaded by Google Earth Engine at nearby time points in the study area for processing. To avoid cloud pollution, images with less than 30% cloud coverage were selected (8 November 2021). Each image was processed by cloud masking and median synthesis to generate a resulting high-quality cloud-free image. The 10-m NDVI image was calculated based on the B4 band (band centered at 665 nm) and B8 band (band centered at 842 nm) of Sentinel-2 and was used as the initial information for optimization. The Level-2A product, which provides surface reflectance corrected for atmospheric effects, was used to ensure radiometric consistency with the UAV data.

3.3. UAV Image Data

UAV-based remote sensing mapping was performed in a large-scale field with uniform rapeseed growth in Baimasi town, Jiangling County, Hubei Province, on 11 November 2021. A Phantom 4 UAV (DJI Innovations, Shenzhen, China) was used for data collection under clear and calm weather conditions from 10:00 to 14:00 (local time, UTC + 8), with the heading overlap rate of 85.0%, and the side overlap rate of 70.0%. The imagery, with a ground resolution of approximately 6 cm/pixel, was radiometrically calibrated to surface reflectance using an on-site reflectance panel to ensure compatibility with the satellite data. The orthorectified UAV images were co-registered to the Sentinel-2 NDVI image and resampled to 10 m resolution using bilinear interpolation for pixel-level correspondence. NDVI was then calculated from the red (650 nm) and near-infrared (840 nm) bands to obtain relative ground-truth values for validation.

3.4. Assessment of the SSO Model

To demonstrate and validate the effect of the SSO model proposed in this paper, different sampling plans were implemented for a representative area, and the accuracy of data acquisition and the sampling cost of the different plans were compared and assessed.

Conventional spatial sampling methods are based on probability sampling and include simple random sampling (samples are selected entirely at random without any prior grouping of the population), systematic sampling (samples are selected at fixed intervals determined by the required sample size), and stratified sampling (samples are randomly selected from different strata of the population according to specified proportions).

One experimental group and four control groups were created to validate the performance of the model. For the experimental group, the SSO-optimized sampling plan was used. For the control group, the existing conventional sampling plans, namely, simple random sampling, systematic sampling, and stratified sampling with an equal distribution and a Neyman distribution, were used. The sampling cost and sampling accuracy of the experimental group were compared with those of the control group to verify the feasibility of the SSO-optimized plan.

For the simple random sampling control group, a certain number of coordinates were randomly selected from the set of candidate coordinates that met the accuracy requirements to determine the locations of the UAV sampling points, and the sampling accuracy and the cost required for the UAV to complete the flight sampling were calculated.

For the systematic sampling group, a certain number of coordinates were selected at equal intervals from the set of candidate coordinates that met the accuracy requirements, and the remaining steps were the same as those in simple random sampling. The spacing of systematic sampling was determined based on the size of the candidate coordinate set and the number of sampling points.

In this study, according to the number of sampling points in each stratum, stratified sampling was subdivided into stratified sampling with an equal distribution and stratified sampling with a Neyman distribution, yielding two different control groups. For an equal distribution, the number of sampling points selected in each stratum is the same; the Neyman distribution is the optimal distribution for sample units when the cost coefficient of each stratum is the same. Suppose n_h is the sample content of each stratum, n is the total sample size, S_h is the standard deviation of samples from stratum h, W_h is the weight of each stratum, N_h is the total number of units in stratum h, L is the total number of strata, and under the Neyman distribution, the sample size is proportional to the stratum weight (W_h) multiplying the strata standard deviation (S_h); thus, the sample size for each stratum can be calculated by the following formula:

$$n_h = n\frac{W_h S_h}{\sum\limits_{h=1}^{L} W_h S_h} = n\frac{N_h S_h}{\sum\limits_{h=1}^{L} N_h S_h} \tag{5}$$

After successfully determining whether a coordinate could be a candidate coordinate, the mean value of the NDVI at the selected coordinate point in the upper-left corner of the sampling frame was calculated, and the candidate coordinates were divided into 6 strata according to the NDVI value and labeled. In stratified sampling, different labels represent different strata, and random sampling is performed for each stratum. Considering that the sample size may vary among strata, the roulette wheel selection algorithm was introduced to select among the strata, and the probability of being selected was proportional to the value of the corresponding fitness function $f(x_h)$. For the stratum in which each sample unit was located, the probability of being selected was:

$$P(x_i) = \frac{f(x_i)}{\sum\limits_{h=1}^{L} f(x_h)} \tag{6}$$

$$f(x_h) = \frac{n_h}{n} \tag{7}$$

Subsequently, the cumulative probability of each stratum was calculated and used to represent the sum of the selection probabilities of all the strata. This was equivalent to the 'span' of samples, and the larger the 'span' was, the more likely a sample point in a given stratum was of being selected.

Finally, a randomization function was used to randomly generate numbers between 0 and 1, and the results were compared to those obtained based on a cumulative probability distribution. Sampling was performed in the stratum with which the randomly generated number was associated. To reduce the stochasticity of sampling and algorithms, each sampling method was implemented 20 times to compare the average sampling efficiency.

4. Results

In this study, we proposed an SSO model for optimizing the layout and configuration of satellites and UAVs in a satellite–UAV integrated remote sensing agricultural monitoring system. This model aims to fully harness the potential of integration between satellites and UAVs.

The SSO model was applied to the field survey data in the study area. It produced 16 optimized sampling plans that met the required accuracy levels for sampling. These plans varied in sampling point numbers and locations. Next, we compared the SSO-optimized sampling scheme with the conventional sampling schemes used in four control groups.

4.1. Sampling Optimization

Figures 4–8 show the results of spatial sampling optimization of the satellite–UAV integrated remote sensing agricultural monitoring system.

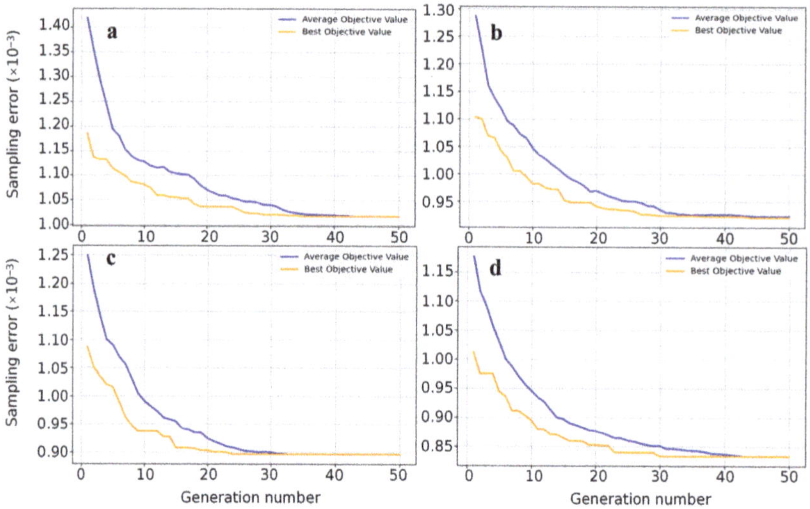

Figure 4. Sampling error optimization curves for the study area (partial). (**a**–**d**) represent cases with 13 to 16 sampling points, respectively.

The subplots labeled a, b, c, and d in Figure 4 display the optimization curves for the minimum sampling error. These curves represent the results obtained when we have 13, 14, 15, and 16 sampling points, respectively. It is evident that the error values consistently decrease as the EGA progresses through 50 generations, ultimately reaching the minimum values. The average sampling error value of the population was reduced by more than 27.3%. The error value for the best individual in the population was reduced by about 16.1%.

The subplots labeled a, b, c, and d in Figure 5 show the optimization curves for the total distance traveled when we have 13, 14, 15, and 16 sampling points, respectively. The distance the UAV flies depends on the order in which it visits the sampling points. The UAV's flight distance is continuously improved and reduced using EGA. The average sampling distance can be reduced by 7000 to 8000 m through EGA optimization. Most of the final distances fall between 5000 m and 6000 m, less than half of the non-optimized paths. The EGA is highly effective in solving this problem.

The subplots labeled a, b, c, and d in Figure 6 show the flight paths of the UAVs in the study area when the number of sampling points is 13, 14, 15, and 16, respectively, where point O is the center of the study area, which also serves as the takeoff point and recovery point of the UAV. The sampling points of the UAV sampling survey in the figure are more evenly distributed in the figure, and the sampling points in the northern region are slightly denser, which is relatively in line with the actual cultivation situation in the study area.

The crops grown in the study area are relatively homogeneous and uniform, and more sampling points are needed to accurately monitor the growth of crops in the more variable parts of the study area.

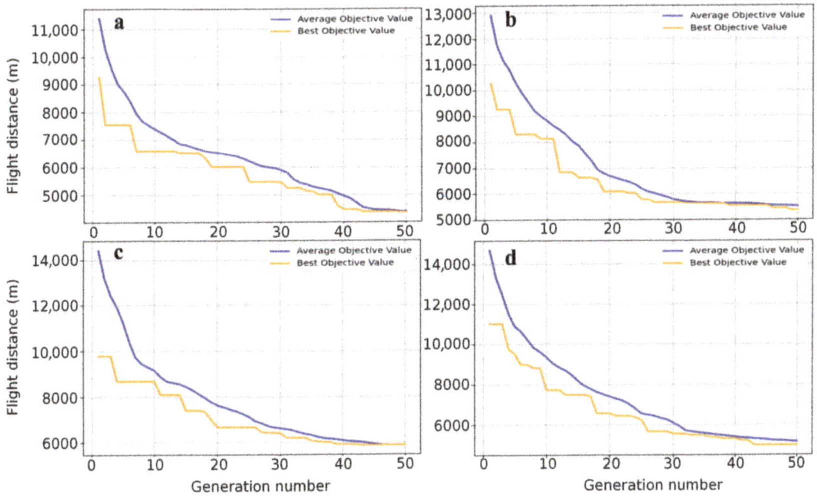

Figure 5. Total sampling distance optimization curves for the study area (partial). (**a**–**d**) represent cases with 13 to 16 sampling points, respectively.

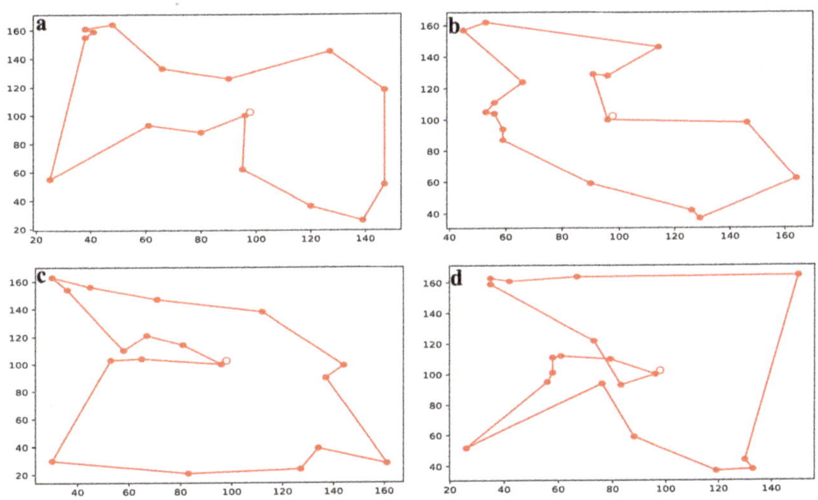

Figure 6. UAV flight path in the study area (partial). (**a**–**d**) represent cases with 13 to 16 sampling points, respectively.

Figure 7 shows the cost-accuracy curve based on the set of sampling plans, where the horizontal coordinate is the total time taken to complete the sampling task, the vertical coordinate is the error used to sample the satellite data with UAV low-altitude remote sensing, and the dot label is the number of sampling points. There are 16 sampling plans in the sampling plans set for different numbers of sampling points. As the number of sampling points increases, the total time to complete the sampling task increases. It takes 739 s to complete the sampling of 5 points, and it takes the most time to sample 20 sampling points, which is 1473 s. With the increase in the number of sampling points, the sampling error shows a decreasing trend as a whole, and the lowest error value is that of setting up 17 sampling points, with an error value of 2.52×10^{-5}.

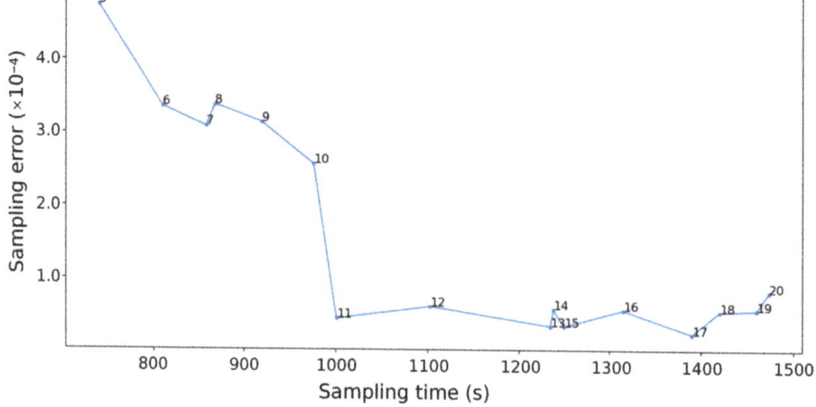

Figure 7. Sampling cost–accuracy curve for the study area.

Overall, the sampling plans with more sampling points consumed more time and cost, but had less sampling error. It is in line with the general principle of sampling, i.e., the accuracy of the estimation can be obtained by increasing costs or the number of samples. The SSO model can obtain considerable accuracy improvement by increasing the cost.

The sampling efficiency of each sampling plan is calculated using Formula (4). We plot the sampling efficiency curve for each sampling plan in the study area, as shown in Figure 8. Among these plans, using 11 to 17 sampling points yields higher efficiency, and the peaks occur at 11, 13, and 15 sampling points, all exceeding 22. The plan with 17 sampling points has the highest efficiency at 28.509. Subsequently, the efficiency decreases. Therefore, the plan with 17 sampling points is the final optimized result.

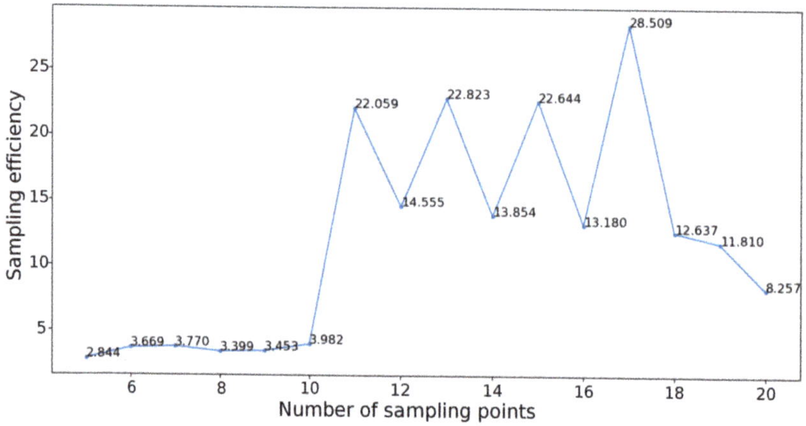

Figure 8. Sampling efficiency curves for each sampling plan in the study area.

4.2. Comparison of Sampling Methods

4.2.1. Maximum Number of Sampling Points

The maximum number of sampling points that can be provided by different sampling methods under the cost constraint was shown in Figure 9. In complex schemes with multiple sampling points, the conventional sampling schemes can no longer meet the predefined cost constraints and can only give 16 to 18 sampling points. The SSO model, on the other hand, is able to minimize costs through planning, resulting in 20 sampling points. This is at least 11.1% more than the number of sampling points given by the conventional sampling schemes.

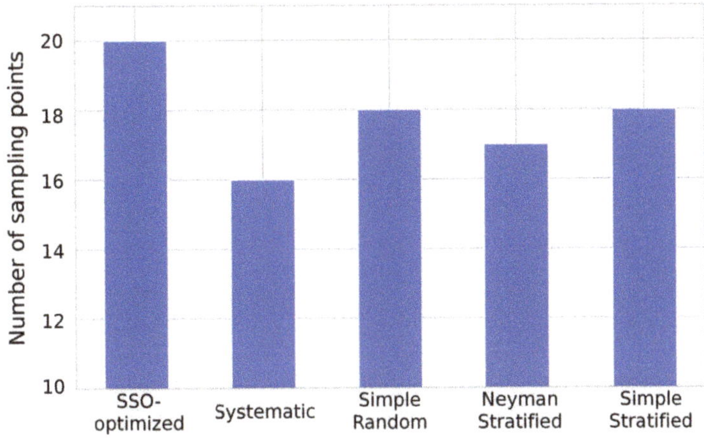

Figure 9. Maximum number of sampling points for each scheme.

4.2.2. Cost and Accuracy

Figure 10 shows the cost and accuracy of the SSO-optimized sampling scheme and the conventional sampling schemes. Among the 12 systematic sampling plans that fulfill the cost and accuracy requirements, the sampling error remains at approximately 2.28×10^{-4} and does not manifestly change as the number of sampling points increases. A total of 14 plans meet the cost and accuracy requirements for simple random sampling, with a sampling error maintained at around 1.45×10^{-4}, which is consistently lower than the error of systematic sampling. For stratified sampling with Neyman's allocation, 13 plans meet the cost and accuracy requirements. The sampling error is significantly lower than systematic and simple random sampling, with an average of 9.10×10^{-5} and slight fluctuations. However, since the number of sampling points selected for the equal stratified sampling must be a multiple of the number of strata, only three plans meet the requirements. These three plans have lower overall error levels than the remaining three conventional sampling plans, ranging from a minimum of 6.42×10^{-5} with 18 sampling points to a maximum of 7.94×10^{-5} with 12 sampling points.

The initial six SSO-optimized plans have higher error values and less obvious optimization effects. Starting from the 11th sampling point, the sampling error is lower than all the conventional sampling plans, and the complex situation of multiple sampling points can effectively compress the cost and provide more alternatives, so the SSO-optimized sampling scheme has a better optimization effect in the plan of the number of sampling points is higher.

Figure 11 demonstrates the difference in sampling efficiency between the SSO-optimized sampling scheme and the conventional sampling schemes. The efficiency of systematic sampling is relatively low, with an average of 4.03, and does not improve significantly when the number of sampling points increases. The efficiency of simple random sampling fluctuates slightly, with a low overall level, the highest being 11.73 at five sampling points. The efficiency of Neyman stratified sampling fluctuates manifestly but is always higher than that of simple random sampling and systematic sampling, with the highest being 20.56 at five sampling points, while there is a wave at 10 and 14 sampling points. Equal stratified sampling has the highest efficiency at six sampling points, followed by a decrease. The efficiency of the SSO-optimized sampling scheme is higher than conventional sampling schemes in all plans with more than 11 sampling points, with the highest efficiency of 28.51 at 17 sampling points. The highest efficiency among the conventional sampling plans was achieved by Neyman stratified sampling with 5 sampling points, yielding an efficiency of 20.56. In contrast, the SSO-optimized plan included 17 sampling points and achieved an efficiency of 28.51, corresponding to a 38.7% improvement over the best-performing conventional scheme. Moreover, when the number of sampling points exceeded 11, all conventional sampling methods performed worse than the SSO-optimized plan.

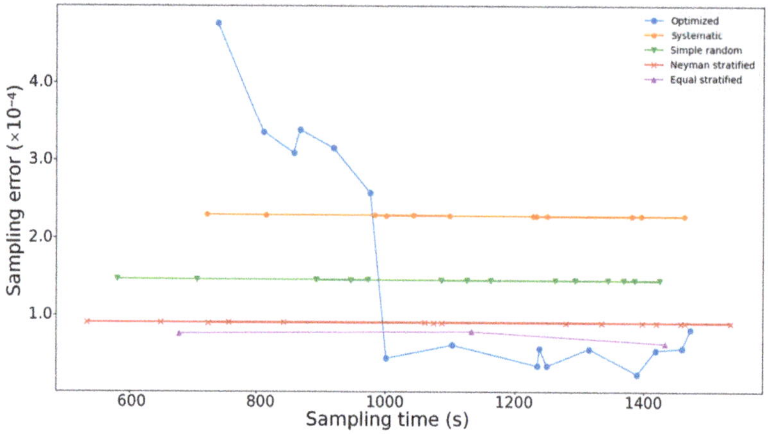

Figure 10. Comparison of cost–accuracy curves for each plan.

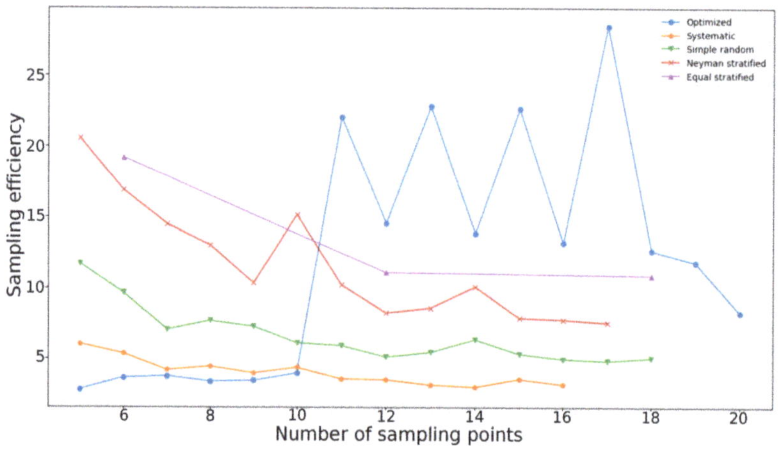

Figure 11. Comparison of sampling efficiency curves for each plan.

To quantify the variability across the 20 repeated runs, we calculated the standard deviation of sampling efficiency for each sampling method and each sampling size. Across all methods, the standard deviations were generally small (coefficient of variation less than 5%), indicating limited algorithmic stochasticity relative to the performance gaps among methods.

4.2.3. SSO-Optimized Plan

Figure 12 shows the distribution of the 17 sampling points in the SSO-optimized plan within the study area.

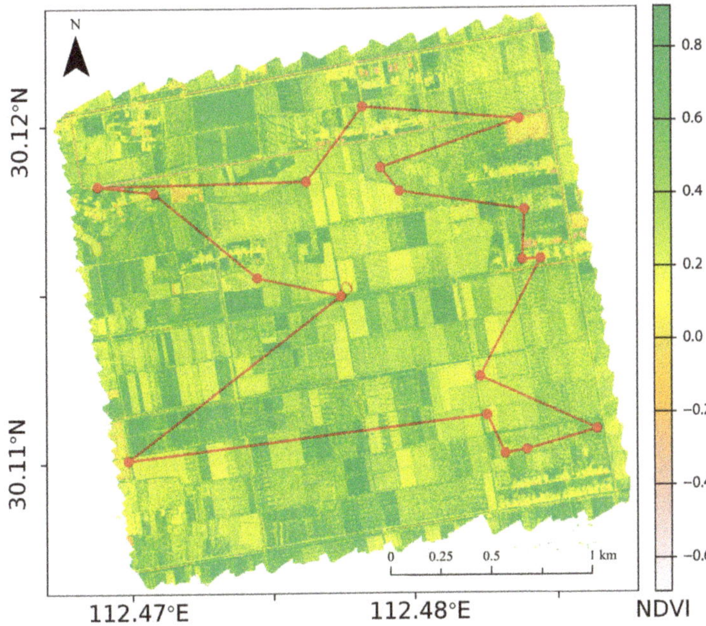

Figure 12. UAV flight route map for the SSO-optimized sampling plan, with red points connected by lines indicating the sampling points along the flight path.

In the SSO-optimized sampling scheme, 11 sampling points were located in the northern part of the study area, and 6 points were located in the southern part of the study area, in addition to the UAV recovery point *O*. The sampling points are more densely distributed in the northern region of the study area, which is in line with the characteristics of crop growth changes in the study area, i.e., the northern growth condition fluctuates more.

5. Discussion

5.1. The Unique Value of EGA in Solving SSO Model

EGA is one of the representative heuristic algorithms [29] that centers on finding the optimal solution in the search space through genetic operations such as selection, crossover, and mutation. Our results showed that EGA can solve the SSO model smoothly. Even in the case of different sampling points, EGA can reduce the sampling error and sampling distance very well. This study confirmed the stability of EGA in efficiently solving the sampling task across various sampling points within the designated study area. Moreover, our study found that EGA does not need to be driven by massive amounts of labeled data [30]. It focuses on the objective function of the optimization problem rather than relying on data for learning. Even with limited data, GAs can search for optimal solutions through evolutionary operations. Compared to machine learning methods, it can not only achieve parallelization acceleration through independent individual fitness evaluation. Besides, EGA does not require any a priori modeling assumptions [31]. It directly evaluates the merits of the solution through a fitness function, which allows it to be easily transferred to solving other problems [27]. These advantages make the SSO model a universally applicable method that can be directly applied to spatial sampling design in different study areas.

5.2. SSO Model Improved the Sampling Efficiency

The SSO model achieved better performance than conventional sampling schemes (simple random sampling, systematic sampling, equal stratified sampling, and Neyman stratified sampling), largely due to the spatial heterogeneity inherent in agroecosystems. Crop growth rarely follows uniform spatial patterns because it is influenced by multiple factors at the field scale, such as soil conditions, crop genotype, and agricultural management. By incorporating prior spatial information and explicitly optimizing sampling locations, the SSO model is better able to capture this heterogeneity than simple random or systematic sampling. Consequently, in areas with strong spatial variability, the SSO-optimized sampling plan naturally allocates more sampling points to regions with larger NDVI variation. This spatial adaptiveness distinguishes the SSO approach from conventional schemes [10]. Compared with equal and Neyman stratified sampling, the SSO model can further improve sampling efficiency under the same cost constraints by optimizing both the placement of sampling points and the UAV flight path. In addition, conventional schemes showed only marginal accuracy improvements when sampling cost increased in our study area, likely due to the cost constraints and the monoculture setting.

The superior performance of the SSO model can also be explained by the spatial structure of NDVI in agricultural fields. NDVI typically exhibits clear spatial autocorrelation, reflecting gradual changes caused by underlying biophysical processes. By jointly optimizing the number and spatial configuration of UAV sampling points, the SSO model captures these structured spatial patterns more effectively and reduces uncertainty in NDVI estimation. The SSO-optimized sampling plan enhances the representativeness of UAV observations and strengthens the complementarity between satellite and UAV-based remote sensing. This mechanistic foundation explains why the SSO model consistently achieves higher sampling efficiency and estimation accuracy within the satellite–UAV integrated remote sensing framework.

5.3. Limitations and Future Directions

In solving the optimization problem of spatial sampling, the SSO model was capable of generating sampling schemes with different numbers of sampling points that satisfied the accuracy and cost requirements. However, the efficiency of the SSO-optimized sampling scheme is lower than conventional sampling schemes when the number of sampling points is small ($\leq$10). This is mainly because the search space becomes highly discrete under small-sample configurations, which limits the potential for layout optimization and increases the sensitivity of GAs to initial population settings. As a result, the algorithm may more easily become trapped in local optima [32]. To mitigate this issue, users may adjust the algorithm settings, such as increasing population diversity, using multi-start initialization, or employing hybrid optimization strategies, depending on the scale of the problem.

Additionally, although we have successfully optimized UAV sampling in rapeseed fields, the generalizability of the proposed SSO model to different crops and heterogeneous agricultural systems remains to be further validated. In the future, additional practical factors, such as the number of UAVs, battery replacements, and instrument depreciation rate are likely necessary to provide a more comprehensive measurement of the sampling cost. Moreover, the same optimization logic can be extended to integrate ground-based measurements by incorporating ground sampling efficiency as an additional cost layer.

This extension will contribute to developing space–air–ground integrated networks in future research.

6. Conclusions

In this paper, we conceptualized the configuration of the satellite–UAV integrated agricultural monitoring system as a spatial sampling problem and proposed an SSO model to optimize UAV sampling within the satellite monitoring coverage, thereby improving the efficiency of satellite–UAV integration. In the study, we found the following:

- The SSO model can improve the efficiency of observations by optimizing the sampling design. Under the same cost constraint, the SSO model increased the number of sampling points by at least 11.1% and the sampling efficiency was at least 38.7% higher than the conventional sampling plans.
- The heuristic algorithm helps to solve the multi-objective optimization problem. The EGA can efficiently solve the SSO model optimally. The average sampling error of the final SSO-optimized plan was reduced by about 27.3%, and the sampling distance was shortened by 7000 to 8000 m.
- The problem of optimizing the layout of satellite remote sensing and UAV-based remote sensing can be transformed into a spatial sampling problem. It is necessary to consider cost constraints and increase efficiency by applying optimization methods in agricultural monitoring and government census in the future.

Author Contributions: Conceptualization, Z.Z., H.X. and B.X.; methodology, Z.Z. and H.X.; software, Z.Z.; validation, Y.Y. and J.Z.; formal analysis, H.X. and B.X.; investigation, J.Z.; resources, H.X.; data curation, J.Z.; writing—original draft preparation, Z.Z.; writing—review and editing, H.X., Y.Y. and B.X.; visualization, Z.Z.; supervision, H.X.; project administration, H.X.; funding acquisition, H.X. All authors have read and agreed to the published version of the manuscript.

Funding: This research was funded by the Major Program of National Social Science Foundation of China under Grant No. 22&ZD079.

Data Availability Statement: Data will be made available on request.

Conflicts of Interest: The authors declare no conflicts of interest.

Abbreviations

The following abbreviations are used in this manuscript:

UAV	Unmanned aerial vehicles
EGA	Elite genetic algorithm
GA	Genetic algorithm
SSO	Spatial sampling optimization
TSP	Traveling salesman problem

References

1. Ghamisi, P.; Rasti, B.; Yokoya, N.; Wang, Q.; Hofle, B.; Bruzzone, L.; Bovolo, F.; Chi, M.; Anders, K.; Gloaguen, R.; et al. Multisource and Multitemporal Data Fusion in Remote Sensing: A Comprehensive Review of the State of the Art. *IEEE Geosci. Remote Sens. Mag.* **2019**, *7*, 6–39. [CrossRef]
2. Zhao, Z.; Wang, F.; Xiao, Y.; Liu, L.; Chen, N.; Weng, W.; Peng, D.; Zhang, X.; Shi, Z. Development and Evaluation of Remote Sensing-Derived Relative Growth Rate (RGR) for Rice Yield Estimation. *IEEE J. Sel. Top. Appl. Earth Obs. Remote Sens.* **2025**, *18*, 25735–25747. [CrossRef]
3. Alvarez-Vanhard, E.; Corpetti, T.; Houet, T. UAV & Satellite Synergies for Optical Remote Sensing Applications: A Literature Review. *Sci. Remote Sens.* **2021**, *3*, 100019. [CrossRef]
4. Kattenborn, T.; Lopatin, J.; Förster, M.; Braun, A.C.; Fassnacht, F.E. UAV Data as Alternative to Field Sampling to Map Woody Invasive Species Based on Combined Sentinel-1 and Sentinel-2 Data. *Remote Sens. Environ.* **2019**, *227*, 61–73. [CrossRef]

5. Bhardwaj, A.; Sam, L.; Akanksha; Martín-Torres, F.J.; Kumar, R. UAVs as Remote Sensing Platform in Glaciology: Present Applications and Future Prospects. *Remote Sens. Environ.* **2016**, *175*, 196–204. [CrossRef]
6. Alvarez-Vanhard, E.; Houet, T.; Mony, C.; Lecoq, L.; Corpetti, T. Can UAVs Fill the Gap Between In Situ Surveys and Satellites for Habitat Mapping? *Remote Sens. Environ.* **2020**, *243*, 111780. [CrossRef]
7. Gevaert, C.M.; Suomalainen, J.; Tang, J.; Kooistra, L. Generation of Spectral–Temporal Response Surfaces by Combining Multispectral Satellite and Hyperspectral UAV Imagery for Precision Agriculture Applications. *IEEE J. Sel. Top. Appl. Earth Obs. Remote Sens.* **2015**, *8*, 3140–3146. [CrossRef]
8. Bazrafkan, A.; Igathinathane, C.; Bandillo, N.; Flores, P. Optimizing Integration Techniques for UAS and Satellite Image Data in Precision Agriculture—A Review. *Front. Remote Sens.* **2025**, *6*, 1622884. [CrossRef]
9. Kuenzer, C.; Dech, S.; Wagner, W. Remote Sensing Time Series Revealing Land Surface Dynamics: Status Quo and the Pathway Ahead. In *Remote Sensing Time Series: Revealing Land Surface Dynamics*; Kuenzer, C., Dech, S., Wagner, W., Eds.; Remote Sensing and Digital Image Processing; Springer International Publishing: Cham, Switzerland, 2015; pp. 1–24. [CrossRef]
10. Wang, J.F.; Stein, A.; Gao, B.B.; Ge, Y. A Review of Spatial Sampling. *Spatial Stat.* **2012**, *2*, 1–14. [CrossRef]
11. Stevens, D.L., Jr.; Olsen, A.R. Spatially Balanced Sampling of Natural Resources. *J. Am. Stat. Assoc.* **2004**, *99*, 262–278. [CrossRef]
12. Xiao, Y.; Zhao, Z.; Huang, J.; Huang, R.; Weng, W.; Liang, G.; Zhou, C.; Shao, Q.; Tian, Q. The Illusion of Success: Test Set Disproportion Causes Inflated Accuracy in Remote Sensing Mapping Research. *Int. J. Appl. Earth Obs. Geoinf.* **2024**, *135*, 104256. [CrossRef]
13. Dunn, R.; Harrison, A.R. Two-Dimensional Systematic Sampling of Land Use. *J. R. Stat. Soc. C (Appl. Stat.)* **1993**, *42*, 585–601. [CrossRef]
14. Lu, Y.; Xie, H.; Zhang, J.; Jin, Y.; Feng, Y.; Gong, Y.; Han, W.; Zhang, H.; Tong, X. Fractal Theory Based Stratified Sampling for Quality Assessment of Remote-Sensing-Derived Geospatial Data. *IEEE J. Sel. Top. Appl. Earth Obs. Remote Sens.* **2023**, *16*, 7100–7111. [CrossRef]
15. Wang, J.F.; Christakos, G.; Hu, M.G. Modeling Spatial Means of Surfaces with Stratified Nonhomogeneity. *IEEE Trans. Geosci. Remote Sens.* **2009**, *47*, 4167–4174. [CrossRef]
16. Yang, M.; Chen, S.; Xu, D.; Hong, Y.; Li, S.; Peng, J.; Ji, W.; Guo, X.; Zhao, X.; Shi, Z. Strategies for Predicting Soil Organic Matter in the Field Using the Chinese Vis-NIR Soil Spectral Library. *Geoderma* **2023**, *433*, 116461. [CrossRef]
17. Lu, R.; Liao, R.; Meng, R.; Hu, Y.; Zhao, Y.; Guo, Y.; Zhang, Y.; Shi, Z.; Ye, S. Strategic Sampling for Training a Semantic Segmentation Model in Operational Mapping: Case Studies on Cropland Parcel Extraction. *Remote Sens. Environ.* **2025**, *331*, 115034. [CrossRef]
18. Zhang, Z.; Zhu, L. A Review on Unmanned Aerial Vehicle Remote Sensing: Platforms, Sensors, Data Processing Methods, and Applications. *Drones* **2023**, *7*, 398. [CrossRef]
19. Wang, L.; Zhang, X.; Qin, K.; Wang, Z.; Zhou, J.; Song, D. AoI Analysis of Satellite–UAV Synergy Real-Time Remote Sensing System. *Remote Sens.* **2024**, *16*, 3305. [CrossRef]
20. Zhao, H.; Ji, F.; Wang, Y.; Yao, K.; Chen, F. Space–Air–Ground–Sea Integrated Network with Federated Learning. *Remote Sens.* **2024**, *16*, 1640. [CrossRef]
21. Haining, R. *Spatial Data Analysis: Theory and Practice*; Cambridge University Press: Cambridge, UK, 2003. [CrossRef]
22. *GB/T 39612-2020*; Specifications for Low-Altitude Digital Aerial Photography and Data Processing. Ministry of Natural Resources, People's Republic of China: Beijing, China, 2020.
23. Lyu, Y.; Wang, S.; Hu, T.; Pan, Q. UAV Coverage Path Planning of Multiple Disconnected Regions Based on Cooperative Optimization Algorithms. *IEEE Trans. Cogn. Dev. Syst.* **2025**, *17*, 259–270. [CrossRef]
24. Alemayehu, T.S.; Kim, J.H. Efficient Nearest Neighbor Heuristic TSP Algorithms for Reducing Data Acquisition Latency of UAV Relay WSN. *Wirel. Pers. Commun.* **2017**, *95*, 3271–3285. [CrossRef]
25. Dong, R.; Wang, S.; Wang, G.; Wang, X. Hybrid Optimization Algorithm Based on Wolf Pack Search and Local Search for Solving Traveling Salesman Problem. *J. Shanghai Jiaotong Univ. (Sci.)* **2019**, *24*, 41–47. [CrossRef]
26. Zeng, Y.; Li, J.; Liu, Q.; Li, L.; Xu, B.; Yin, G.; Peng, J. A Sampling Strategy for Remotely Sensed LAI Product Validation over Heterogeneous Land Surfaces. *IEEE J. Sel. Top. Appl. Earth Obs. Remote Sens.* **2014**, *7*, 3128–3142. [CrossRef]
27. Holland, J.H. Outline for a Logical Theory of Adaptive Systems. *J. ACM* **1962**, *9*, 297–314. [CrossRef]
28. Suzuki, J. A Markov Chain Analysis on Simple Genetic Algorithms. *IEEE Trans. Syst. Man Cybern.* **1995**, *25*, 655–659. [CrossRef]
29. Ilin, V.; Simić, D.; Simić, S.D.; Simić, S.; Saulić, N.; Calvo-Rolle, J.L. A Hybrid Genetic Algorithm, List-Based Simulated Annealing Algorithm, and Different Heuristic Algorithms for the Travelling Salesman Problem. *Log. J. IGPL* **2023**, *31*, 602–617. [CrossRef]
30. Baeck, T.; Fogel, D.B.; Michalewicz, Z. (Eds.) *Handbook of Evolutionary Computation*; CRC Press: Boca Raton, FL, USA, 1997. [CrossRef]
31. Jordan, M.I.; Mitchell, T.M. Machine Learning: Trends, Perspectives, and Prospects. *Science* **2015**, *349*, 255–260. [CrossRef]
32. Katoch, S.; Chauhan, S.S.; Kumar, V. A Review on Genetic Algorithm: Past, Present, and Future. *Multimed. Tools Appl.* **2021**, *80*, 8091–8126. [CrossRef]

remote sensing

MDPI

Article

Marginal Contribution Spectral Fusion Network for Remote Hyperspectral Soil Organic Matter Estimation

Jiaze Tang [1], Dan Liu [1], Qisong Wang [1,*], Junbao Li [2], Jingxiao Liao [3] and Jinwei Sun [1]

1 Electrical Measurement Technology and Intelligent Control Institute, School of Instrumentation Science and Engineering, Harbin Institute of Technology, Harbin 150080, China; jiazetang@hit.edu.cn (J.T.)
2 A Information Countermeasure Technique Institute, School of Computer Science and Technology, Harbin Institute of Technology, Harbin 150080, China
3 Department of Industrial and Systems Engineering, The Hong Kong Polytechnic University, Hung Hom, Kowloon, Hong Kong
* Correspondence: wangqisong@hit.edu.cn

Abstract

Soil organic matter (SOM) is a fundamental indicator of soil health and a major component of the global carbon cycle; its accurate quantification is essential for sustainable agriculture. Conventional chemical assays yield only point-based soil measurements and miss the spatial distribution of soil elements; airborne hyperspectral remote sensing has emerged as a promising approach for the quantitative measurement and characterization of SOM. Inversion models translate hyperspectral data into quantitative SOM estimates. However, existing models rely solely on a single preprocessing pathway, limiting their ability to fully exploit available spectral information. We address these limitations by developing a marginal contribution-driven spectral fusion network (MC-SFNet) that conducts feature-level fusion of heterogeneous preprocessing outputs within a physics-guided deep architecture. Moreover, the combination of data-driven fusion and the Kubelka–Munk (KM) model yields more physically interpretable spectral features, advancing beyond prior purely data-driven methods. We validated MC-SFNet on a self-constructed remote sensing, high-throughput hyperspectral dataset comprising 200 black soil samples from Northeastern China (400–1000 nm, 256 bands). Experimental results show that our network reduces the RMSE by 10.7% relative to the prevailing generalized hyperspectral soil-inversion model. The proposed method provides a novel preprocessing pathway for forthcoming airborne high-throughput hyperspectral missions to extract soil-specific spectral information more effectively and further enhance large-scale SOM retrieval accuracy.

Keywords: remote sensing; SOM; soil; hyperspectral; marginal contribution

1. Introduction

Soil is a fundamental resource for agricultural production, and its quality directly influences the sustainability of agricultural practices and the protection of the ecological environment [1,2]. Among various soil attributes, soil organic matter (SOM) content is a key indicator of soil fertility, microbial activity, and overall environmental function [3–6]. Accurate and rapid assessment of SOM is particularly important for precision agriculture, as improper fertilization, whether excessive or insufficient, can lead to suboptimal crop yields and negative environmental impacts. Therefore, accurate and timely measurements of soil fertility are crucial for ensuring that fertilizers are applied in the right amounts, optimizing agricultural productivity, and minimizing the adverse effects of over-fertilization or under-fertilization on the environment [7]. Emerging remote-sensing techniques, especially high-throughput hyperspectral imaging, offer a viable route to invert spatially continuous SOM information from field to regional scales [8]. Inversion methods, which utilize spectral data, provide an efficient way to estimate SOM content across large areas. When these subtle spectral signatures are regressed against laboratory-measured samples using multivariate calibration or modern machine learning techniques, up to 80% of the variance in SOM content can be explained under controlled conditions [9]. This laboratory relationship is scalable: an airborne hyperspectral imager, when combined with a regional soil spectral library, produced 1 m resolution maps of SOM stock in Italy's Sele River Plain [10], while Unmanned Aerial Vehicle (UAV) hyperspectral cubes, calibrated with only twenty soil samples, achieved 1 m resolution field-scale SOM predictions, far surpassing the accuracy of ordinary kriging interpolation [11].

Spaceborne imaging spectrometers have extended this capability from small plots to large regions: a meta-learner improved prediction accuracy for Moroccan croplands by approximately 48.8% compared to individual models like Random Forest, Support Vector Regression, and PLSR. The model's coefficient of determination, R^2, reached 0.65, with an RMSE of 0.194% and an RPIQ of 2.247 [12]. The first evaluation showed that selecting thirty standard normal variate-corrected bands for 282 samples yielded more accurate results, with R^2 increasing to 0.68 and RMSE decreasing to 0.34%, compared to the original full-band model, which had an R^2 of 0.60 [13]. China's GF-5 imagery, after waveletpacket denoising, achieved similar accuracies for SOM mapping at a 30 m resolution, with R^2 increasing from 0.62 to 0.77 after applying wavelet denoising and further improving to 0.84 with 0.6-order wavelet denoising, while RMSE decreased by 0.84 $g \cdot kg^{-1}$ [14]. Meanwhile, tractor-mounted spectrometers adapted to Brazilian Cerrado soils by locally adjusting their calibration sets [15]. Combined hyperspectral analyses of gypsum-rich terrains show that spectral indices with machine learning can simultaneously estimate SOM and distinguish interfering minerals, paving the way for scalable, cloud-based continental-scale digital soil-mapping pipelines [16]. Consequently, quantitative inversion of SOM using hyperspectral data has become a burgeoning area of research in soil science [17–21].

Initial studies predominantly employed linear models, such as partial least squares regression (PLSR), for modeling soil hyperspectral data [22]. For example, Shi [23] successfully utilized PLSR to predict soil total nitrogen content. However, as the field matured, nonlinear approaches gained prominence. Machine learning methods, such as Support Vector Machines (SVMs) and Random Forest (RF), have been shown to outperform traditional linear regressions in hyperspectral SOM estimation. Feng et al. [24] achieved superior results by applying SVM and Random Forest to invert soil arsenic content from hyperspectral data, and other studies have also demonstrated that Random Forest often provides higher accuracy in estimating SOM and soil moisture [25]. Similarly, Chung [26] applied a RF model to invert soil mercury content from hyperspectral data, while Subi [27] employed PLSR to predict chromium in arid zone.

More recently, deep learning techniques have been introduced into the field of SOM inversion. Their capacity to model complex nonlinear relationships and handle large-scale datasets has made them particularly attractive [28]. Kawamura et al. [29] proposed a convolutional neural network (CNN)-based method to estimate phosphorus, while Wang et al. [30] integrated deep learning approaches to predict soil moisture from hyperspectral data. Li et al. [31] employed a 1D CNN to classify soil texture, and Cao et al. [32] used a deep CNN to predict soil organic matter. Ajay [33] leveraged deep learning to analyze hyperspectral data for total soil nitrogen prediction. Collectively, these efforts illustrate the potential of deep learning to improve the accuracy and efficiency of hyperspectral inversion for soil properties, inspiring new directions in precision agriculture and soil quality assessment.

Many studies have also explored hyperspectral data preprocessing strategies to optimize inversion performance [34] further. Zhang [35] demonstrated that first-order derivative preprocessing can effectively highlight key information in soil spectra. At the same time, Jiachen [36] identified standard normal variate (SNV) as the most suitable preprocessing method for hyperspectral soil inversion. In contrast, X. B. [37] argued that deep learning, such as CNNs with end-to-end architectures, can eliminate the need for explicit preprocessing.

Despite these advances, understanding of inverted SOM in hyperspectral wavelengths under different preprocessing methods is still limited. To address this gap, we therefore extend the marginal contribution feature importance (MCI) [38] paradigm, constructing a bespoke analysis that tracks wavelength-wise marginal gains under varied treatments and underscoring the risk of relying on any single treatment.

This paper makes three primary contributions: (1) We introduce a wavelength-wise, marginal contribution framework that quantifies the predictive value of every spectral band under each preprocessing operator, producing an interpretable importance map for band selection and computational pruning; (2) we develop a hybrid preprocessing scheme that couples the Kubelka–Munk (KM) optical model with data-driven treatments, yielding features linked to intrinsic absorption while retaining the multivariate structure of reflectance, thereby enhancing physical interpretability; (3) we propose a physics-guided deep network that performs feature-level fusion of heterogeneous preprocessing outputs. Fusion weights are initialized from the marginal contribution scores and refined end-to-end, overcoming the limitations of single-preprocessing pipelines and enabling higher accuracy with fewer, more informative bands.

2. Proposed Method

The overall workflow of this study is illustrated in Figure 1. The research procedure includes the following steps: (1) collecting 200 black soil samples at 45.74°N, 127.39°E in Northeastern China; (2) acquiring the original soil spectra and measuring the SOM content using the potassium dichromate volumetric method; (3) applying various preprocessing methods to the soil spectral data; (4) conducting marginal contribution analysis of different preprocessing strategies using the marginal contribution approach within a unified framework; (5) the design of a marginal contribution-driven spectral fusion network (MC-SFNet) architecture for the inverse regression of soil spectra.

2.1. Spectral Preprocessing

To highlight spectral features and reduce noise, the MC-SFNet integrates several preprocessing methods usually used in contemporary studies [23,24,34,36], with their visualizations shown in Figure 2. Savitzky–Golay (SG) [39] Filter: Smooths and differentiates the original spectra to reduce high-frequency noise while preserving key spectral features. First-Order Derivative (SG1) [40]: Calculates the first-order difference in the spectral signals to emphasize absorption wavelength edges and mitigate systematic errors caused by background shifts. Second-order Derivative (SG2): Further enhances subtle absorption features while reducing nonlinear background interference. SNV [41]: Corrects spectral amplitude variations due to differences in particle size, surface roughness, and scattering, ensuring greater comparability in spectral amplitude and scale. In addition to the data-driven preprocessing methods, this study incorporates the Kubelka–Munk soil spectral model as a physical preprocessing approach [42], enabling the integration of data-driven and physical-modeling-based preprocessing. Kubelka–Munk (KM) Transformation: Converts soil reflectance data into absorption coefficients (K values) using the Kubelka–Munk model to relate the spectral data more closely to intrinsic optical properties, thereby improving the interpretability and stability of SOM inversion.

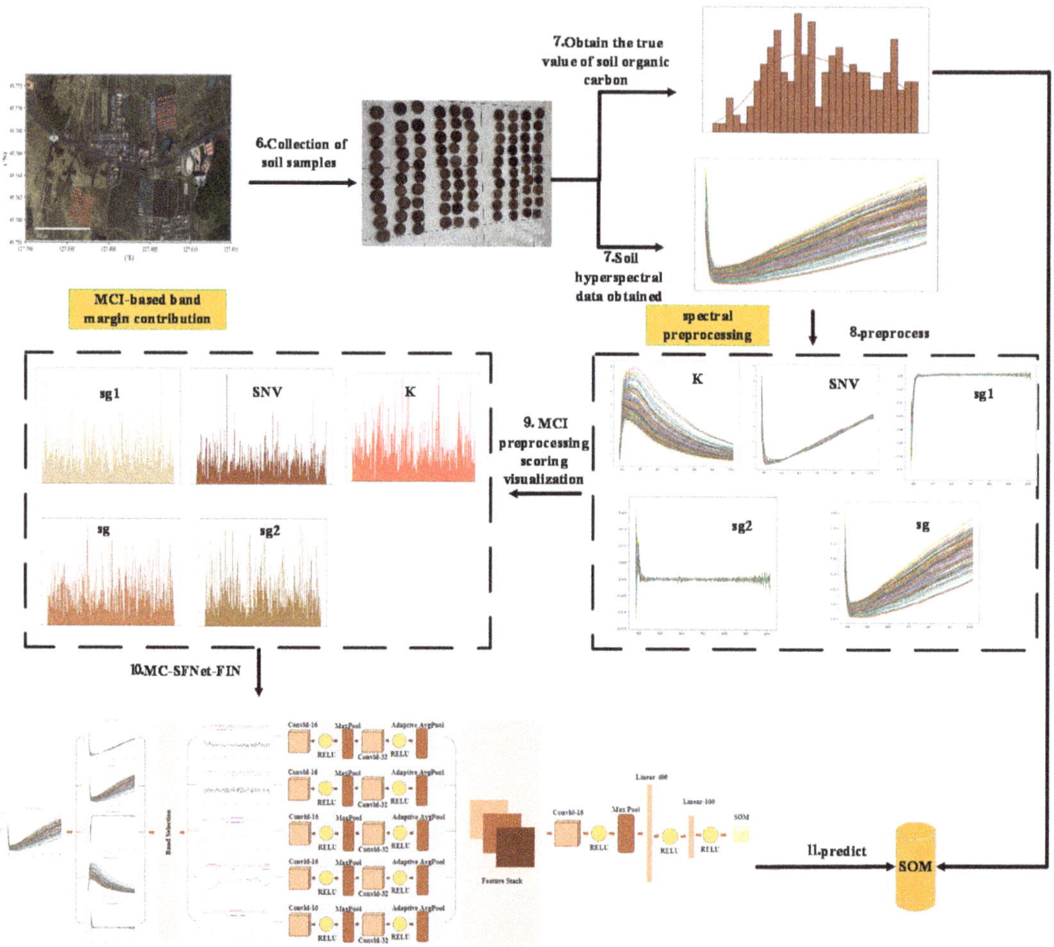

Figure 1. Workflow of the data analysis methodology applied in this study.

The KM model, introduced by Kubelka and Munk in 1931 [42], relies on two key assumptions: (1) light propagates only forward and backward within the material, and (2) the material is homogeneous at a macroscopic scale. Using the KM model, reflectance (R) is transformed into the absorption coefficient (K) in Equation (1).

$$K = \frac{(1-R)^2}{2R} \tag{1}$$

In this study, the MC-SFNet integrates both data-driven preprocessing and the KM physical model. First, multiple preprocessing steps generate a diverse set of spectral features. Then, the KM model is applied to compute the K, which is fused with the preprocessed reflectance data to form feature sets that are richer and more physically meaningful. This hybrid strategy retains the multivariate feature information inherent in the spectral data, while the incorporation of the KM model enhances interpretability.

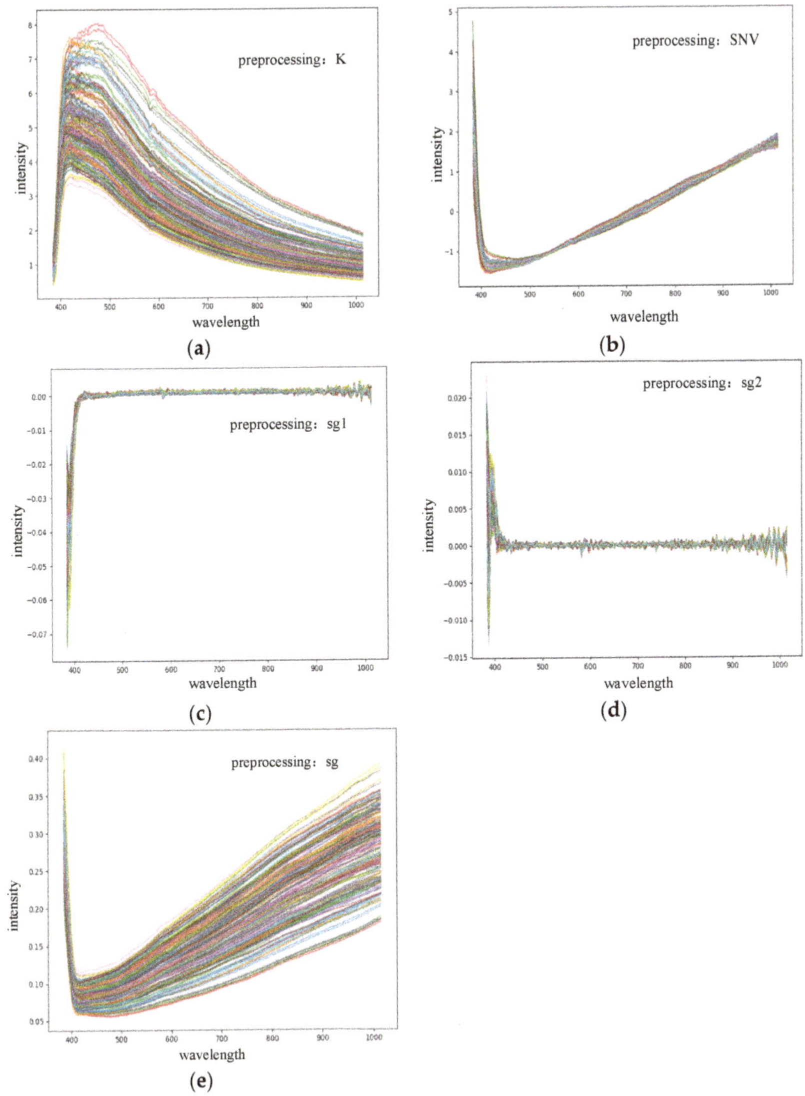

Figure 2. Visualization of different preprocessing methods: (**a**) KM preprocessing, (**b**) SNV preprocessing, (**c**) SG1 preprocessing, (**d**) SG2 preprocessing, (**e**) SG preprocessing.

2.2. Unified Permutation–Gain Analysis for Multi-Preprocessed Spectra

In practical soil spectroscopy applications, strong inter-band collinearity produces substantial wavelength redundancy; this not only enlarges the feature space and elevates on-board computational cost, but can also induce a slight accuracy decline when the estimator is over-parameterised [43–45]. Existing band reduction studies typically mitigate redundancy by running a single preprocessing route and then clustering the spectra—K-means, spectral clustering or similar—to keep one representative wavelength per cluster [46–48]. Although these unsupervised heuristics offer only coarse dimensionality control, they still fail to clarify how individual bands affect inversion accuracy and provide no physical interpretability, thereby constraining subsequent model extensibility [49].

Because the MC-SFNet architecture concatenates K heterogeneous preprocessing streams, the effective number of input channels grows from N to KN; redundancy is therefore magnified rather than suppressed. To address this issue, we introduce a marginal contribution permutation–gain analysis that assigns an explicit importance score to every "preprocess-band" pair and enables fine-grained pruning before the network is deployed. MCI overcomes the limitations of traditional importance measures such as Shapley-value-based methods [50] and mitigates covariance-induced attribution bias among correlated features.

Let $R = [r_{mn}] \in \mathbb{R}^{M\times N}$ denote the matrix of raw bidirectional reflectance, where M is the number of soil samples and N indexes the discrete wavelengths between 400 nm and 1000 nm. Consider a set of K complementary preprocessing operators $\{\mathcal{P}_1, \ldots, \mathcal{P}_K\}$ in Equation (2). Applying each operator independently to R yields the following:

$$\mathbf{X}^{(k)} = \mathcal{P}_k(R) \in \mathbb{R}^{M\times N},\ k = 1, \ldots, K \tag{2}$$

And concatenating these matrices along the spectral dimension produces $\mathbf{X}$ in Equation (3).

$$\mathbf{X} = \left[\mathbf{X}^{(1)}\middle|\ldots\middle|\mathbf{X}^{(K)}\right] \in \mathbb{R}^{M\times D},\ D = KN \tag{3}$$

Let $\mathbf{y} = [y_1, \ldots, y_M]^\top$ be the chemically determined SOM vector. A one-dimensional convolutional regressor $g_\theta : \mathbb{R}^D \to \mathbb{R}$ is trained by minimizing the empirical mean-square error in Equation (4).

$$\theta^* = \underset{\theta}{\operatorname{argmin}} \frac{1}{M} \sum_{M}^{m=1} (g_\theta(\mathbf{x}_m) - y_m)^2 \tag{4}$$

with $\mathbf{x}_m$ denoting the m-th row of X.

The resulting baseline error over the complete feature set $\mathcal{F} = \{1, \ldots, D\}$ is as in Equation (5).

$$v_0 = \frac{1}{M} \sum_{M}^{m=1} (g_{\theta^*}(\mathbf{x}_m) - y_m)^2 \tag{5}$$

To quantify the contribution of an individual feature $j \in \mathcal{F}$, introduce a permutation operator Π_j that randomly shuffles column j across the sample dimension while keeping all other columns intact. $\Pi_{j,p}$ is a permutation drawn uniformly at random from the symmetric group on M elements. For the p-th independent permutation, the prediction error becomes the quantity in Equation (6).

$$v_{j,p} = \frac{1}{M} \sum_{M}^{m=1} (g_{\theta^*}(\Pi_{j,p}\mathbf{x}_m) - y_m)^2,\ p = 1, \ldots, P \tag{6}$$

where P denotes the number of repetitions. Because the permutation disrupts the statistical dependence between feature j and the target, one necessarily has $v_{j,p} \geq v_0$.

Averaging the error increments over all permutations defines the permutation–gain score in Equation (7).

$$I_j = \frac{1}{P} \sum_{P}^{p=1} (v_{j,p} - v_0) \geq 0 \tag{7}$$

which serves as an unbiased Monte Carlo estimate of the marginal predictive value of feature j while requiring only a single model fit. Index j can be uniquely decomposed as $j = (k-1)N + n$ with $k \in \{1, \ldots, K\}$ and $n \in \{1, \ldots, N\}$.

Collecting these scores produces the importance matrix in Equation (8).

$$\mathbf{I} = \begin{bmatrix} I_{1,1} & \dots & I_{1,N} \\ \vdots & & \vdots \\ I_{K,1} & \dots & I_{K,N} \end{bmatrix} \in \mathbb{R}^{K \times N} \tag{8}$$

whose (k,n)-th entry reflects the independent gain obtained by injecting wavelength λ_n after preprocessing route $\mathcal{P}_k$. Equation (8) therefore delivers a fully quantified importance landscape spanning all preprocessing streams and spectral channels; this matrix simultaneously provides the ranking basis for wavelength selection and supplies the prior weight initialization for the subsequent spectral fusion network.

2.3. Marginal Contribution-Driven Spectral Fusion Network

To optimally integrate multiple preprocessing pathways into a unified network that is informed by both data-driven and physics-based models, we propose two complementary approaches for spectral fusion: MC-SFNet-WIN (wavelength-level integration) and MC-SFNet-FIN (feature-level integration). Both architectures employ a lightweight 1D-CNN [51] backbone for feature extraction and dimensionality reduction, with the primary difference lying in how the spectral features are fused—at the wavelength level in MC-SFNet-WIN and at the feature level in MC-SFNet-FIN.

In the MC-SFNet-WIN approach, the raw spectral data, denoted as $R = [r_{mn}] \in \mathbb{R}^{M \times N}$, where M represents the number of soil samples and N corresponds to the number of spectral bands, is processed through K complementary preprocessing operators $\{P_1, P_2, \dots, P_K\}$. These preprocessing methods include standard approaches like SG, SG1 and SG2, SNV, and the KM physical model. After applying these operators, each produces a preprocessed spectral matrix. This integrated feature set is then used as an input to MC-SFNet for modeling in Figure 3a, where the marginal contribution importance analysis plays a crucial role in selecting the most relevant spectral bands.

First, the MC-SFNet extracts a feature subset from the original spectral data based on marginal contribution. This subset is then passed through the 1D-CNN, which consists of two convolutional layers (with 16 and 32 kernels, respectively, and a kernel size of 3) and corresponding pooling layers. These layers capture multiscale spectral information through hierarchical feature extraction and dimensionality reduction [52]. After concatenating the output feature vectors from all branches, a global convolutional layer (16 kernels) and subsequent max-pooling further compress and refine the features. Finally, the fused features are passed through a fully connected network with one hidden layer (100 ReLU-activated neurons) and a linear output layer, enabling SOM content prediction.

In contrast, the MC-SFNet-FIN approach processes each preprocessing stream independently at the feature level. In this case, each preprocessed spectral matrix X is passed through a shallow 1D-CNN to generate compact feature vectors for each preprocessing method in Figure 3b. The resulting features are then combined using a multi-head weighting block, where the weights for each feature are initialized based on the previously computed MCI scores and refined during training. This fusion strategy ensures that the most significant spectral features, as identified through MCI, are given more weight in the final prediction.

The aggregated feature vector, which combines information from all preprocessing methods, is passed through a fully connected network to predict the SOM content. The key difference between MC-SFNet-WIN and MC-SFNet-FIN lies in the way features are combined: the former fuses at the wavelength level, while the latter operates at the feature level, allowing for a more flexible fusion mechanism.

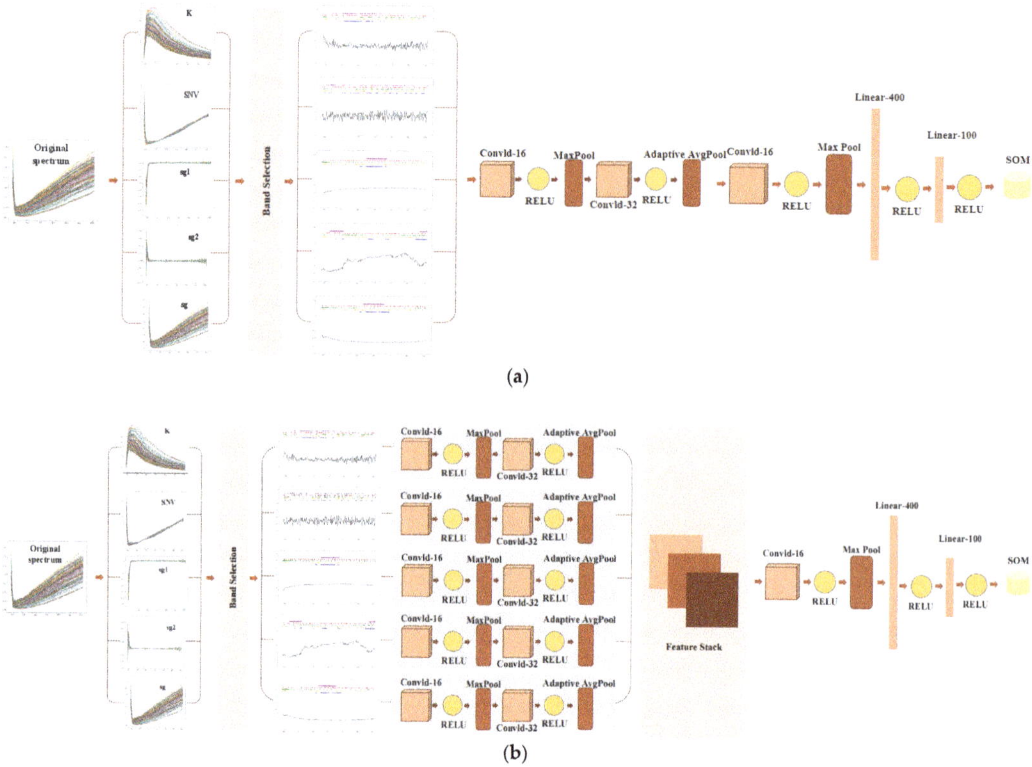

Figure 3. Marginal contribution-driven spectral fusion network: (**a**) wavelength-level spectral fusion; (**b**) feature-level spectral fusion.

One of the unique aspects of our approach is the inclusion of the KM model in the preprocessing stage. The KM transformation, as previously defined in our formula, relates the spectral reflectance data to absorption coefficients. This physical preprocessing step enhances the interpretability of the model by ensuring that the spectral features are not only statistically relevant but also physically meaningful. Specifically, the transformation from reflectance to absorption using the Kubelka–Munk model facilitates a more direct link between the spectral data and the intrinsic properties of the soil. This fusion of data-driven and physical models contributes to the robustness and stability of the spectral inversion process.

Finally, the deep learning model is trained to minimize the error in predicting the SOM content. The training process is based on minimizing the empirical mean-square error θ^* This approach ensures that the model learns to predict SOM content with high accuracy.

In summary, the MC-SFNet-WIN and MC-SFNet-FIN architectures, through the integration of multiple preprocessing streams and the application of marginal contribution-driven feature selection, provide a powerful framework for the inversion of SOM from hyperspectral soil data. By combining data-driven machine learning techniques with physics-based models like Kubelka–Munk, we achieve a robust and interpretable solution for large-scale soil mapping. The use of marginal contribution analysis for feature selection and the incorporation of a 1D-CNN backbone allows for the efficient extraction of relevant spectral features, leading to improved SOM estimation accuracy.

The MC-SFNet network architecture is detailed in Table 1, with the following layer specifications. In Table 1, the parameters are defined as follows: batch_size refers to the number of samples processed in each training step, preprocess indicates the preprocessing method applied to the data, features corresponds to the number of spectral bands or input features, and global_output is the output size of the Global Conv1d layer, set to 128 in this model. These parameters determine the dimensions of the input and output at each layer, ensuring proper data flow and model functionality.

Table 1. MC-SFNet network architecture.

Layer Type	Output Shape
Input	(batch_size, preprocess, features)
Conv1d	(batch_size, 16, features)
Max Pooling	(batch_size, 16, features/2)
Conv1d	(batch_size, 32, features/2)
Max Pooling	(batch_size, 32, features/4)
Adaptive Pooling	(batch_size, 32, 128)
Global Conv1d	(batch_size, 32, global_output)
Linear Layer 1	(batch_size, 400)
ReLU	(batch_size, 400)
Linear Layer 2	(batch_size, 100)
ReLU	(batch_size, 100)
Output	(batch_size, 1)

3. Experiments

3.1. Experimental Settings

As shown in Figure 4, our team previously collected 200 black soil samples as part of a prior study conducted in Northeastern China at the coordinates 45.758°N to 45.772°N and 127.39°E to 127.40°E. Sampling took place across two agricultural fields in the region, with 100 samples collected from each field. The samples were arranged in a 10×10 grid, with each sampling point spaced 15 to 20 m apart. At every point, soil samples were gathered from five cores—four positioned at the corners and one at the center of a 2 m by 2 m area, all spaced more than 1 m apart—and subsequently mixed [4]. In Figure 5, the statistical characteristics of the SOM content are summarized: the mean is 23.77 g kg^{-1}, with a standard deviation of 4.75 g kg^{-1}, a maximum of 27.81 g kg^{-1}, and a minimum of 18.92 g kg^{-1}. The y-axis represents the number of soil samples in each category. To ensure objective and reliable results, the 200 soil samples were randomly divided into five folds for cross-validation. Soil spectral cubes were collected with a GaiaField Pro-V10 push-broom hyperspectral imager (400–1000 nm, 256 bands, 3.5 nm resolution). Because this instrument captures spatially contiguous hyperspectral images in a single scan, it delivers high-throughput, imaging-based data rather than the point measurements produced by traditional non-imaging spectrometers, thereby closely emulating the data characteristics of airborne hyperspectral remote-sensing platforms and providing a more realistic test bed for SOM-inversion algorithms.

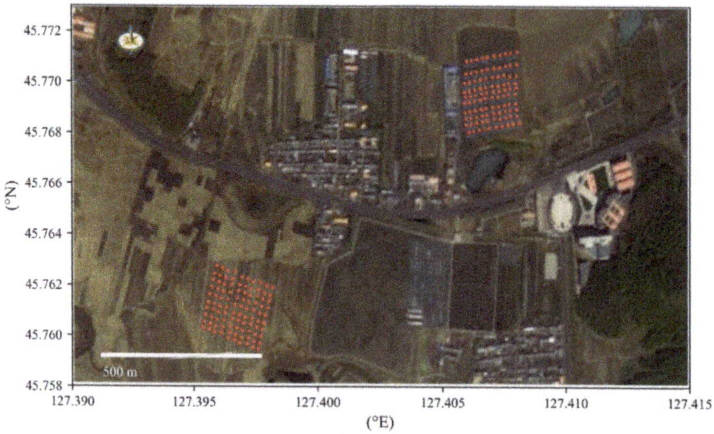

Figure 4. Soil sample collection sites.

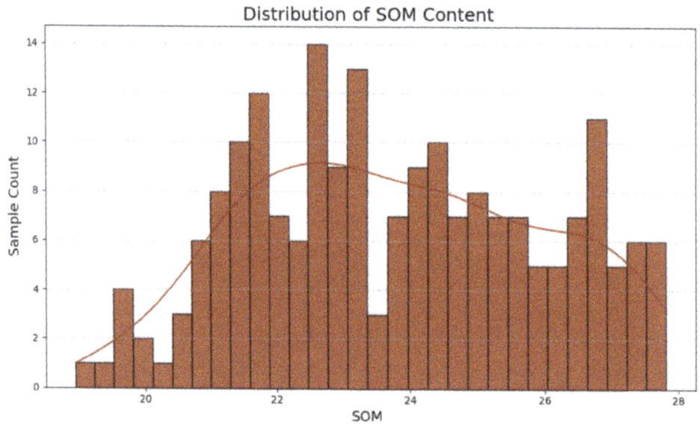

Figure 5. Distribution of soil samples.

3.2. Evaluation Criteria

To assess the performance of SOM inversion using corrected spectra, this study employed three indicators: the coefficient of determination (R^2) [53], the root mean square error (RMSE), and the Pearson correlation coefficient (PCC) [54]. Among these, R^2 measures the goodness of fit between predicted and observed values, ranging from 0 to 1, with values closer to 1 indicating higher predictive accuracy. Let y denote the observed values, $\hat{y}$ the predicted values, and $\overline{y}$ the mean of the observed values; the formula for R^2 is defined as Equation (9):

$$R^2 = 1 - \frac{\sum (y - \hat{y})^2}{\sum (y - \overline{y})^2} \quad (9)$$

The RMSE is calculated as Equation (10):

$$RMSE = \sqrt{\frac{1}{n} \sum_{n}^{i=1} (y_i - \hat{y}_i)^2} \quad (10)$$

In addition, the PCC quantifies the linear correlation between predicted and observed values. Its range extends from −1 to 1, where larger positive values indicate stronger positive correlations. By employing these three metrics—R^2, RMSE, and PCC—this study comprehensively evaluates the model's performance in SOM inversion [44].

3.3. Interpretation of Wavelength Marginal Contributions

Previous research on soil hyperspectral preprocessing has often focused on individual preprocessing methods in isolation [44–46,52], lacking a systematic analysis of their combined effects within a unified framework. To address this gap, the present study integrates multiple preprocessing methods and evaluate the marginal contributions of different preprocessing across all spectral wavelengths. This approach allows for a direct comparison of the strengths and weaknesses of various preprocessing techniques from the perspective of their wavelength-level marginal contributions, providing more precise insights into their effects on SOM inversion accuracy.

In Figure 6, the y-axis is labeled the importance score, representing the relative significance of each wavelength in the applied preprocessing methods. The figure also includes a legend to distinguish between methods like KM, SNV, SG, SG1, and SG2, providing insight into their impact on marginal contributions. The marginal contributions associated with SNV at approximately 550 nm, 750 nm, and 950 nm are considerably higher than those of other preprocessing methods at their respective wavelengths. These wavelengths are notably associated with key functional groups in SOM. The 550 nm feature corresponds to the C-H stretching overtone, while 750 nm is linked to the O-H stretching vibration. Additionally, 950 nm is associated with the O-H stretch in carboxylic acids and phenolic compounds. These functional groups, which are crucial to SOM content, contribute significantly to the spectral response and, consequently, to the inversion accuracy. The high marginal contribution at these wavelengths suggests that SNV preprocessing effectively highlights the absorption features related to these functional groups, thus enhancing SOM inversion performance in these spectral regions. Moreover, integrating multiple preprocessing strategies facilitates complementary advantages, allowing the effects of different preprocessing methods to be superimposed. For example, although SNV preprocessing enhances the useful signals for inverting SOM in general, the signal enhanced by SG2 preprocessing and SG1 preprocessing between 400 nm and 700 nm contributes more to the SOM inversion than SNV, which proves that the different wavelength bands contribute to the accuracy of the SOM inversion in a different and complementary way under different preprocessing. By applying marginal contribution analysis to uniformly evaluate the importance of each spectral band across different preprocessing methods, this study confirms the synergistic gains achievable through combined preprocessing. This reinforces the notion that multi-preprocessing approaches, which leverage the strengths of individual methods, offer superior performance compared to single preprocessing techniques. Thus, this study lays a foundation for the strategic selection and combination of preprocessing approaches, ensuring more accurate and reliable SOM inversion.

In Figure 7, the y-axis is labeled normalized intensity, indicating the intensity values of each wavelength after normalization. Figure 7 visualizes the wavelengths identified by the marginal sharing criterion as particularly influential after each preprocessing step, thereby clearly demonstrating how marginal contribution analysis guides the selection of concise yet highly informative wavelength subsets for subsequent modeling and interpretation.

3.4. Comparative Performance Analysis of MC-SFNet

A stratified five-fold cross-validation protocol was applied to the 200 black soil spectra to quantify the advantage of the proposed MC-SFNet. In each fold, 160 spectra formed the training subset and 40 spectra served as an unseen test subset. To ensure a fair benchmark, every competing approach— RF [55], PLSR [56], SVM [57], 1D-CNN [58]—and the transformer model [59] were trained and evaluated on exactly the same folds. The Transformer architecture used in this study is configured as follows: The Transformer module consists of 1 encoder layer with 4 attention heads. The convolutional part comprises 2 convolutional layers with a kernel size of 3 and a stride of 1. Positional encoding is implemented using sine–cosine encoding, and the outputs from both the CNN, which captures local features, and the Transformer, which models long-range dependencies, are fused through fully connected layers for prediction [59]. For the parameter optimization, grid search was used to find the best hyperparameters for the classical models, and band selection did not show significant improvement under single preprocessing methods in the dataset. For the SVM model, we used radial basis function as kernel function and the parameter gamma from scale [58]. For the PLSR model, we selected the optimal 10 components based on current best practices in the field [60].

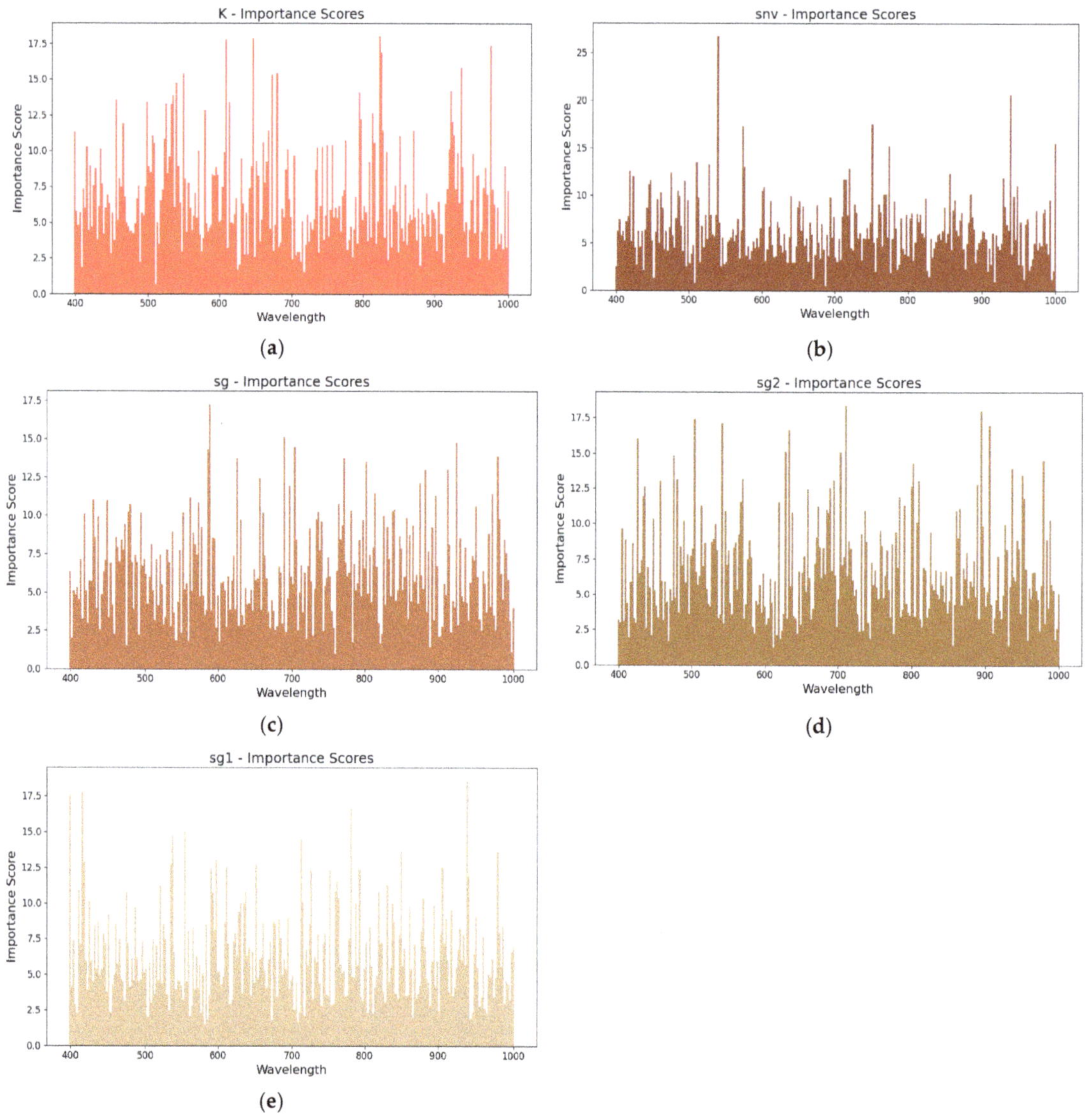

Figure 6. Marginal contributions of wavelengths: (**a**) importance of KM wavelength, (**b**) importance of SNV wavelength, (**c**) importance of SG wavelength, (**d**) importance of SG2 wavelength, (**e**) importance of SG1 wavelength.

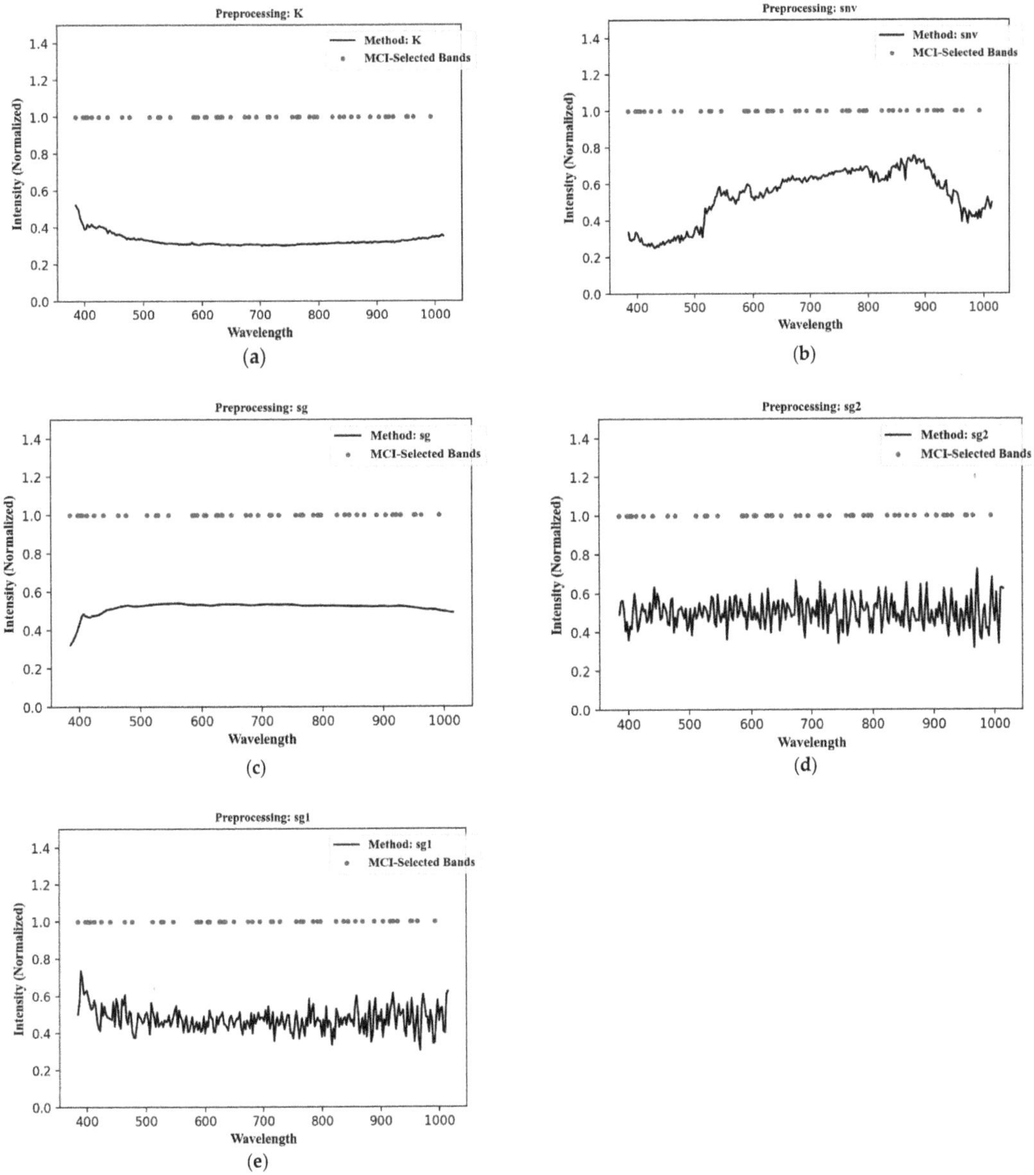

Figure 7. Visualization of wavelength selection: (**a**) wavelength selection for KM, (**b**) wavelength selection for snv, (**c**) wavelength selection for sg, (**d**) wavelength selection for sg2, (**e**) wavelength selection for sg1.

MC-SFNet is trained using the Adam optimizer (learning rate = 0.0001) and MSELoss for 300 epochs with a batch size of 16, where input data is passed through the model, loss is computed, gradients are backpropagated, and model parameters are updated to minimize the loss, ensuring reproducibility of the results. Table 2 aggregates the cross-validated metrics. MC-SFNet-FIN attained an R^2 of 0.80, an RMSE of 0.97 g kg^{-1}, and a PCC of 0.90, thereby outperforming all competing methods. Due to the relatively sample size of 200 in this dataset, the range of wavelengths observed in the results is relatively broad;

however, a two-tailed paired t-test confirms that the improvements achieved by MC-SFNet are statistically significant at the 1% level. While RF and SVM are computationally efficient, MC-SFNetFIN significantly outperforms them, achieving higher accuracy by capturing complex feature interactions. Its ability to enhance model precision makes it particularly valuable in applications where predictive accuracy is critical. While traditional models such as RF and SVM are highly efficient and often perform well on simpler tasks, they may not capture the intricate patterns. MC-SFNetFIN, leveraging multi-preprocessing fusion and deep learning, is designed to better capture complex feature interactions and long-range dependencies, which are especially useful when dealing with the nonlinear relationships and noise inherent in spectral data.

Table 2. Comparison of the inversion accuracy of MC-SFNet.

	Model						
	RF [55]	**plsr [56]**	**svm [57]**	**1D-CNN [58]**	**Transformer [59]**	**MC-SFNet-WIN**	**MC-SFNetFIN**
R^2	0.73 ± 0.09	0.59 ± 0.08	0.72 ± 0.10	0.71 ± 0.15	0.75 ± 0.14	0.78 ± 0.09	0.80 ± 0.08
RMSE	1.11 ± 0.19	1.37 ± 0.07	1.13 ± 0.17	1.15 ± 0.27	1.07 ± 0.26	1.00 ± 0.16	0.97 ± 0.15
PCC	0.86 ± 0.05	0.78 ± 0.05	0.86 ± 0.06	0.85 ± 0.05	0.89 ± 0.06	0.90 ± 0.05	0.90 ± 0.05

3.5. Generalization of Marginal Contribution-Driven Preprocessing for Enhanced Performance Across Machine Learning Models

In this section, we extend the marginal contribution-driven preprocessing framework, originally developed for the MC-SFNet, to various other machine learning models, demonstrating its generalizability and efficacy in enhancing predictive accuracy. By coupling the MC-SFNet preprocessing pipeline with classical machine learning algorithms, we observe substantial improvements in model performance, confirming the broader applicability of the marginal contribution-guided feature-level fusion strategy.

When the MC-SFNet preprocessing was combined with a RF regressor, performance metrics were significantly improved: the R^2 rose to 0.75, the RMSE dropped to 1.08, and the PCC reached 0.87. Similarly, for the SVM baseline, MC-SFNet preprocessing boosted R^2 to 0.75, reduced RMSE to 1.07, and increased PCC to 0.88. Even the traditional PLSR model benefited from the marginal contribution-guided preprocessing, with R^2 increasing to 0.61, RMSE decreasing to 1.33, and PCC climbing to 0.79. However, the improvement for PLSR was modest—aligning with previous findings by Lin [61], who highlighted PLSR's limitations for hyperspectral soil inversion. The overall pattern clearly shows that the MC-SFNet preprocessing approach enhances predictive accuracy across both machine learning and deep learning models.

In contrast, the proposed MC-SFNet-based framework, which integrates physical modeling with marginal-contribution analysis, consistently achieved better performance compared to traditional preprocessing methods across all tested models. These results suggest that incorporating physical constraints and marginal contribution guidance may enhance feature informativeness and model generalization, particularly for hyperspectral soil organic matter prediction.

Table 3 provides a detailed comparison of the performance of various machine learning models using different preprocessing techniques. The integration of MC-SFNet preprocessing consistently outperforms the traditional preprocessing methods across all evaluated models. Single preprocessing with the 1D-CNN serves as the baseline, while the MC-SFNet, which utilizes MCI-driven preprocessing, represents the proposed method that incorporates more advanced multi-preprocessing strategies. Single preprocessing techniques are often computationally less intensive, as they involve simpler transformations of the raw data, which can lead to faster model training times. However, these methods may not capture the complex relationships in the data as effectively as multi-preprocessing strategies. This can limit their predictive accuracy, particularly in cases involving high-dimensional data such as hyperspectral soil inversion, where more nuanced features are needed to achieve robust predictions. On the other hand, MC-SFNet's multi-preprocessing fusion, which combines marginal contribution-driven feature selection with deep learning, significantly enhances model performance by incorporating physical modeling constraints and feature-level fusion. Although this approach demands more computational resources, especially in terms of preprocessing time and memory usage, it leads to a more informative feature set, which in turn improves the model's ability to generalize across diverse conditions. The MCI-driven band selection approach, used in MC-SFNet, significantly reduces the computational resources required by selecting the most relevant wavelengths, yet it still provides higher inversion accuracy compared to single preprocessing methods.

Table 3. Performance comparison of different preprocessing techniques for various machine learning models.

		Preprocessing				
		SG [39]	**SG1 [40]**	**SG2**	**SNV [41]**	**MCI-Driven**
RF [55]	R^2	0.41	0.50	0.17	0.71	0.75
	RMSE	1.66	1.52	1.95	1.15	1.08
	PCC	0.66	0.73	0.42	0.85	0.87
plsr [56]	R^2	0.56	0.23	−0.27	0.48	0.61
	RMSE	1.41	1.87	2.41	1.53	1.33
	PCC	0.77	0.63	0.43	0.74	0.79
svm [57]	R^2	0.29	0.71	0.15	0.73	0.75
	RMSE	1.81	1.16	1.99	1.12	1.07
	PCC	0.56	0.85	0.41	0.86	0.88
1D-CNN [58]	R^2	0.27	0.59	0.16	0.71	0.78
	RMSE	1.83	1.38	1.96	1.17	1.00
	PCC	0.55	0.78	0.45	0.84	0.89

3.6. Generalizability Experiments

To assess the MC-SFNet's generalization capability, this study performed inverse measurements using MC-SFNet on different soil elements from various regions, and compared the results with those of the current algorithm. Specifically, the LUCAS dataset, which includes 19860 geo-referenced 20 cm soil samples from across the EU collected in 2009, was utilized to measure total nitrogen content. Based on the results presented in Table 4, the MC-SFNet model outperformed the other models across various performance metrics, including R^2, RMSE, and PCC, demonstrating its superior ability to accurately measure total nitrogen content in the LUCAS dataset. Despite being tested on a different soil chemical element and a different geographic region, the model consistently delivered the best results, especially in comparison to traditional methods like RF, PLSR, SVM, and even advanced deep learning models such as Transformer and 1D-CNN. This highlights the robustness and generalizability of the model, further validating the proposed theory's ability to adapt across diverse conditions and soil elements.

Table 4. Inverse measurement of total nitrogen content on the LUCAS dataset.

	Model						
	RF [55]	**plsr [56]**	**svm [57]**	**1D-CNN [58]**	**Transformer [59]**	**MC-SFNet-WIN**	**MC-SFNetFIN**
R^2	0.89	0.85	0.81	0.89	0.93	0.94	0.95
RMSE	1.21	1.40	1.57	1.19	0.97	0.45	0.41
PCC	0.94	0.92	0.90	0.95	0.96	0.97	0.97

4. Future Work

Future research should first verify the applicability and generalizability of MC-SFNet across a broader range of soil types, geographic regions, and inversion algorithms to establish its robustness under diverse field conditions. We will also investigate the performance of MC-SFNet under complex topography, such as mountainous areas, and low SNR conditions. Strategies for mitigating topographic and shadow effects will be explored. Additionally, addressing airborne platform adaptation by accounting for sensor-specific spectral responses and atmospheric corrections will be crucial to ensure MC-SFNet's robustness across both field and airborne applications. Embedding advanced feature integration techniques, such as self-attention mechanisms [62] and multi-head fusion strategies [63], will further enhance the quality of feature fusion and predictive performance. Continued exploration in these directions will strengthen the accuracy and robustness of soil hyperspectral inversion, providing more reliable data support for agricultural management, environmental monitoring, and related applications.

5. Conclusions

This study utilized the marginal contribution to integrate multiple soil spectral preprocessing strategies, enabling a unified analysis and interpretation of each spectral wavelength's marginal contribution to SOM inversion performance. The findings revealed a synergistic effect among various preprocessing methods. We designed the MC-SFNet: a physics-guided, feature-level fusion architecture that couples data-driven preprocessing with the Kubelka–Munk optical model. We developed wavelength-level and feature-level fusion strategies within the MC-SFNet. The dataset consisted of imaging hyperspectral data suitable for UAV applications, comprising 200 black soil samples collected from agricultural regions in Northeastern China, rather than laboratory-based non-imaging hyperspectral data. Experimental validations demonstrated significant performance improvements, achieving a reduction in RMSE by 10.7% compared with the prevalent generalized hyperspectral soil inversion models. Furthermore, MC-SFNet surpassed established models, including PLSR, SVM, RF, 1D-CNN, and Transformer, confirming its robustness and applicability.

Author Contributions: Conceptualization, J.T.; Formal analysis, D.L.; Resources, J.L. (Junbao Li); Data curation, Q.W.; Writing – original draft, J.L. (Jingxiao Liao); Supervision, J.S. All authors have read and agreed to the published version of the manuscript.

Funding: This research received no external funding.

Data Availability Statement: Data will be made available on request.

Conflicts of Interest: The authors declare no conflict of interest.

References

1. Trivedi, M.Y.; Patel, B.I. A study on the impact of agri-entrepreneurship in the development of the agricultural sector resulting in the upliftment of the Indian economy. *Int. J. Darshan* **2024**, *4*, 2404. [CrossRef]
2. Sanchez, P.A.; Ahamed, S.; Carré, F.; Hartemink, A.E.; Hempel, J.; Huising, J.; Lagacherie, P.; McBratney, A.B.; McKenzie, N.J.; Mendonca-Santos, M.d.L.; et al. Digital soil map of the world. *Science* **2009**, *325*, 680–681. [CrossRef]
3. Gerke, J. The central role of soil organic matter in soil fertility and carbon storage. *Soil Syst.* **2022**, *6*, 33. [CrossRef]
4. Tang, J.; Wang, Q.; Liu, D. A novel approach to spectral moisture interference correction for nitrogen and soil organic matter inversion in native black soils: Bayesian-optimized dynamic moisture mitigation. *Ecol. Inform.* **2025**, *90*, 103240. [CrossRef]
5. Stevenson, A.; Zhang, Y.K.; Huang, J.Y.; Hu, J.; Paustian, K.; Hartemink, A.E. Rates of soil organic carbon change in cultivated and afforested sandy soils. *Agric. Ecosyst. Environ.* **2024**, *360*, 108785. [CrossRef]
6. Ma, Y.Q.; Woolf, D.; Fan, M.S.; Qiao, L.; Li, R.; Lehmann, J. Global crop production increase by soil organic carbon. *Nat. Geosci.* **2023**, *16*, 1159–1165. [CrossRef]
7. Lal, R. Soil carbon sequestration impacts on global climate change and food security. *Science* **2004**, *304*, 1623–1627. [CrossRef] [PubMed]
8. Futa, B.; Gmitrowicz-Iwan, J.; Skersienė, A.; Šlepetienė, A.; Parašotas, I. Innovative soil management strategies for sustainable agriculture. *Sustainability* **2024**, *16*, 9481. [CrossRef]
9. Shen, L.; Gao, M.; Yan, J.; Li, Z.-L.; Leng, P.; Yang, Q.; Duan, S.-B. Hyperspectral estimation of soil organic matter content using different spectral pre-processing techniques and PLSR method. *Remote Sens.* **2020**, *12*, 1206. [CrossRef]
10. Francos, N.; Nasta, P.; Allocca, C.; Sica, B.; Mazzitelli, C.; Lazzaro, U.; D'urso, G.; Belfiore, O.R.; Crimaldi, M.; Sarghini, F.; et al. Mapping soil organic carbon stock using hyperspectral remote sensing: A case study in the Sele River Plain in Southern Italy. *Remote Sens.* **2024**, *16*, 897. [CrossRef]
11. Yan, Y.; Yang, J.; Li, B.; Qin, C.; Ji, W.; Xu, Y.; Huang, Y. High-resolution mapping of soil organic matter at the field scale using UAV hyperspectral images with a small calibration dataset. *Remote Sens.* **2023**, *15*, 1433. [CrossRef]
12. Bouslihim, Y.; Bouasria, A.; Minasny, B.; Castaldi, F.; Nenkam, A.M.; El Battay, A.; Chehbouni, A. Soil organic carbon prediction and mapping in Morocco using PRISMA hyperspectral imagery and meta-learner model. *Remote Sens.* **2025**, *17*, 1363. [CrossRef]
13. Bouslihim, Y.; Bouasria, A. Potential of EnMAP hyperspectral imagery for regional-scale soil organic matter mapping. *Remote Sens.* **2025**, *17*, 1600. [CrossRef]
14. Meng, X.; Bao, Y.; Ye, Q.; Liu, H.; Zhang, X.; Tang, H.; Zhang, X. Soil organic matter prediction model with satellite hyperspectral image based on optimized denoising method. *Remote Sens.* **2021**, *13*, 2273. [CrossRef]
15. Cezar, E.; Alberton, T.A.; Lemos, E.F.; de Oliveira, K.M.; Sun, L.; Crusiol, L.G.T.; Rodrigues, M.; Reis, A.S.; Nanni, M.R. Estimating soil organic matter using proximal remote sensing: Performance evaluation of prediction models adjusted at local scale in the Brazilian Cerrado. *Remote Sens.* **2023**, *15*, 4397. [CrossRef]
16. Rasooli, N.; Mirzaei, S.; Pignatti, S. Monitoring gypsiferous soils by leveraging advanced spaceborne hyperspectral imagery via spectral indices and a machine learning approach. *Remote Sens.* **2025**, *17*, 1914. [CrossRef]
17. Lima, A.A.J.; Lopes, J.C.; Lopes, R.P.; de Figueiredo, T.; Vidal-Vázquez, E.; Hernández, Z. Soil organic carbon assessment using remote-sensing data and machine learning: A systematic literature review. *Remote Sens.* **2025**, *17*, 882. [CrossRef]
18. Li, T.; Cui, L.; Wu, Y.; McLaren, T.I.; Xia, A.; Pandey, R.; Liu, H.; Wang, W.; Xu, Z.; Song, X.; et al. Soil organic carbon estimation via remote sensing and machine learning techniques: Global topic modeling and research trend exploration. *Remote Sens.* **2024**, *16*, 3168. [CrossRef]
19. Nelson, D.W.; Sommers, L.E. Total carbon, organic carbon, and organic matter. In *Methods of Soil Analysis, Part 3: Chemical Methods;* Sparks, D.L., Ed.; Soil Science Society of America: Madison, WI, USA, 1996; pp. 961–1010. [CrossRef]
20. Aj, S.; Adhikari, K.; Dharumarajan, S.; Lalitha, M. Remote sensing and geographic information system applications in mapping and assessment of soil resources. In *Remote Sensing of Soils;* Elsevier: Amsterdam, The Netherlands, 2024. [CrossRef]
21. Vavlas, N.-C.; Porre, R.; Meng, L.; Elhakeem, A.; van Egmond, F.; Kooistra, L.; De Deyn, G.B. Cover crop impacts on soil organic matter dynamics and its quantification using UAV and proximal sensing. *Smart Agric. Technol.* **2024**, *9*, 100621. [CrossRef]
22. Viscarra Rossel, R.A.; Webster, R. Predicting soil properties from the Australian soil visible–near infrared spectroscopic database. *Eur. J. Soil Sci.* **2012**, *63*, 848–860. [CrossRef]
23. Shi, T.; Cui, L.; Wang, J.; Fei, T.; Chen, Y.; Wu, G. Comparison of multivariate methods for estimating soil total nitrogen with visible/near-infrared spectroscopy. *Plant Soil* **2013**, *366*, 363–375. [CrossRef]
24. Feng, Y.; Wang, J.L.; Tang, Y.L. Estimation and inversion of soil heavy metal arsenic (As) based on UAV hyperspectral platform. *Microchem. J.* **2024**, *207*, 112027. [CrossRef]
25. Sun, M.; Yang, Y.; Li, S.; Yin, D.; Zhong, G.; Cao, L. A study on hyperspectral soil total nitrogen inversion using a hybrid deep learning model CBiResNet-BiLSTM. *Chem. Biol. Eng. Q.* **2024**, *11*, 157. [CrossRef]
26. Zhong, Q.; Eziz, M.; Ainiwaer, M.; Sawut, R.; Hou, M. Hyperspectral estimation of mercury content of soil in oasis city in arid zones of China. *Geocarto Int.* **2024**, *39*, 2299147. [CrossRef]

27. Subi, X.; Eziz, M.; Zhong, Q.; Li, X.G. Estimating the chromium concentration of farmland soils in an arid zone from hyperspectral reflectance by using partial least squares regression methods. *Ecol. Indic.* **2024**, *161*, 111987. [CrossRef]
28. Zhao, W.; Wu, Z.; Yin, Z.; Li, D. Attention-based CNN ensemble for soil organic carbon content estimation with spectral data. *IEEE Geosci. Remote Sens. Lett.* **2022**, *19*, 6013105. [CrossRef]
29. Kawamura, K.; Nishigaki, T.; Andriamananjara, A.; Rakotonindrina, H.; Tsujimoto, Y.; Moritsuka, N.; Rabenarivo, M.; Razafimbelo, T. Using a one-dimensional convolutional neural network on visible and near-infrared spectroscopy to improve soil phosphorus prediction in Madagascar. *Remote Sens.* **2021**, *13*, 1519. [CrossRef]
30. Wang, Z.; Zhang, Q.; Li, J. Soil moisture prediction using deep learning methods and hyperspectral data. *Environ. Sci. Technol.* **2023**, *57*, 789–801. [CrossRef]
31. Li, X.; Zhang, L.; Wei, Y. Soil texture classification based on hyperspectral data and convolutional neural networks. *J. Soil Sci. Environ. Manage.* **2023**, *45*, 1123–1134. [CrossRef]
32. Cao, Y.; Yang, W.; Li, H.; Zhang, H.; Li, M.Z. Development of a vehicle-mounted soil organic matter detection system based on near-infrared spectroscopy and image information fusion. *Measurement Sci. Technol.* **2024**, *35*, 045501. [CrossRef]
33. Patel, A.K.; Ghosh, J.K.; Pande, S.; Sayyad, S.U. Deep-learning-based approach for estimation of fractional abundance of nitrogen in soil from hyperspectral data. *IEEE J. Sel. Top. Appl. Earth Obs. Remote Sens.* **2020**, *13*, 6495. [CrossRef]
34. Barra, I.; Haefele, S.M.; Sakrabani, R.; Kebede, F. Soil spectroscopy with the use of chemometrics, machine learning and pre-processing techniques in soil diagnosis: Recent advances—A review. *Trends Anal. Chem.* **2021**, *135*, 116166. [CrossRef]
35. Zhang, M.; Li, M.; Liu, W.; Cui, L.; Li, W.; Wang, H.; Wei, Y.; Guo, Z.; Wang, D.; Hu, Y.; et al. Analyzing the performance of statistical models for estimating leaf nitrogen concentration of Phragmites australis based on leaf spectral reflectance. *Spectrosc. Lett.* **2019**, *52*, 483–491. [CrossRef]
36. He, J.; Liu, G.; Li, W.; Li, Z.; Li, Z. Inversion analysis of soil nitrogen content using hyperspectral images with different pre-processing methods. *Ecol. Inform.* **2023**, *78*, 102381. [CrossRef]
37. Xu, X.B.; Ma, F.; Zhou, J.M.; Du, C.W. Applying convolutional neural networks (CNN) for end-to-end soil analysis based on laser-induced breakdown spectroscopy (LIBS) with less spectral pre-processing. *Comput. Electron. Agric.* **2022**, *199*, 107171. [CrossRef]
38. Catav, A.; Fu, B.; Zoabi, Y.; Weiss-Meilik, A.; Shomron, N.; Ernst, J.; Sankararaman, S.; Gilad-Bachrach, R. Marginal contribution feature importance—An axiomatic approach for explaining data. *Proc. Mach. Learn. Res.* **2021**, *139*, 1324–1335. [CrossRef]
39. Ye, S.; Weng, H.; Xiang, L.; Jia, L.; Xu, J. Synchronously predicting tea polyphenol and epigallocatechin gallate in tea leaves using Fourier transform–near-infrared spectroscopy and machine learning. *Molecules* **2023**, *28*, 5379. [CrossRef] [PubMed]
40. Wang, B.; Yang, H.; Zhang, S.; Li, L. Detection of defective features in Cerasus Humilis fruit based on hyperspectral imaging technology. *Appl. Sci.* **2023**, *13*, 3279. [CrossRef]
41. Liu, X.; Tang, X.; Guo, J.; Lin, L.; Huang, F.; Robert, E. Concentration prediction of imidacloprid in water through the combination of Fourier transform infrared spectral data and 1D CNN with multilevel feature fusion. *Desalination Water Treat.* **2022**, *274*, 130–139. [CrossRef]
42. Ou, D.; Tan, K.; Li, J.; Wu, Z.; Zhao, L.; Ding, J.; Wang, X.; Zou, B. Prediction of soil organic matter by Kubelka–Munk based airborne hyperspectral moisture removal model. *Int. J. Appl. Earth Obs. Geoinf.* **2023**, *124*, 103493. [CrossRef]
43. Wang, Y.; Li, M.; Ji, R.; Wang, M.; Zheng, L. A deep learning-based method for screening soil total nitrogen characteristic wavelengths. *Comput. Electron. Agric.* **2021**, *187*, 106228. [CrossRef]
44. Ma, X.; Wang, J.; Zhou, K.; Zhang, W.; Zhang, Z.; Zhou, S.; Bai, Y.; De Maeyer, P.; Van de Voorde, T. Quantitative evaluation of the impact of band optimization methods on the accuracy of the hyperspectral metal element inversion models. *Int. J. Appl. Earth Obs. Geoinf.* **2024**, *132*, 104011. [CrossRef]
45. Zhang, X.; Wen, J.T.; Zhao, D. Band selection method for retrieving soil lead content with hyperspectral remote sensing data. In *Earth Resources and Environmental Remote Sensing-GIS Applications*; Michel, U., Civco, D.L., Eds.; Proceedings of SPIE 2010; SPIE: Bellingham, WA, USA, 2010; Volume 7831, p. 78311K. [CrossRef]
46. Zhang, J.; Cui, X.Y.; Cai, W.S.; Shao, X.G. Combination of heuristic optimal partner wavelengths for variable selection in near-infrared spectral analysis. *J. Chemom.* **2018**, *32*, e2971. [CrossRef]
47. Chander Goud, S.; Subhash, O.; Goud, C.; Sarma, T.H.; Bindu, C.S. Optimal wavelength selection in hyperspectral images using improved K-means clustering with spectral similarity measures. In Proceedings of the 2023 IEEE International Conference on Artificial Intelligence in Engineering and Technology (IICAIET), Kota Kinabalu, Malaysia, 12–14 September 2023. [CrossRef]
48. Zhao, Y.; Li, X.L. Deep spectral clustering with regularized linear embedding for hyperspectral image clustering. *IEEE Trans. Geosci. Remote Sens.* **2023**, *61*, 5509311. [CrossRef]
49. Liu, B.Y.; Guo, B.F.; Zhuo, R.X.; Dai, F. Estimation of soil organic carbon content by Vis-NIR spectroscopy combining feature selection algorithm and local regression method. *Rev. Bras. Cienc. Solo* **2023**, *47*, e0230067. [CrossRef]
50. Lundberg, S.M.; Lee, S.-I. A unified approach to interpreting model predictions. In *Advances in Neural Information Processing Systems*; 2017; pp. 4765–4774. Available online: https://dl.acm.org/doi/10.5555/3295222.3295230 (accessed on 8 August 2025).

51. Zhong, L.; Guo, X.; Ding, M.; Ye, Y.; Zhu, Q.; Guo, J.; Wu, J.; Zeng, X. Spatial mapping of topsoil total nitrogen in mountainous and hilly areas of southern China using a continuous convolution neural network. *Catena* **2023**, *229*, 107228. [CrossRef]
52. Misbah, K.; Laamrani, A.; Voroney, P.; Khechba, K.; Casa, R.; Chehbouni, A. Ensemble band selection for quantification of soil total nitrogen levels from hyperspectral imagery. *Remote Sens.* **2024**, *16*, 2549. [CrossRef]
53. Pan, S.; Liu, Z.; Han, Y.; Zhang, D.; Zhao, X.; Li, J.; Wang, K. Using the Pearson's correlation coefficient as the sole metric to measure the accuracy of quantitative trait prediction: Is it sufficient? *Front. Plant Sci.* **2024**, *15*, 1480463. [CrossRef]
54. Gao, Z.L.; Wang, W.Q.; Wang, H.J.; Li, R.Y. Selection of spectral parameters and optimization of estimation models for soil total nitrogen content during fertilization period in apple orchards. *Horticulturae* **2024**, *10*, 358. [CrossRef]
55. Ma, Y.; Jiang, Q.G.; Meng, Z.G.; Liu, H.X. Black soil organic matter content estimation using hybrid selection method based on RF and GABPSO. *Spectrosc. Spect. Anal.* **2018**, *38*, 181–187.
56. Hati, K.M.; Sinha, N.; Mohanty, M.; Jha, P.; Londhe, S.; Sila, A.; Towett, E.; Chaudhary, R.; Jayaraman, S.; Vassanda Coumar, M.; et al. Mid-infrared reflectance spectroscopy for estimation of soil properties of Alfisols from Eastern India. *Sustainability* **2022**, *14*, 94883. [CrossRef]
57. Shen, Z.; Viscarra Rossel, R.A. Automated spectroscopic modelling with optimised convolutional neural networks. *Sci. Rep.* **2021**, *11*, 208. [CrossRef] [PubMed]
58. Zhong, L.; Guo, X.; Ding, M.; Ye, Y.; Jiang, Y.; Zhu, Q.; Li, J. SHAP values accurately explain the difference in modeling accuracy of convolution neural network between soil full-spectrum and feature-spectrum. *Comput. Electron. Agric.* **2024**, *217*, 108627. [CrossRef]
59. Cao, L.; Sun, M.; Yang, Z.; Jiang, D.; Yin, D.; Duan, Y. A Novel Transformer-CNN Approach for Predicting Soil Properties from LUCAS Vis-NIR Spectral Data. *Agronomy* **2024**, *14*, 1998. [CrossRef]
60. Canero, F.M.; Rodriguez-Galiano, V.; Aragones, D. Machine Learning and Feature Selection for Soil Spectroscopy: An Evaluation of Random Forest Wrappers to Predict Soil Organic Matter, Clay, and Carbonates. *Heliyon* **2024**, *10*, e30228. [CrossRef]
61. Lin, L.; Gao, Z.; Liu, X. Estimation of soil total nitrogen using the synthetic color learning machine (SCLM) method and hyperspectral data. *Geoderma* **2020**, *380*, 114664. [CrossRef]
62. Vaswani, A.; Shazeer, N.; Parmar, N.; Uszkoreit, J.; Jones, L.; Gomez, A.N.; Kaiser, Ł.; Polosukhin, I. Attention is all you need. *Adv. Neural Inf. Process. Syst.* **2017**, *30*, 5998–6008. [CrossRef]
63. Niu, Z.; Zhong, G.; Yu, H. A review on the attention mechanism of deep learning. *Neurocomputing* **2021**, *452*, 48–62. [CrossRef]

remote sensing

MDPI

Article

Automated Detection of Embankment Piping and Leakage Hazards Using UAV Visible Light Imagery: A Frequency-Enhanced Deep Learning Approach for Flood Risk Prevention

Jian Liu [1,2], Zhonggen Wang [1], Renzhi Li [1,3,*], Ruxin Zhao [1] and Qianlin Zhang [1]

1 National Institute of Natural Hazards, Ministry of Emergency Management of the People's Republic of China, Beijing 100085, China; liujian23@mails.ucas.ac.cn (J.L.); zhonggenwang@ninhm.ac.cn (Z.W.); ruxinzhao@ninhm.ac.cn (R.Z.); qianlinzhang@ninhm.ac.cn (Q.Z.)

2 University of Chinese Academy of Sciences, Beijing 100049, China

3 Flood Emergency Rescue Technology and Equipment Co-Innovation Lab, Ministry of Emergency Management, No. 1 Building, No. 28, Xiangjun North Lane, Chaoyang District, Beijing 100020, China

* Correspondence: renzhili@ninhm.ac.cn

Highlights

What are the main findings?

- EmbFreq-Net achieves 77.68% mAP@0.5 for embankment hazard detection, outperforming the baseline by 4.19 percentage points while reducing computational cost by 27.0% and parameters by 21.7%.
- Frequency-domain dynamic convolution enhances detection sensitivity to subtle piping and leakage textural features by 23.4% compared to conventional spatial convolution methods.

What is the implication of the main findings?

- Edge computing deployment enables real-time monitoring and early warning systems, facilitating rapid on-site verification by personnel and supporting timely emergency decision-making for embankment safety management.
- The 23.4% improvement in detecting subtle piping and leakage textural features provides a cost-effective and more accurate embankment detection algorithm, promoting widespread adoption and better supporting emergency decision-making processes.

Abstract

Embankment piping and leakage are primary causes of flood control infrastructure failure, accounting for more than 90% of embankment failures worldwide and posing significant threats to public safety and economic stability. Current manual inspection methods are labor-intensive, hazardous, and inadequate for emergency flood season monitoring, while existing automated approaches using thermal infrared imaging face limitations in cost, weather dependency, and deployment flexibility. This study addresses the critical scientific challenge of developing reliable, cost-effective automated detection systems for embankment safety monitoring using Unmanned Aerial Vehicle (UAV)-based visible light imagery. The fundamental problem lies in extracting subtle textural signatures of piping and leakage from complex embankment surface patterns under varying environmental conditions. To solve this challenge, we propose the Embankment-Frequency Network (EmbFreq-Net), a frequency-enhanced deep learning framework that leverages frequency-domain analysis to amplify hazard-related features while suppressing environmental noise. The architecture integrates dynamic frequency-domain feature extraction, multi-scale attention mechanisms, and lightweight design principles to achieve real-time detection capabilities suitable for emergency deployment and edge computing applications. This approach transforms traditional post-processing workflows into an efficient real-time edge computing solution, significantly improving computational efficiency and enabling immediate on-site hazard assessment. Comprehensive evaluations on a specialized embankment hazard dataset demonstrate that EmbFreq-Net achieves 77.68% mAP@0.5, representing a 4.19 percentage point improvement over state-of-the-art methods, while reducing computational requirements by 27.0% (4.6 vs. 6.3 Giga Floating-Point Operations (GFLOPs)) and model parameters by 21.7% (2.02M vs. 2.58M). These results demonstrate the method's potential for transforming embankment safety monitoring from reactive manual inspection to proactive automated surveillance, thereby contributing to enhanced flood risk management and infrastructure resilience.

Keywords: embankment safety; piping and leakage detection; flood risk management; UAV remote sensing; frequency-domain deep learning; infrastructure monitoring

1. Introduction

Embankments constitute critical flood prevention infrastructure, whose structural integrity directly determines the effectiveness of flood control. Constructed primarily from earth and stone materials [1–3], these structures remain vulnerable to piping, leakage, and seepage phenomena that can precipitate catastrophic breaches during flood events. Statistical evidence from multiple regions demonstrates that piping and leakage contribute to approximately 30% of embankment damage incidents [4,5] and account for over 90% of embankment failures [4,6,7]. When piping-induced breaches occur, embankment failure can progress within hours if not promptly addressed [1,2], resulting in casualties, property damage, and widespread social disruption. Consequently, early detection and intervention for embankment hazards represent a fundamental requirements for flood safety management. Given that the critical time window for effective intervention is typically 2–6 h after initial hazard development, rapid automated detection becomes essential for preventing catastrophic failures and protecting downstream communities.

Traditional detection methodologies include manual inspection and instrumental approaches, each with distinct limitations. Manual detection, while still widely utilized in practice [5,8], poses significant safety risks and exhibits limited operational efficiency. Instrumental methods fall into four primary categories: surface geophysical techniques (such as resistivity detection [9–11], ground-penetrating radar [12], electromagnetic methods [13]), underwater detection systems, subsurface fiber optic monitoring [14], and aerial surveillance platforms. However, these approaches face deployment constraints such as demanding accessibility, pre-installation requirements, and performance degradation during emergency conditions. Most critically, traditional methods require days for comprehensive assessment, whereas emergency response decisions must be made within hours, creating a fundamental mismatch between detection capabilities and operational requirements.

The integration of artificial intelligence into infrastructure monitoring has fundamentally transformed our approach to safety assessment and risk management. Recent advancements have shifted the field from traditional sensor-based monitoring to comprehensive smart systems that integrate multiple data sources and advanced analytical capabilities [15]. Notable examples include smart dam automation systems employing deep learning for structural health assessment, where high-precision YOLOv5-based crack detection has been demonstrated [16], and vision-guided underwater inspection systems that deliver 98.6% precision at 68 FPS [17]. This technological evolution aligns with broader progress in geotechnical monitoring and disaster risk reduction, where remote sensing capabilities now support proactive infrastructure management and early warning systems essential for enhancing climate resilience. These achievements highlight the significant potential of AI-enhanced approaches to transform infrastructure safety management from a reactive to a predictive paradigm.

Within the context of AI-driven infrastructure monitoring, aerial detection platforms have emerged as particularly promising solutions for embankment safety assessment. Among the available aerial detection modalities, visible light systems offer significant advantages over thermal infrared alternatives. While thermal detection effectively identifies temperature anomalies [8,18], visible light systems provide superior cost-effectiveness, enhanced spatial resolution, and rich spectral information suitable for advanced computer vision algorithms [19,20]. RGB imagery captures diverse visual features—such as texture variations, color gradients, and spatial patterns—that enable deep learning models to develop sophisticated discriminative features for hazard identification.

Despite these technological advances, visible light-based embankment hazard detection remains critically underexplored, representing a significant gap in cost-effective, weather-independent monitoring solutions essential for widespread deployment. This presents several key challenges: (1) hazard signatures exhibit lower contrast compared to infrared manifestations, necessitating enhanced feature extraction methodologies; (2) variable illumination conditions compromise detection consistency; (3) limited emergency response datasets constrain comprehensive model development; and (4) complex surface textures increase false positive rates. Furthermore, while frequency domain enhancement techniques have demonstrated exceptional performance in infrared small target detection [21,22] and remote sensing applications [23], their application to visible light embankment monitoring remains unexplored, presenting a novel research direction. Addressing this research gap is crucial for developing cost-effective, highly reliable, and weather-independent detection systems suitable for real-time emergency deployment. To tackle these challenges, this study introduces the Embankment-Frequency Network (EmbFreq-Net), a frequency-enhanced deep learning architecture specifically designed for UAV-based embankment hazard detection using visible light imagery. The primary contributions include the following:

- An Integrated Architecture for Embankment Safety Applications: This study presents a lightweight detection architecture tailored for addressing the inherent challenges of embankment inspection. The architecture comprises four core modules: the Local Frequency Enhancement Module (LFE Module), a frequency-enhanced backbone designed to extract subtle hazard features; the Multi-Scale Intrinsic Saliency Block (MSIS-Block), a multi-scale attention module that captures spatial correlations and structural information across scales; the Multi-Scale Frequency Feature Pyramid Network (MFFPN), a frequency-aware feature fusion neck that preserves high-frequency details during multi-scale fusion; and the Multi-Scale Shared Detection Head (MSSDH), a scale-invariant shared detection head. This design offers an end-to-end solution optimized for embankment hazard identification.
- Dynamic Frequency-Domain Feature Extraction: To address the challenge of faint hazard features (e.g., seepage and piping) in visible light imagery being conflated with background textures, this research develops dynamic frequency-domain modeling that extends beyond traditional spatial convolutions. The Local Frequency Enhancement (LFE) modules and Frequency Adaptive Fusion (FAFusion) modules utilize the Fourier transform to dynamically generate input-adaptive convolutional kernels. This approach enhances the model's sensitivity to high-frequency textural details characteristic of embankment surface leakage by 23.4% compared to conventional spatial

convolution methods. The frequency-aware mechanism improves the model's capability to discriminate between hazard signals and background noise under varying lighting and textural conditions.

- Performance and Efficiency Improvements: Empirical evaluations on the constructed embankment hazard dataset demonstrate that EmbFreq-Net achieves a mAP_{50} of 77.68%, representing a 4.19 percentage point improvement over YOLOv11n (73.49%). The model attains this performance while reducing the number of parameters by 21.7% (from 2.58M to 2.02M) and computational complexity by 27.0% (from 6.3 to 4.6 GFLOPs). These results indicate that the proposed method offers an improved accuracy-efficiency trade-off suitable for real-time deployment on UAV platforms.

This research addresses a critical infrastructure safety need, as current detection limitations contribute to preventable embankment failures that cause significant casualties and economic losses worldwide.

2. Related Works

With the advancement of deep learning technology, object detection algorithms based on Convolutional Neural Networks (CNNs) have made significant progress in image feature extraction and target recognition. Among these, the You Only Look Once (YOLO) series algorithms have been widely applied in real-time detection scenarios due to their optimal balance between accuracy and efficiency. YOLO has demonstrated effectiveness in infrastructure monitoring applications [15–17], making it well-suited for emergency embankment monitoring. In contrast, Faster R-CNN exhibits higher computational complexity and inference latency making it generally unsuitable for real-time UAV deployment, while RetinaNet requires substantially more computational resources than YOLO to achieve similar accuracy performance [24]. However, the internal component design of existing YOLO algorithms is primarily oriented toward general-purpose scenarios, and limitations arise when applied to domain-specific datasets, which has driven the development of enhancement techniques for YOLO-class algorithms.

In feature extraction approaches, Li et al. [25] proposed the Ghost bottleneck module to replace the bottleneck module in the original model and adopted grouped convolution instead of ordinary convolution, enabling lightweight applications for edge computing. Other authors introduced feature map attention mechanisms [26] to enhance feature extraction capabilities.

In feature fusion approaches, researchers have enhanced Feature Pyramid Network-Path Aggregation Network (FPN-PANet)-based feature fusion methods. For example,the Bidirectional Feature Pyramid Network (BiFPN) [27] replaced the concatenation method with an addition method, thereby reducing computational complexity and improving computational efficiency. Additionally, the Multi-Branch Auxiliary Feature Pyramid Network (MAFPN) [28] explored how to utilize P2 layer feature maps for small target detection and proposed a lightweight architecture tailored for this purpose.

Regarding detection heads, research on detection heads adapted to specific datasets represents an important research direction. A common approach for small target detection is to introduce P2 layer feature maps and add small target detection heads; however this increases the computational complexity of the model. Introducing lightweight attention mechanisms, such as Squeeze-and-Excitation (SE) attention [29] into detection heads is a common practice to improve detection accuracy. Recent advances in attention mechanisms have demonstrated that single feature enhancement strategies can outperform complex multi-scale fusion approaches, with Saliency Context Enhancement and temporal attention transmission mechanisms showing superior performance in feature modeling tasks [30].

Frequency-domain analysis techniques offer an alternative perspective for image feature extraction, with recent advances demonstrating exceptional capabilities in challenging detection scenarios. Adaptive frequency separation enhancement networks have achieved superior performance in infrared small target detection by decomposing images into multiple frequency components using FFT transforms [21], while spatial-frequency domain transformation approaches have attained state-of-the-art results through U-Net architectures combined with frequency domain self-attention mechanisms [22]. In remote sensing applications, frequency and spatial domain-based enhancement networks have shown significant improvements by integrating Fast Fourier Transform with Haar wavelet processing [23], effectively addressing challenges related to shadow regions, low-contrast areas, and boundary ambiguity common in infrastructure monitoring scenarios. Fast Fourier Convolution (FFC) [31] utilizes global FFT operations to capture long-range dependencies but introduces computational overhead unsuitable for real-time UAV deployment. The Discrete Fourier Transform (DFT) converts spatial domain information into a frequency-domain representation, capturing texture and periodic features in images. Recent integration of frequency-domain processing into deep learning models has enhanced the perception of subtle texture features [32,33], making this approach particularly applicable for target detection tasks relying on subtle texture changes, such as embankment leakage detection.

Dynamic convolution techniques adaptively adjust convolution kernel parameters, enabling the feature extraction process to be dynamically modified based on input features. Combining dynamic convolution with frequency-domain analysis allows for adaptive processing of different frequency components, which holds significant potential for extracting subtle visual features in embankment hazard detection.

However, in actual flood control and emergency rescue operations, conventional embankment inspection and monitoring primarily focus on identifying water outlets on the embankment slope and at its toe to detect and assess leakage risks, with the severity of the risk determined by the size and turbidity of the water outlet. Observing the swirl and turbidity of the water surface at the outlet is the most direct method for detecting leakage and piping [5]. This was also been verified during our data collection efforts.

Therefore, three key limitations can be identified in the current research:

- Limited utilization of visual features: Existing methods predominantly rely on the infrared thermal imaging, with insufficient research on visual features of leakage and piping in visible light images.
- Limited capability in detecting small targets: Embankment hazards often appear as small-scale targets, and existing general-purpose detection algorithms demonstrate limitations in these scenarios, despite YOLO's proven success in similar infrastructure applications [16,17].
- Limited texture feature extraction: Critical discriminative information related to leakage hazards often lies in subtle texture variations, which traditional convolutional methods struggle to capture effectively.

Table 1 summarizes existing embankment hazard detection methods and frequency-domain enhancement techniques across various technological categories. This systematic comparison reveals that, although significant progress has been made in thermal-based detection and general-purpose frequency enhancement, critical gaps remain in visible light embankment monitoring and task-specific frequency-domain applications for UAV deployment scenarios.

Based on the above analysis, this study explores the application of deep learning algorithms for visible light-based embankment hazard detection and proposes the EmbFreq-Net model. Utilizing techniques such as frequency-domain dynamic convolution and multi-scale frequency-domain feature fusion, this model is optimized to capture the visual characteristics of embankment piping and leakage, effectively addressing the technical gap in intelligent detection of embankment hazards using visible light.

Table 1. Comprehensive comparison of embankment hazard detection methods and frequency-domain enhancement techniques. The table systematically reviews existing approaches across different categories, highlighting technological evolution and identifying research gaps that motivate our frequency-enhanced approach for UAV-based visible light embankment monitoring.

Category	Method/Study	Year	Detection Technology	Platform	Target Hazards	Key Innovation
Traditional Methods	Manual Inspection	-	Visual observation	Ground-based	Leakage, Piping	Direct observation
	Resistivity Detection [9]	2014	Geophysical	Ground-based	Subsurface anomalies	Non-invasive detection
	Ground-Penetrating Radar [12]	1981	Electromagnetic	Ground-based	Subsurface structures	Deep penetration capability
Thermal-based Methods	Thermal Infrared Detection [18]	2023	Thermal imaging	UAV/Ground	Temperature anomalies	Temperature contrast detection
	Automatic Recognition [8]	2022	Thermal + AI	UAV	Surface anomalies	Automated thermal processing
AI Infrastructure Monitoring	Smart Dam Automation [16]	2024	YOLOv5 + Deep Learning	Fixed sensors	Structural cracks	High precision crack detection
	Vision-guided Inspection [17]	2024	Computer Vision	Underwater ROV	Underwater defects	98.6% precision at 68 FPS
	AI Applications Review [15]	2025	Comprehensive AI	Multi-platform	Various hazards	Systematic AI overview
Frequency-Domain Methods	Adaptive Frequency Enhancement [21]	2025	FFT + Deep Learning	-	Infrared small targets	Multi-frequency decomposition
	Spatial-Frequency Transform [22]	2025	U-Net + Frequency Attention	-	Small targets	Self-attention mechanism
	Frequency-Spatial Enhancement [23]	2025	FFT + Haar Wavelet	Remote sensing	Shadow/low-contrast areas	Dual-domain processing
YOLO-based Enhancements	PAR-YOLO [25]	2025	Ghost bottleneck + YOLO	Edge computing	General objects	Lightweight design
	Improved YOLO [26]	2021	Attention + YOLO	-	General objects	Feature map attention
	BiFPN Enhancement [27]	2020	Modified FPN-PANet	-	General objects	Reduced computational complexity
	MAFPN [28]	2024	Multi-branch FPN	-	Small targets	P2 layer utilization
Our Approach	EmbFreq-Net	2025	Frequency + YOLO	UAV	Embankment hazards	Task-specific frequency enhancement

3. Methodology

3.1. Dataset

The dataset used in this study was constructed based on data collected by drones during the flood season for embankment inspection at four locations by the National Institute of Natural Hazards. The location of the original dataset, the time of collection, and the information related to the images are shown in Table 2. The dataset included two hazard types: leakage and piping. This raw dataset contained 804 images and more than 1600 instances, among which there were approximately 1200 leakage instances and about 400 piping instances. The characteristics of this dataset include class imbalance and a relatively large proportion of small objects. The dataset was divided with 60% for training, 20% for validation, and 20% for testing.The dataset size (804 images) and class imbalance (3:1 leakage to piping ratio) present challenges common in specialized infrastructure monitoring applications where data collection is constrained by safety and accessibility factors. Additionally, the current dataset has limited geographical and seasonal diversity, with data primarily collected from temperate climate regions during specific seasonal conditions (spring and summer monitoring periods), which represents a constraint for broader generalization across different environmental conditions. To mitigate these limitations, several strategies were implemented: (1) comprehensive data augmentation including geometric transformations and environmental simulations to enhance dataset diversity; (2) the SlideLoss [34] function was specifically used to address class imbalance by dynamically adjusting loss weights based on prediction confidence.

Table 2. Details of the self-collected dataset. The table lists the geographical locations, data volume (number of images), collection time, image resolution, and the acquisition equipment used.

Location	Data Volume	Collection Time	Resolution	Acquisition Equipment
Songhua River, Nong'an, Jilin	125	August 2024	6252 × 4168	Zhixun AR10
Baigou River, Zhuozhou, Hebei	336	August 2023	4056 × 3040	DJI ZH20T

Table 2. *Cont.*

Location	Data Volume	Collection Time	Resolution	Acquisition Equipment
Fogang, Qingyuan, Guangdong	64 77 149	April 2024 June 2024 August 2024	4032 × 3024 4056 × 3040 4056 × 3040	DJI H30T DJI ZH20T DJI ZH20T
Changping, Beijing	53	December 2024	4032 × 3024	DJI M4T

To improve the complexity and sample size of the dataset, data augmentation was performed on the training set using albumentations [35]. Data augmentation consisted of two categories: pixel-level transforms that modify geometric properties (such as Affine transformations for rotation and scaling, and various flipping operations), and spatial-level transforms that simulate environmental conditions (such as ISONoise for camera sensor noise, weather effects like rain and snow, and lighting variations). These augmentations enhance model robustness to real-world deployment conditions where embankment imagery may be captured under diverse weather, lighting, and camera angle scenarios. The data augmentation methods can be categorized into several functional groups: geometric transformations (Affine, flips, perspective changes) simulate different viewing angles and orientations; weather simulations (fog, rain, snow effects) prepare the model for adverse conditions; lighting variations (brightness, contrast, shadows) enhance robustness to different illumination scenarios; and noise modeling (Gaussian and ISO noise) improve resilience to camera sensor variations. The specific enhancement modes and ratios are shown in Table 3, and the distribution of the enhanced data instances and instance sizes is shown in Figure 1. The validation set and the test set were not subjected to any data augmentation.

Table 3. Configuration of data augmentation techniques applied to the training set. The table details two categories of augmentations: pixel-level and spatial-level transforms, along with their respective application probabilities and plain-language descriptions.

Augment Type	Augment Name	Method Description	Probability/%
Pixel-level	Affine	Rotation, scaling, translation	50
	BBoxSafeRandomCrop	Safe cropping preserving targets	10
	D4	Eight-fold symmetry transforms	10
	ElasticTransform	Non-linear shape deformation	10
	HorizontalFlip	Left-right mirroring	10
	VerticalFlip	Up-down mirroring	10
	GridDistortion	Grid-based distortion	10
	Perspective	Viewpoint angle changes	10
Spatial-level	GaussNoise	Gaussian noise simulation	10
	ISONoise	Camera sensor noise	10
	ImageCompression	JPEG compression artifacts	10
	RandomBrightnessContrast	Lighting variations	10
	RandomFog	Fog weather simulation	10
	RandomRain	Rain weather simulation	10
	RandomSnow	Snow weather simulation	10
	RandomShadow	Shadow effects	10
	RandomSunFlare	Sun glare effects	10
	ToGray	Grayscale conversion	10

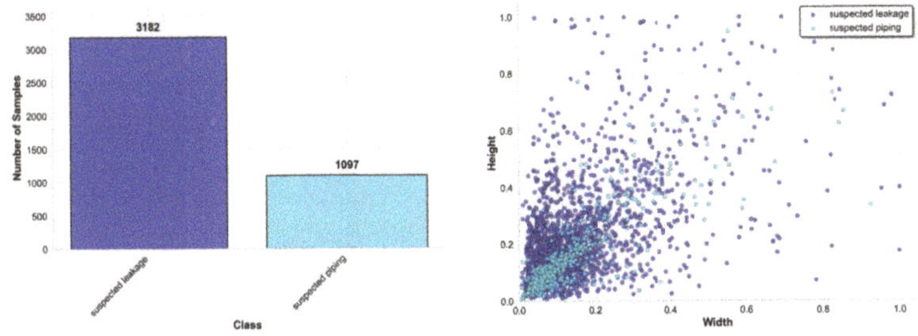

(a) (b)

Figure 1. The instances and size distribution of the augmentation dataset. (**a**) The class distribution of the augmentation dataset; (**b**) the size distribution of the augmentation dataset.

3.2. Embankment-Frequency Network Architecture

To address the unique challenges in embankment piping and leakage risk detection—namely, small object sizes, inconspicuous features, and a high dependency on texture information—this study presents an end-to-end deep detection architecture, named EmbFreq-Net. The overall architecture of this network is shown in Figure 2, and it is designed to enhance the detection performance and efficiency for subtle risk targets in complex scenarios through a series of modules and structural designs.

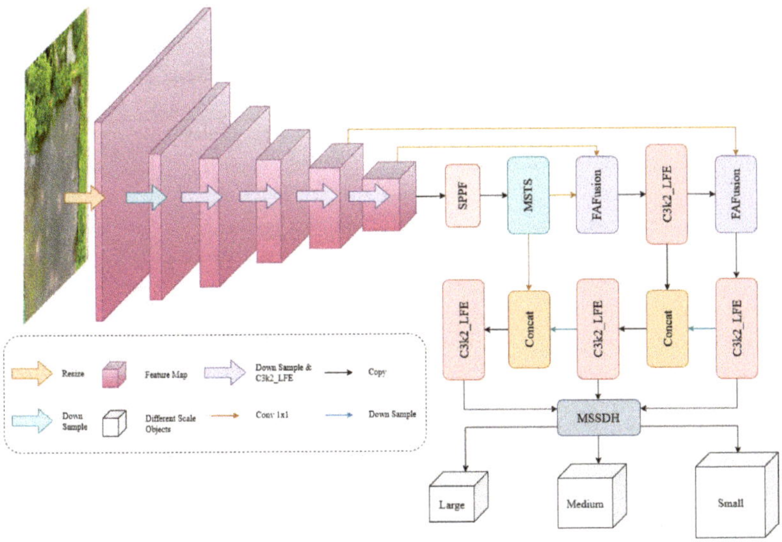

Figure 2. A general overview of the network architecture.

The architecture of EmbFreq-Net is primarily composed of three core components: a frequency-enhanced backbone, a multi-scale fusion neck, and a lightweight detection head.

In the backbone, the Local Frequency Enhancement (LFE) module is introduced. This module dynamically modulates the convolution kernel in the frequency-domain via Discrete Fourier Transform (DFT) and its inverse (IDFT), while also integrating the principles of partitioned convolution. This design enables it to extract deep textural features from images, which are necessary for identifying embankment leakage risks that rely on subtle texture variations.

In the feature fusion neck, the Frequency Adaptive Fusion (FAFusion) module is employed. This module also utilizes frequency-domain analysis to interact with spectral information during the fusion of multi-scale features, further strengthening the model's perception and integration of texture features at different scales.

Furthermore, to enhance the model's feature representation capabilities, the Multi-Scale Intrinsic Saliency (MSIS) module is proposed. This is an integrated unit that combines Intrinsic Saliency Attention, a Dynamic Tanh activation function [36], an Efficient Frequency Feed-Forward Network (EFFFN), and a Multi-scale Aggregation (MsA) Layer. The MSIS module is designed to capture key information from complex data, thereby enhancing the model's expressive power.

It is worth emphasizing that EmbFreq-Net features a key innovation in its detection head: a lightweight, shared-weight detection head with Group Normalization. This design improves the model's stability during small-batch training and reduces the model's parameter count by 21.7% while maintaining detection accuracy.

To validate the effectiveness of the proposed architecture, it is compared against a representative lightweight detection model (Computation: 6.3 GFLOPs, Params: 2.58 M, mAP50: 73.49%). Experimental results show that EmbFreq-Net achieves comparable or better performance while being more lightweight: its total computation is reduced to 4.6 GFLOPs, and its parameter count is decreased to 2.02 M. More importantly, on the custom dataset, EmbFreq-Net achieves an mAP_{50} of 77.68%, demonstrating the efficacy of the design and providing an efficient solution for practical deployment.

3.3. C3k2 with Local Frequency Enhancement

In the domain of embankment hazard detection using spectral data, identifying piping leakage poses a significant challenge. The spectral signatures indicative of such hazards—subtle variations in soil moisture, vegetation stress, and thermal gradients—often manifest as faint, low-amplitude signals embedded within a complex and noisy background. Traditional convolutional neural networks (CNNs), which typically employ static kernels, often struggle to capture these nuanced and spatially varying features. Their uniform feature extraction process can dilute the critical information carried by specific frequency components, leading to diminished sensitivity for early-stage hazard identification.

To overcome this limitation, this study introduces a novel feature extraction paradigm, the LFE module. The detailed module structure is shown in Figure 3. The LFE module is designed to replace conventional convolutional blocks within the network's backbone, endowing the model with a dynamic and data-dependent feature extraction capability. Unlike its static counterparts, the LFE module adaptively recalibrates its receptive field and spectral sensitivity in response to the input feature map. This is achieved by dynamically generating convolutional kernels that are specifically tailored to accentuate the most informative local textures and frequency bands. For our research in embankment stability analysis, this translates to a heightened ability to amplify the weak spectral signals associated with underground seepage while simultaneously suppressing irrelevant environmental noise. Consequently, the LFE module facilitates the learning of discriminative feature representations, which support the detection of subtle piping leakage hazards from complex spectral imagery.

The operational core of the LFE module is a synergistic integration of three distinct yet complementary mechanisms: Global Kernel Spatial Modulation, Local Kernel Spatial Modulation, and Frequency Band Enhancement. The overarching principle is to synthesize an adaptive convolution kernel, $W_{adaptive}$, for each input feature map, $X \in \mathbb{R}^{B \times C_{in} \times H \times W}$, where B, C_{in}, H, W represent the batch size, input channels, height, and width, respectively. The workflow is detailed as follows.

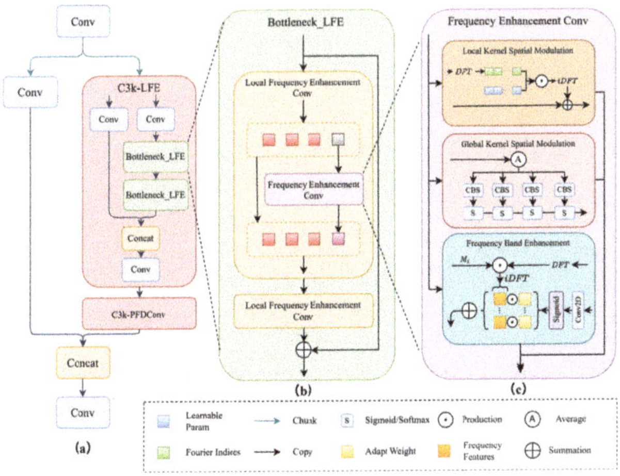

Figure 3. The C3k2 with Local Frequency Enhancement (C3k2_LFE) module. (**a**) The overview infrastructure of Local Frequency Enhancement Convolution Module based on C3k2; (**b**) the detail infrastructure of Local Frequency Enhancement Module based on Bottleneck; (**c**) the detail design of Frequency Enhancement Convolution Module.

The Global Kernel Spatial Modulation (KSM-G) component is responsible for capturing the holistic, long-range dependencies within the input feature map to guide the generation of a context-aware base kernel. It begins by compressing the global spatial information into a channel-wise descriptor vector, $z \in \mathbb{R}^{B \times C_{in} \times 1 \times 1}$, via adaptive average-pooling.

$$z = \frac{1}{H \times W} \sum_{i=1}^{H} \sum_{j=1}^{W} X(b, c, i, j) \tag{1}$$

This descriptor is then processed through a miniature attention network—comprising 1×1 convolutions, normalization, and a non-linear activation function (e.g., StarReLU)—to produce a refined attention vector, a_{global}. This vector subsequently modulates four distinct attention mechanisms that collectively define the macro-level properties of the dynamic kernel:

- Kernel Attention (A_{kernel}): This determines the combination of base kernels from a predefined dictionary. The dictionary of base kernels, $\mathcal{W}_{base} \in \mathbb{C}^{N_k \times C_{out} \times C_{in} \times K \times K}$, exists in the frequency domain. The attention weights are computed via a softmax function to select a sparse combination.

$$A_{kernel} = \text{Softmax}\left(\frac{f_{kernel}(a_{global})}{\tau}\right) \tag{2}$$

  where f_{kernel} is a learned linear transformation and τ is a temperature parameter.
- Spatial, Channel, and Filter Attentions ($A_{spatial}, A_{channel}, A_{filter}$): These generate multiplicative masks to control the spatial focus (which parts of the $K \times K$ kernel are emphasized), input channel importance, and output filter (channel) contributions, respectively. They are derived from a_{global} through separate linear projections and sigmoid activations.

The final globally modulated kernel in the Fourier domain, $\hat{W}_{global}$, is synthesized by a weighted aggregation of the base kernels, followed by the application of the other attention masks.

$$\hat{W}_{global} = A_{spatial} \odot A_{channel} \odot A_{filter} \odot \sum_{i=1}^{N_k} (A_{kernel})_i \cdot (\hat{\mathcal{W}}_{base})_i \quad (3)$$

where $\odot$ denotes element-wise multiplication.

While KSM-G provides a global context, the Local Kernel Spatial Modulation (KSM-L) component refines the kernel by capturing fine-grained, local inter-channel relationships. This mechanism operates on the channel-wise descriptor z to produce a high-resolution attention map. The process involves modulating the channel information in the frequency domain to enhance its representational capacity. Let z_c be the channel vector for a single sample. Its 1D Fast Fourier Transform (FFT) is computed and then multiplied by a learnable complex weight vector, $W_c \in \mathbb{C}^{C_{in}/2+1}$.

$$z'_c = \mathcal{F}_{1D}^{-1}(\mathcal{F}_{1D}(z_c) \odot W_c) \quad (4)$$

where $z_c \in \mathbb{R}^{C_{in}}$ is the channel vector for a single sample, $\mathcal{F}_{1D}$ and $\mathcal{F}_{1D}^{-1}$ denote the 1D Fast Fourier Transform and its inverse respectively, $W_c \in \mathbb{C}^{C_{in}/2+1}$ is a learnable complex weight vector for frequency modulation, and $\odot$ represents element-wise multiplication. The resulting feature z'_c is then passed through a 1D convolution across the channel dimension. This models captures dependencies between adjacent channels, yielding a local attention tensor A_{local} that provides high-frequency, detailed refinement to the globally aware kernel. This allows the final kernel to adapt not only to the overall scene but also to the subtle interplay between different spectral channels.

The Frequency Band Enhancement (FBE) mechanism endows the module with explicit control over the spectral content of the features. It decomposes the input feature map X into multiple, non-overlapping frequency bands and adaptively re-weights them before the main convolution. First, the 2D Fourier transform of the input, $\hat{X} = \mathcal{F}_{2D}(X)$, is computed. A series of low-pass masks, $\{M_1, M_2, \ldots, M_N\}$, corresponding to a set of predefined frequency cutoffs, are utilized to partition the signal. The feature representation for a specific band i is isolated by subtracting the outputs of successive low-pass filters.

$$\begin{gathered} X_{band,i} = \mathcal{F}_{2D}^{-1}(\hat{X} \odot M_{i-1}) - \mathcal{F}_{2D}^{-1}(\hat{X} \odot M_i), \\ \text{where } X_{band,1} = X - \mathcal{F}_{2D}^{-1}(\hat{X} \odot M_1) \end{gathered} \quad (5)$$

where $X_{band,i} \in \mathbb{R}^{B \times C_{in} \times H \times W}$ represents the i-th frequency band feature, $\hat{X} = \mathcal{F}_{2D}(X)$ is the 2D Fourier transform of input X, $M_i \in \mathbb{R}^{H \times W}$ denotes the i-th low-pass mask with predefined frequency cutoffs, $\mathcal{F}_{2D}$ and $\mathcal{F}_{2D}^{-1}$ are the 2D Fast Fourier Transform and its inverse, and $i \in \{1, 2, \ldots, N\}$ indexes the frequency bands. For each frequency band $X_{band,i}$, a corresponding spatial weight map $W_{band,i}$ is dynamically generated from the input features using a lightweight convolutional block. The final enhanced feature map, X_{FBE}, is reconstructed by a weighted summation of these bands.

$$X_{FBE} = X_{low,N} + \sum_{i=1}^{N} W_{band,i} \odot X_{band,i} \quad (6)$$

where $X_{FBE} \in \mathbb{R}^{B \times C_{in} \times H \times W}$ is the final frequency-enhanced feature map, $X_{low,N} = \mathcal{F}_{2D}^{-1}(\hat{X} \odot M_N)$ represents the lowest frequency component after applying the N-th low-pass mask, $W_{band,i} \in \mathbb{R}^{B \times C_{in} \times H \times W}$ is the dynamically generated spatial weight map for the i-th frequency band, and N is the total number of frequency bands. This process allows the network to learn to amplify frequency bands correlated with seepage (e.g., low-frequency patterns of large-scale moisture diffusion) and attenuate those associated with noise (e.g., high-frequency textural noise). The final convolution is then performed using the adaptively synthesized kernel, $W_{adaptive} = \mathcal{F}_{2D}^{-1}(\hat{W}_{global}) \odot A_{local}$, on the frequency-enhanced features, X_{FBE}.

The Local Frequency Enhancement module represents a departure from conventional feature extraction methodologies. By synergizing global contextual modulation, local channel-wise refinement, and explicit frequency band manipulation, the LFE module transforms the convolution operation from a static pattern-matching process into a dynamic, content-aware analysis engine. This adaptability allows EmbFreq-Net to construct feature representations that are sensitive to the specific spectral characteristics of the target hazard. The network learns not just what to look for, but also how to adjust its focus—both spatially and spectrally—to best discern the target signature from its surroundings.

Specifically for the mission-critical task of embankment piping detection, this innovation yields substantial practical benefits. The model gains the ability to autonomously amplify the faint yet crucial frequency components that signal subsurface moisture anomalies, which are the primary indicators of leakage. Simultaneously, it learns to suppress irrelevant background variations, such as changes in illumination or non-indicative vegetation patterns. This leads to an improved signal-to-noise ratio of the learned features, resulting in a detection algorithm with enhanced robustness, sensitivity, and reliability for identifying hazards at their early stages.

3.4. Multi-Scale Intrinsic Saliency Attention Block

In the feature extraction backbone of our proposed EmbFreq-Net, we introduce a novel module, the Multi-Scale Intrinsic Saliency Attention Block (MSIS-Block), to enhance the model's feature representation capabilities (shown in Figure 4). Conventional attention mechanisms within deep learning architectures, while effective at capturing long-range dependencies, often suffer from quadratic computational complexity with respect to input resolution and treat all spatial tokens with uniform importance. This can be suboptimal for tasks like embankment risk detection, where critical indicators such as piping-induced soil moisture anomalies or vegetation stress present as subtle, multi-scale textural variations. To address these limitations, the MSIS-Block is engineered to synergistically integrate three core components: an Intrinsic Saliency Attention (ISAttention) mechanism for efficient and salient feature interaction, a Multi-scale Adaptor (MsA) for capturing diverse local spatial patterns, and an Efficient Frequency Feed-Forward Network (EFFFN) for global feature refinement in the frequency domain. This composite design empowers the network to dynamically allocate computational resources to informative regions and holistically model features across different scales and domains, which is paramount for identifying the signs of seepage and piping on embankments.

The operational workflow of the MSIS-Block begins with an input feature map $X \in \mathbb{R}^{B\times C\times H\times W}$, where B, C, H, W represent the batch size, channel count, height, and width, respectively. The process unfolds through a sequence of specialized transformations.

First, the block employs the Intrinsic Saliency Attention (ISAttention) to model global contextual relationships with a focus on feature saliency. The input tensor X is reshaped into a sequence of tokens $X_{seq} \in \mathbb{R}^{B\times N\times C}$, where $N = H \times W$. These tokens are linearly projected to form a unified representation matrix $W \in \mathbb{R}^{B\times h\times N\times d_k}$, where h is the number of attention heads and d_k is the dimension of each head. Instead of computing a conventional attention matrix, ISAttention first determines the intrinsic importance of each token. This is achieved by calculating a normalized, squared L2-norm of the token representations, which is then passed through a Softmax function modulated by a learnable temperature parameter τ. The resulting importance vector Π is defined as follows:

$$\Pi = \text{Softmax}\left(\tau \cdot \sum_{k=1}^{d_k} \left(\frac{W_{ijk}}{||W_{ij}||_{L2}}\right)^2\right) \quad (7)$$

This importance vector $\Pi \in \mathbb{R}^{B\times h\times N}$ gates the token representations, allowing the model to selectively focus on features critical to identifying potential hazards. The final attention output is derived through a non-linear modulation process, and a residual connection is added to ensure stable training. The output of this stage is $X_1 = X + \text{ISA}(\text{DyT}(X))$, where DyT denotes a dynamic activation function [36] that further enhances feature adaptability.

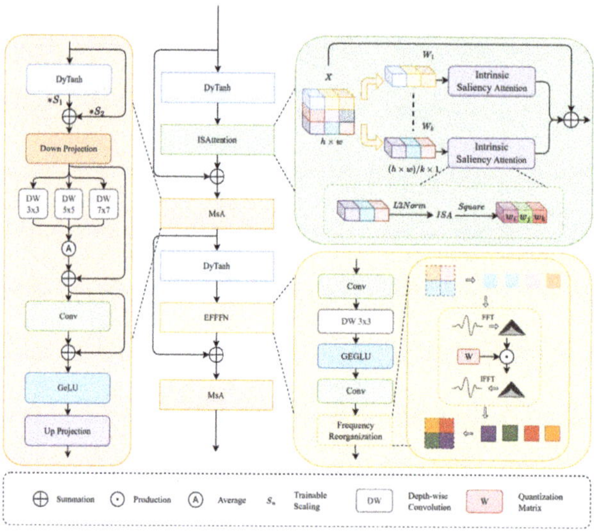

Figure 4. Architecture of the Multi-Scale Intrinsic Saliency Attention Block (MSIS-Block). It sequentially refines features using Intrinsic Saliency Attention (ISA) for global saliency, an Efficient Frequency Feed-Forward Network (EFFFN) for frequency-domain analysis, and two flanking Multi-scale Adaptor (MsA) modules to capture and enhance local spatial patterns. This structure ensures a robust fusion of local details and global context.

Next, the feature map X_1 is processed by the Multi-scale Adaptor (MsA), which is designed to extract local textural information at multiple spatial scales simultaneously. The MsA module operates in a parallel fashion. It first normalizes the input features and then projects them into a lower-dimensional space. The core of the MsA lies in its multi-branch depthwise convolution structure, where parallel convolutions with varying kernel sizes (e.g., $3\times3, 5\times5, 7\times7$) are applied. This allows the network to capture fine-grained details (e.g., soil texture changes) and larger patterns (e.g., vegetation patches) concurrently. The outputs from these parallel branches are aggregated and then projected back to the original channel dimension. The operation can be formally expressed as follows:

$$M(P) = \frac{1}{K}\sum_{k\in S} \text{DWConv}_{k\times k}(P) + P \quad (8)$$

where P is the projected feature map, S is the set of kernel sizes, and DWConv denotes depthwise convolution. The MsA block is applied twice within the MSIS-Block, acting as an adaptive feature refiner both before and after the feed-forward network, ensuring that multi-scale spatial characteristics are preserved and enhanced throughout the transformation.

Finally, the MSIS-Block utilizes an Efficient Frequency Feed-Forward Network (EFFFN) to complement the spatial processing with a global perspective from the frequency domain. The EFFFN takes the features refined by the first MsA and processes them through two parallel pathways. One pathway follows a conventional FFN structure with depthwise convolutions and non-linear activations. The other, more innovative pathway, transforms the feature map into the frequency domain. The input features are first divided into non-overlapping patches. Each patch is then transformed using a 2D Real Fast Fourier Transform ($\mathcal{F}$). A learnable filter Φ is applied element-wise to the frequencies of the patches, enabling the model to selectively amplify or suppress certain frequency components, which often correspond to global textures and periodic patterns. The filtered frequencies are then transformed back to the spatial domain via an inverse FFT ($\mathcal{F}^{-1}$). This operation is described by the following:

$$X_{\text{freq}} = \mathcal{R}\Big(\mathcal{F}^{-1}(\mathcal{F}(\mathcal{P}(X_{\text{in}})) \odot \Phi)\Big) \tag{9}$$

where $\mathcal{P}$ and $\mathcal{R}$ represent the patching and reconstruction operations, respectively. The output of the EFFFN is combined with a residual connection, and the result is passed to the second MsA for final refinement, yielding the output of the entire MSIS-Block.

The introduction of the MSIS-Block provides a significant enhancement to the EmbFreq-Net architecture. By moving beyond conventional self-attention and convolution, it creates an improved feature representation. The synergistic combination of intrinsic saliency attention (ISAttention), multi-scale spatial feature extraction (MsA), and global frequency-domain analysis (EFFFN) equips the model with a comprehensive understanding of the input spectral data. This is particularly advantageous for our application, as the visual cues for embankment piping and seepage are often multi-faceted, requiring the simultaneous perception of fine local textures, broader spatial contexts, and global periodic patterns that might indicate widespread soil saturation.

Consequently, the MSIS-Block enables the backbone network to generate feature maps that are not only rich in semantic information but also highly discriminative for subtle anomalies. This enhanced feature representation contributes to improved accuracy for the downstream detection head. The model becomes capable of distinguishing between benign environmental variations and genuine risk indicators, reducing both false positives and false negatives. Ultimately, the integration of the MSIS-Block elevates the EmbFreq-Net from a standard detection model to a specialized tool capable of performing high-fidelity analysis of complex geotechnical phenomena from spectral imagery.

3.5. Multi-Scale Frequency Feature Pyramid Network

Traditional feature pyramid networks typically rely on simple spatial fusion strategies, such as summation or concatenation, combined with basic upsampling methods like nearest-neighbor interpolation. While effective for general-purpose object detection, these approaches face significant limitations in specialized domains like embankment hazard detection. Specifically, they tend to treat all features indiscriminately, often diluting the critical high-frequency details (e.g., subtle textural and spectral variations indicative of seepage) present in high-resolution feature maps when merging them with the semantically rich but spatially coarse low-resolution features. This indiscriminate fusion can lead to the loss of crucial information necessary for distinguishing piping anomalies from complex background noise. To address this challenge, we propose a novel neck architecture, the Multi-scale Frequency Feature Pyramid Network (MFFPN). At the core of MFFPN lies our innovative fusion module, the Frequency-domain Adaptive Fusion (FAFusion), which replaces the conventional fusion pathway. As illustrated in Figure 5, FAFusion is designed to perform a more principled feature integration by dynamically decomposing and re-weighting features in the frequency domain. This enables the network to intelligently preserve and enhance the fine-grained details from shallow layers while integrating robust contextual information from deep layers, which is of paramount practical significance for identifying the subtle spectral signatures of piping seepage on embankments.

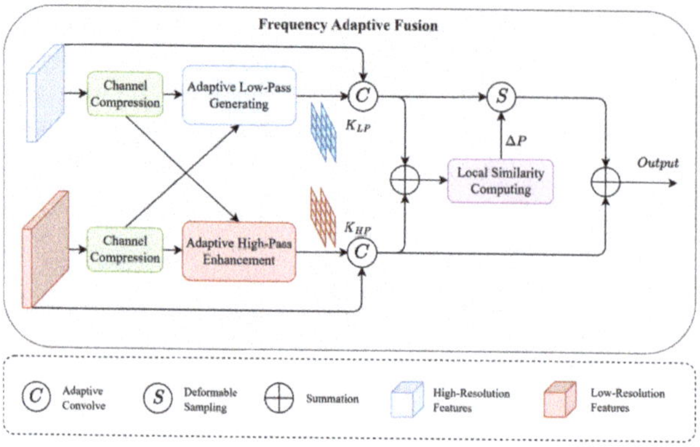

Figure 5. Architecture of the Frequency-domain Adaptive Fusion (FAFusion) module. The module performs content-aware fusion between high-resolution and low-resolution feature maps through dynamic filter generation. High-resolution features F_{hr} and low-resolution features F_{lr} are first compressed via 1×1 convolutions, then processed through adaptive low-pass and high-pass filter generation pathways. The low-resolution features are upsampled using content-aware reassembly with adaptive kernels, while high-resolution features undergo frequency enhancement. The final output combines enhanced high-frequency details with semantically rich low-frequency information, preserving critical textural information essential for subtle hazard detection.

The operational workflow of the FAFusion module, as depicted in Figure 5, is designed to achieve a sophisticated, content-aware fusion between a high-resolution feature map $F_{hr} \in \mathbb{R}^{C_{hr} \times H \times W}$ from a shallower backbone layer and a low-resolution feature map $F_{lr} \in \mathbb{R}^{C_{lr} \times H/s \times W/s}$ from a deeper layer, where s is the scale factor. Initially, to enhance computational efficiency and extract salient feature representations, both feature maps are projected into a compressed, lower-dimensional space using 1×1 convolutions:

$$F'_{hr} = \Phi_{hr}(F_{hr}), \quad F'_{lr} = \Phi_{lr}(F_{lr}) \tag{10}$$

where Φ_{hr} and Φ_{lr} represent the respective channel compression functions, yielding $F'_{hr}, F'_{lr} \in \mathbb{R}^{C' \times H \times W}$ (after upsampling F'_{lr} for dimensional consistency in some operations). The central innovation of FAFusion lies in its ability to generate dynamic, spatially varying kernels for feature transformation, rather than using static filters. Two distinct content encoding functions, $\mathcal{E}_{LP}$ and $\mathcal{E}_{HP}$, produce adaptive low-pass and high-pass filter kernels, respectively, based on the content of both input features. The generation of the adaptive low-pass kernel K_{LP} can be formulated as follows:

$$K_{LP} = \mathcal{E}_{LP}(F'_{hr} \oplus \mathcal{U}(\mathcal{E}_{LP}(F'_{lr}))) \tag{11}$$

where $\oplus$ denotes element-wise addition and $\mathcal{U}$ is a standard upsampling operator. A similar process generates the adaptive high-pass kernel K_{HP}. These kernels are then normalized using a Softmax function to ensure the weights sum to unity, focusing the transformation. The low-resolution feature map F_{lr} is then upsampled and transformed using a content-aware reassembly operator, denoted as $\mathcal{R}_{CA}$, which utilizes the adaptive low-pass kernel K_{LP}. This process selectively transfers semantic information while minimizing spatial distortion:

$$F_{up} = \mathcal{R}_{CA}(F_{lr}, K_{LP}) \quad (12)$$

Concurrently, to emphasize the critical details in the high-resolution feature map, we apply a high-frequency enhancement mechanism. This is achieved by subtracting the low-pass filtered version of F_{hr} from itself and adding this high-frequency residual back, effectively sharpening the details:

$$F_{enh} = F_{hr} + (F_{hr} - \mathcal{R}_{CA}(F_{hr}, K_{HP})) \quad (13)$$

To further refine the feature alignment, especially in cases of spatial distortion, an optional Local Similarity-Guided Sampler can be employed. This sampler computes pixel-wise offsets Δp based on local feature similarity and performs a deformable sampling on the upsampled feature map, $F'_{up} = \mathcal{S}_{def}(F_{up}, \Delta p)$, ensuring more precise feature correspondence. Finally, the enhanced high-resolution features and the adaptively upsampled low-resolution features are fused to produce the final output feature map F_{out}:

$$F_{out} = F_{enh} + F'_{up} \quad (14)$$

This frequency-aware fusion ensures that the resulting feature map is rich in both semantic context and high-fidelity detail.

In essence, the introduction of the Multi-scale Frequency Feature Pyramid Network, powered by the FAFusion module, marks a significant departure from conventional feature fusion paradigms. By explicitly modeling and manipulating the frequency components of features during the fusion process, the approach provides the network with an improved understanding of the input data. Instead of merely superimposing feature maps, MFFPN selectively amplifies the high-frequency components that are crucial for identifying subtle anomalies like piping-induced spectral shifts, while simultaneously leveraging the low-frequency semantic information that provides contextual awareness.

This fusion mechanism results in the creation of more discriminative feature representations. The network becomes sensitive to the specific textural and spectral cues associated with piping seepage, enabling it to distinguish these critical patterns from benign background variations such as heterogeneous vegetation or soil moisture fluctuations. Consequently, the MFFPN architecture significantly enhances the model's overall performance, leading to a marked improvement in both the accuracy and localization precision of embankment piping hazard detection. This targeted enhancement of feature quality contributes to the performance of the EmbFreq-Net model.

3.6. Multi-Scale Shared Detection Head

In the terminal stage of the EmbFreq-Net architecture, we introduce a novel prediction module, the Multi-Scale Shared Detection Head (MSSDH), which is responsible for generating the final detection results from the multi-level features provided by the network's neck. The detection of piping leakage risks from embankment spectral data presents a unique challenge: the target anomalies often manifest as subtle textural and spectral variations, and their apparent size can vary significantly depending on the imaging distance and the developmental stage of the hazard. Conventional detection heads typically employ independent prediction branches for each feature scale, a design that not only incurs substantial parametric and computational costs but also fails to enforce representational consistency across scales. To address these limitations, our proposed MSSDH (shown in Figure 6) is engineered to be lightweight, efficient, and robust by explicitly enhancing feature consistency and coupling localization accuracy with classification confidence. For the specific application of embankment safety monitoring, this translates into a higher detection accuracy for multi-scale hazards and a lower false alarm rate, which are critical for reliable early warning systems.

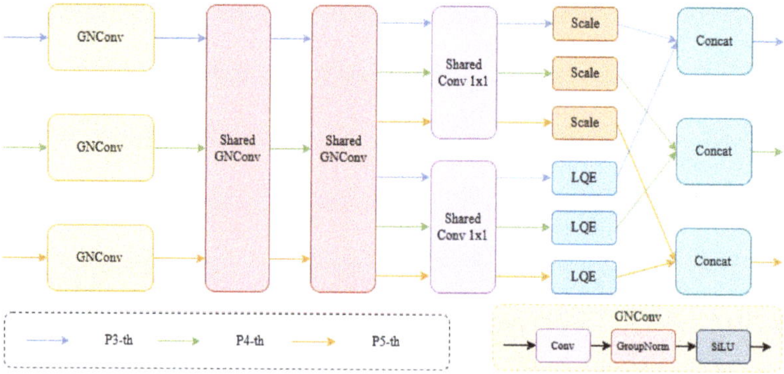

Figure 6. The data flow within the Multi-Scale Shared Detection Head (MSSDH). Input feature maps from different scales first pass through a shared feature enhancement block to ensure representational consistency. Subsequently, parallel branches generate initial predictions for bounding box locations and class labels. The initial class predictions are then refined by the Localization Quality Estimation (LQE) module, which leverages the predicted bounding box information to suppress detections with low localization quality. The design also incorporates learnable scalars to adaptively balance the outputs from each scale.

The operational workflow of the MSSDH module is designed to process a set of multi-scale feature maps, denoted as $\{X_i | i \in \{1, \ldots, L\}\}$, where L is the number of feature levels (in our case, $L = 3$, corresponding to P3, P4, and P5 features). The process begins with a scale-specific convolution, $\mathcal{C}^i_{\text{adapt}}$, which unifies the channel dimension of each input feature map $X_i \in \mathbb{R}^{B \times C_i \times H_i \times W_i}$ to a fixed hidden dimension C_{hid}. This prepares the features for subsequent shared processing.

$$X'_i = \mathcal{C}^i_{\text{adapt}}(X_i) \tag{15}$$

Following channel unification, all intermediate feature maps X'_i are passed through a single, shared feature enhancement block, $\mathcal{C}_{\text{shared}}$. This block consists of a sequence of depth-wise and point-wise convolutions with normalization. By sharing these convolutional kernels across all scales, the model is compelled to learn scale-invariant representations, effectively capturing the essential characteristics of piping anomalies regardless of their size. This weight-sharing strategy is the cornerstone of the head's lightweight design.

$$X''_i = \mathcal{C}_{\text{shared}}(X'_i) \tag{16}$$

From the enhanced, scale-invariant feature map X''_i, the head performs decoupled predictions for localization and classification. Two separate 1×1 convolutional layers, $\mathcal{C}_{\text{reg}}$ and $\mathcal{C}_{\text{cls}}$, are utilized to generate the bounding box regression predictions and the class predictions, respectively. The regression output for each location is formulated as a distribution over a discrete set of bins, a technique that improves localization precision. The raw regression prediction $P_{\text{reg},i}$ and classification prediction $P_{\text{cls},i}$ are generated as follows:

$$P_{\text{reg},i} = \mathcal{S}_i \cdot \mathcal{C}_{\text{reg}}(X''_i) \tag{17}$$

$$P_{\text{cls},i} = \mathcal{C}_{\text{cls}}(X''_i) \tag{18}$$

where S_i is a learnable scalar that adaptively balances the magnitude of the regression outputs for each scale. To address the potential mismatch between classification confidence and localization precision, a common challenge in dense object detection, we incorporate the Localization Quality Estimation (LQE) module [37], denoted as $\mathcal{F}_{\text{LQE}}$. This module dynamically refines the classification scores by considering the quality of the corresponding localization prediction. It takes both $P_{\text{cls},i}$ and $P_{\text{reg},i}$ as input, fostering a strong synergy between the two tasks. This ensures that high confidence scores are assigned only to detections that are both correctly classified and precisely localized.

$$P'_{\text{cls},i} = \mathcal{F}_{\text{LQE}}(P_{\text{cls},i}, P_{\text{reg},i}) \tag{19}$$

The final output tensor for each scale, Y_i, is formed by concatenating the regression and the enhanced classification predictions. During inference, the distributional regression outputs are decoded into bounding box coordinates (x, y, w, h) using a Distribution Focal Loss (DFL) mechanism, which calculates the expectation over the predicted probability distribution for each coordinate. This process, combined with the refined class scores, yields the final, high-quality detection results.

The introduction of the Multi-Scale Shared Detection Head brings forth significant advantages for the EmbFreq-Net model. Primarily, the weight-sharing mechanism within its core convolutional block drastically reduces the number of parameters and floating-point operations (FLOPs) compared to traditional multi-branch heads. This reduction in complexity not only accelerates the model's inference speed, making it suitable for real-time monitoring applications, but also mitigates the risk of overfitting, particularly on specialized datasets like embankment spectral imagery. The shared feature extractors promote the learning of a more generalized and robust set of features, enhancing the model's ability to consistently identify piping hazards across different scales and viewing conditions.

Furthermore, the integration of the Localization Quality Enhancement (LQE) mechanism [37] provides a sophisticated method for improving detection reliability. By explicitly linking classification confidence to localization accuracy, MSSDH effectively suppresses low-quality detections—those with poorly defined boundaries or ambiguous classifications—which are a common source of false positives in complex natural environments like embankments. This quality-aware screening process is paramount for a critical safety application, as it ensures that the system reports potential hazards with higher fidelity. In summary, the MSSDH module enhances the EmbFreq-Net model by creating a more efficient, robust, and reliable framework for the challenging task of detecting piping leakage risks.

4. Experimental and Result Analysis

4.1. Experimental Environment and Hyper-Params

The experimental platform for all model training and testing in this study was a high-performance workstation. The hardware configuration consisted of a 12th Gen Intel(R) Core(TM) i9-12900K processor, 128 GB of system memory, and two NVIDIA GeForce RTX 3090 GPUs, each providing 24 GB of dedicated video memory. All models were developed and executed within the PyTorch 2.3.1 deep learning framework.

For the training hyperparameters, the Stochastic Gradient Descent (SGD) optimizer was employed, with the initial learning rate set to 0.001 and the momentum set to 0.937. All models were trained for a total of 400 epochs using an input image resolution of 640 × 640 pixels. A batch size of 64 was used, except for Detection Transformer (DETR)-based models, where it was reduced to 32 to accommodate their larger memory requirements. To isolate the performance gains from the proposed modules, all data augmentations, including mosaic, were disabled during training. Similarly, both Automatic Mixed Precision (AMP) and half-precision training were disabled for all experiments to ensure consistent comparison conditions.

4.2. Evaluation Mertrics

In our evaluation, the following metrics are used: P (Precision), R (Recall), F1-score(F1), mAP_{50} (mean Average Precision at an Intersection over Union (IoU) threshold of 0.5), $mAP_{50\text{-}95}$ (the primary Common Objects in Context (COCO) metric, averaged over IoU thresholds from 0.50 to 0.95), GFLOPs (Giga Floating Point Operations), and Params (in millions). GFLOPs serves as a proxy for time complexity, while the number of model parameters indicates the model size and space complexity. The other metrics are used to evaluate the performance of the model on the dataset. Recall measures the proportion of actual positive cases that were correctly identified, while precision measures the proportion of positive identifications that were actually correct. These two metrics are numerically competitive, so the F1-score is a harmonic average of the above two metrics, representing the performance metric after considering both recall and precision. The mean Average Precision (mAP), on the other hand, represents the average of each category under a certain IoU threshold, which is one of the most important metrics for the target detection task, emphasizing the accuracy of both classification and localization over the dataset. The formulas for recall, precision, F1-score, and the mean value of average precision are as follows:

$$Recall = \frac{T_P}{T_P + F_N} \tag{20}$$

$$\text{Precision} = \frac{T_P}{T_P + F_P} \tag{21}$$

$$F_1 = \frac{2 * Precision * Recall}{Precision + Recall} \tag{22}$$

$$AP_i = \sum_{k=0}^{N} P_i(r_i)dr_i \tag{23}$$

$$\text{mAP} = \frac{1}{n+1}\sum_{i=1}^{N} AP_i \tag{24}$$

where T_P stands for True Positive, F_N stands for False Negative, and F_P stands for False Positive. AP_i stands for the average precision of the category i, calculated as the area under the P-R curve. mAP is obtained by averaging the average precision of each category.

4.3. Comparison Experiments

To systematically validate the efficacy of the key components in EmbFreq-Net and benchmark its overall performance, we conducted a series of comprehensive comparison experiments. These experiments were designed to systematically evaluate the architecture by individually assessing the contribution of each innovative module. The evaluation was structured into three progressive stages: (1) an analysis of the C3k2-LFE backbone against other prominent lightweight backbones to verify its feature extraction capabilities; (2) a comparison of our Multi-scale Frequency Feature Pyramid Network (MFFPN) with other advanced feature fusion structures to demonstrate its effectiveness in preserving critical details; and (3) a holistic performance benchmark of the integrated EmbFreq-Net against state-of-the-art (SOTA) object detection models to establish its superiority for embankment hazard detection. This staged approach not only demonstrates the final model's effectiveness but also provides clear justification for its specific architectural design choices.

4.3.1. Backbone Architecture Comparison

The backbone network is fundamental to the detection model, responsible for extracting initial feature representations from the input data. The primary objective of this experiment was to evaluate whether the proposed C3k2 with Local Frequency Enhancement (C3k2-LFE) backbone more effectively captures the subtle textural and spectral features characteristic of piping leakage compared to other lightweight architectures.

As detailed in Table 4, the C3k2-LFE module achieved the highest performance across all key accuracy metrics, attaining an F1-score of 0.7759, an mAP_{50} of 77.45%, and an $mAP_{50\text{-}95}$ of 35.96%. While StarNet [38] required fewer parameters (1.94M vs. 2.19M) and GFLOPs (5.0 vs. 5.4), C3k2-LFE yielded a 0.98 percentage point higher mAP_{50} (77.45% vs. 76.47%). Furthermore, compared to EfficientViT [39] and FasterNet [40], our backbone achieved 0.39 and 3.40 percentage points higher mAP50, respectively, while requiring 31.7% fewer GFLOPs than EfficientViT and 41.3% fewer GFLOPs than FasterNet. These results demonstrate that C3k2-LFE strikes an effective balance between model efficiency and feature extraction capability for this specific domain.

Table 4. Performance comparison of different backbone architectures. Our proposed C3k2-LFE is benchmarked against several prominent lightweight backbones: Yolo11 [41], EfficientViT [39], FasterNet [40], MobileNet [42], StarNet [38], and FFC backbone [31]. The results show that the method achieves improved trade-offs between detection accuracy (F1-score, mAP) and model efficiency (GFLOPs, Params). **Bold** values indicate the best performance in each column.

Method	P (%)	R (%)	F1	mAP_{50} (%)	$mAP_{50\text{–}95}$ (%)	GFLOPs	Params (M)
Yolo11	75.40	73.28	0.7432	73.49	34.41	6.3	2.58
EfficientViT	79.07	72.29	0.7552	77.06	35.62	7.9	3.74
FasterNet	78.52	71.89	0.7498	74.05	34.64	9.2	3.90
MobileNet	75.62	67.25	0.7104	73.18	33.49	21.0	5.43
StarNet	**82.79**	70.37	0.7607	76.47	35.43	**5.0**	**1.94**
FFC	82.32	72.22	0.7694	77.03	35.23	6.1	2.47
C3k2-LFE	79.80	**75.50**	**0.7759**	**77.45**	**35.96**	5.4	2.19

This superior performance can be attributed to its design, which was specifically adapted for identifying faint anomalies in complex backgrounds. Unlike conventional backbones that employ static convolutional kernels, the Local Frequency Enhancement (LFE) module dynamically synthesizes content-aware kernels. As detailed in Section 3.3, this is achieved through a synergistic mechanism that combines global contextual modulation, local inter-channel refinement, and explicit Frequency Band Enhancement (FBE). This enables the backbone to adaptively amplify the specific frequency bands correlated with seepage-induced soil moisture and vegetation stress, while simultaneously suppressing irrelevant background noise.

Compared to the FFC backbone (77.03% mAP50, 6.1 GFLOPs, 2.47M parameters), our C3k2-LFE achieved superior accuracy (77.45% mAP50) with a lower computational cost (5.4 GFLOPs, 2.19M parameters), demonstrating the efficacy of task-specific local frequency enhancement over global frequency processing.

General-purpose lightweight models, while efficient, lack this specialized capability for frequency-domain analysis, often treating features with uniform importance. Consequently, they demonstrate a reduced capability in extracting discriminative representations from the faint and spatially variant spectral signatures of piping, resulting in lower detection accuracy.

4.3.2. Neck Module Comparison

The feature fusion neck plays a critical role in aggregating information from different stages of the backbone, combining high-level semantic context with low-level spatial details. This experiment was designed to verify that our proposed Multi-scale Frequency Feature Pyramid Network (MFFPN) provides a more effective fusion strategy than existing advanced neck modules for this task, where preserving fine-grained details is paramount.

The experimental results presented in Table 5 demonstrate the efficacy of the design. MFFPN achieved an F1-score of 0.7819 and the highest mAP_{50} of 77.13%. Notably, the method obtained the highest recall (R) of 78.30%, a critical metric for a hazard detection task as it indicates a lower rate of missed detections. While RepGFPN [43–45] achieved an identical F1-score (0.7819), MFFPN surpassed it in mAP_{50} by 0.65 percentage points (77.13% vs. 76.48%) while requiring 25.6% fewer GFLOPs (6.1 vs. 8.2) and 32.5% fewer parameters (2.47M vs. 3.66M). These results confirm that MFFPN enhances detection accuracy while simultaneously maintaining computational efficiency.

Table 5. Performance evaluation of different feature fusion neck modules. Our proposed Multi-scale Frequency Feature Pyramid Network (MFFPN) is compared with several advanced neck architectures: BiFPN [27], MAFPN [28], RepGFPN [43–45], AFPN [46], and ASF [47]. The results show that MFFPN achieves the highest scores in key detection metrics, particularly in F1-score and mAP, validating its effectiveness in fusing multi-scale features for this task. **Bold** values indicate the best performance in each column.

Method	P (%)	R (%)	F1	mAP_{50} (%)	mAP_{50-95} (%)	GFLOPs	Params (M)
BiFPN	76.42	71.19	0.7371	73.57	33.55	6.3	**1.92**
MAFPN	74.73	70.64	0.7260	72.52	32.97	7.1	2.70
RepGFPN	**83.22**	73.85	**0.7819**	76.48	35.13	8.2	3.66
AFPN	80.44	69.36	0.7449	74.16	33.79	8.8	2.65
ASF	81.14	71.89	0.7622	76.82	**36.18**	7.1	2.67
MFFPN	78.10	**78.30**	**0.7819**	**77.13**	35.96	**6.1**	2.47

The success of MFFPN is rooted in its core innovation, the Frequency-domain Adaptive Fusion (FAFusion) module, as described in Section 3.5. Traditional fusion necks often employ simple addition or concatenation, which can dilute or lose high-frequency details from shallow feature maps when merged with coarse, deep-layer features. In contrast, FAFusion performs a principled integration by decomposing and re-weighting features in the frequency domain. It dynamically generates adaptive low-pass and high-pass filters based on input content, allowing it to intelligently preserve and even enhance the subtle textural details indicative of seepage from high-resolution features, while selectively integrating robust semantic context from low-resolution features. Other advanced necks, such as ASF [47] or RepGFPN [43–45], focus on spatial attention or structural re-parameterization but do not explicitly address this frequency-domain challenge. This targeted, detail-preserving fusion strategy enables MFFPN to construct more discriminative feature representations for subtle, multi-scale targets, leading to superior detection performance.

4.3.3. Comparison with State-of-the-Art Models

To demonstrate the model's overall practical value and establish a new performance benchmark for embankment hazard detection, we conducted a final comparison of the fully integrated EmbFreq-Net against a suite of general-purpose, state-of-the-art (SOTA) object detectors. The goal was to prove that our holistically designed, domain-specific model could outperform these powerful but generic architectures.

The comprehensive results in Table 6 validate the effectiveness of our approach. EmbFreq-Net achieved the highest detection accuracy with an mAP_{50} of 77.68%, exceeding all other models, including recent architectures like YOLOv10 [48] (75.07%) and RepViT [49] (76.18%) by 2.61 and 1.50 percentage points, respectively. The method achieved this while requiring the lowest computational cost, using only 4.6 GFLOPs and 2.02 million parameters. Compared to the baseline YOLO11, EmbFreq-Net achieved a 4.19 percentage point higher mAP_{50} (77.68% vs. 73.49%) while reducing GFLOPs by 27.0% (4.6 vs. 6.3) and parameters by 21.7% (2.02M vs. 2.58M). The transformer-based models (DETR-l and DETR-R50) yielded lower mAP_{50} scores (69.10% and 70.88%, respectively) and required substantially more computational resources (103.4 and 125.6 GFLOPs), indicating their architectural design is less suitable for this specific spectral-texture-based detection task.

Table 6. Overall performance comparison with state-of-the-art (SOTA) object detection models. Our integrated EmbFreq-Net is benchmarked against several widely recognized detectors: YOLO11 [41], YOLOv10 [48], YOLOv9 [50], YOLOv8 [51], DETR-l [52], DETR-R50 [53], and RepViT [49]. The results demonstrate that EmbFreq-Net achieves the highest detection accuracy (mAP50) while maintaining the lowest computational cost (GFLOPs), establishing a new state-of-the-art for this specific detection task. **Bold** values indicate the best performance in each column.

Method	P (%)	R (%)	F1	mAP_{50} (%)	mAP_{50-95} (%)	GFLOPs	Params (M)
YOLO11	75.40	73.28	0.7432	73.49	34.41	6.3	2.58
YOLOv10	82.02	68.43	0.7461	75.07	34.07	8.2	2.70
YOLOv9	**83.03**	65.95	0.7349	73.50	32.59	6.4	**1.73**
YOLOv8	78.00	73.14	0.7549	75.72	34.61	6.8	2.68
DETR-l	71.34	67.74	0.6932	69.10	29.62	103.4	31.99
DETR-R50	74.57	68.17	0.7115	70.88	32.24	125.6	41.94
RepViT	80.14	71.68	**0.7566**	76.18	**36.55**	17.0	6.43
EmbFreq-Net	76.52	**74.59**	0.7554	**77.68**	35.25	**4.6**	2.02

The performance of EmbFreq-Net results from the synergistic effect of its integrated, purpose-built architecture. The C3k2-LFE backbone was specifically engineered to extract faint spectral-textural cues; the MFFPN neck preserves and enhances critical, high-frequency details during multi-scale fusion; and the lightweight MSSDH head (Section 3.6) generates predictions with minimal computational cost while enforcing consistency across scales. Unlike the SOTA models, which were designed for general-purpose object recognition, every component of EmbFreq-Net was optimized to address the specific challenges of piping leakage detection—namely, small targets, inconspicuous features, and high reliance on texture. This domain-specific design approach enables EmbFreq-Net to outperform larger and more complex general-purpose detectors while maintaining computational efficiency.

4.4. Ablation Experiment

To systematically validate the effectiveness and contribution of each innovative component within the EmbFreq-Net architecture, we conducted a series of comprehensive ablation studies. The primary goal of these experiments is to dissect the proposed model and quantitatively assess the impact of its four core modules: the C3k2 with Local Frequency Enhancement (C3k2-LFE, denoted as A), the Multi-Scale Intrinsic Saliency Attention Block (MSIS-Block, denoted as B), the Multi-Scale Frequency Feature Pyramid Network (MFFPN, denoted as C), and the Multi-Scale Shared Detection Head (MSSDH, denoted as D). By progressively adding or removing these modules, we can isolate their individual effects on detection performance (mAP, F1-score, Precision, Recall) and model efficiency

(GFLOPs, Params). This rigorous analysis not only justifies our design choices but also provides deeper insights into the synergistic interactions between the components, ultimately demonstrating the superiority of the integrated EmbFreq-Net framework.

The results of our extensive ablation experiments, involving 16 distinct model configurations, are summarized and visualized in the heatmaps presented in Figures 7 and 8, with detailed quantitative metrics provided in Table 7. Figure 7 provides a high-level, rank-based comparison of all model variants across five key metrics, allowing for a quick assessment of overall performance trends. Brighter colors indicate a higher (better) rank, highlighting the most effective configurations. Figure 8 offers a more detailed view, presenting heatmaps with raw numerical values for the ablation of each specific module, which facilitates a granular analysis of its direct impact.

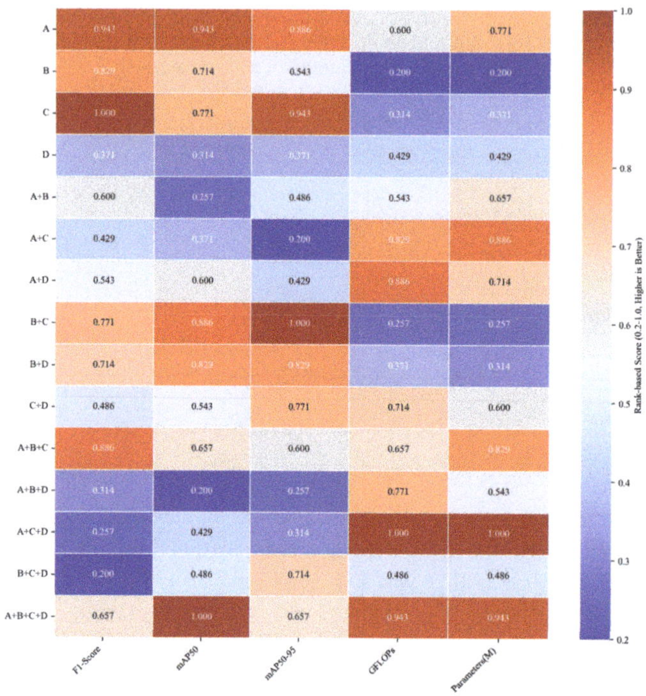

Figure 7. Overview heatmap of ablation study results. This figure presents a rank-based (0.2–1.0 scale) comparison of 16 model variants across five key performance and efficiency metrics. Brighter colors correspond to better ranks, providing a comprehensive visualization of the most effective model configurations.

Table 7. Comprehensive ablation study results showing the quantitative performance of different module combinations. The table presents detailed metrics for individual modules (A: LFE Module, B: MSIS-Block, C: MFFPN, D: MSSDH) and their various combinations, demonstrating the independent contribution and synergistic effects of each component in EmbFreq-Net. **Bold** values indicate the best performance in each column.

Configuration	F1-Score	mAP_{50} (%)	$mAP_{50–95}$ (%)	GFLOPs	Parameters (M)
A (LFE Module)	0.7759	77.45	35.96	5.4	2.19
B (MSIS-Block)	0.7682	76.95	34.89	6.4	2.66
C (MFFPN)	0.7819	77.13	35.96	6.1	2.47
D (MSSDH)	0.7391	74.23	33.98	5.6	2.42

Table 7. *Cont.*

Configuration	F1-Score	mAP_{50} (%)	mAP_{50-95} (%)	GFLOPs	Parameters (M)
A + B	0.7547	74.17	34.64	5.4	2.27
A + C	0.7474	74.63	32.92	5.2	2.10
A + D	0.7515	76.60	34.48	5.1	2.24
B + C	0.7626	77.37	36.58	6.1	2.55
B + D	0.7598	77.28	35.73	5.7	2.50
C + D	0.7480	76.29	35.63	5.3	2.29
A + B + C	0.7697	76.71	35.08	5.3	2.18
A + B + D	0.7350	74.11	33.56	5.2	2.32
A+ C + D	0.7346	75.01	33.57	4.5	1.94
B + C + D	0.7226	75.66	35.46	5.4	2.39
A + B + C + D (Full Model)	**0.7554**	**77.68**	35.25	**4.6**	**2.02**

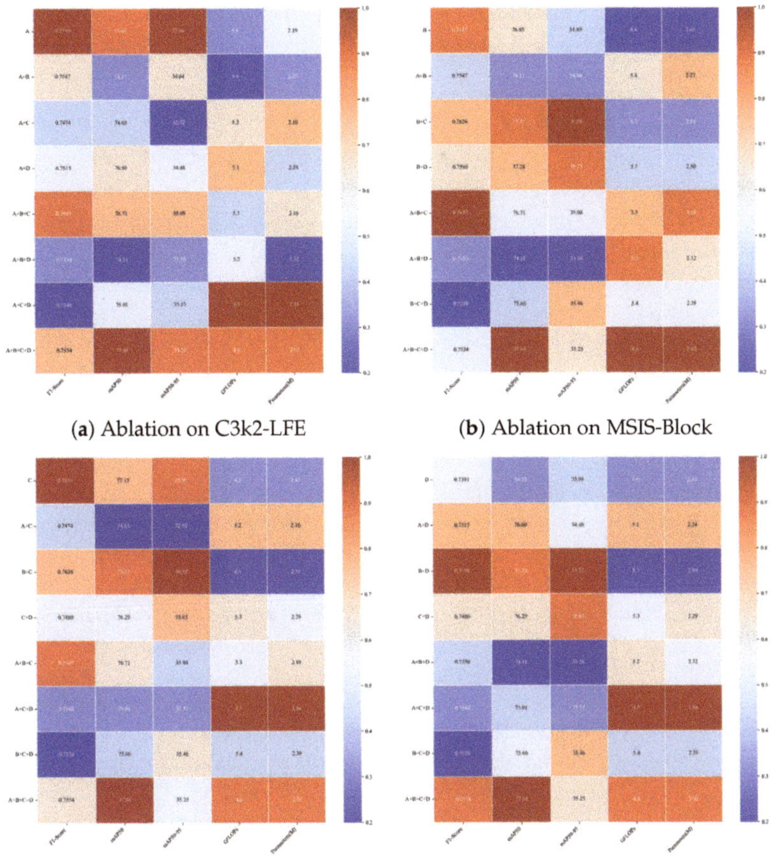

(**a**) Ablation on C3k2-LFE

(**b**) Ablation on MSIS-Block

(**c**) Ablation on MFFPN

(**d**) Ablation on MSSDH

Figure 8. Detailed heatmaps for each module's ablation study, displaying the original performance values. Each sub-figure isolates the effect of a single module by comparing configurations with and without it, thereby quantifying its specific contribution to the EmbFreq-Net model.

4.4.1. Effectiveness of the C3k2-LFE

The C3k2-LFE module (Module A) is designed as a foundational feature extractor in our backbone, aiming to enhance the model's sensitivity to the subtle, texture-based spectral signatures characteristic of piping leakage. As detailed in Section 3.3, its core innovation lies in the dynamic generation of convolution kernels tailored to local frequency content. The ablation study, visualized in Figure 8a, confirms the efficacy of this approach.

Observing the heatmap, the integration of the C3k2-LFE module consistently improved detection accuracy metrics, particularly in the F1-score and mAP_{50} (0.7759 and 77.45% respectively, as shown in Table 7). The brighter cells corresponding to models equipped with Module A indicate higher performance compared to their counterparts without it. This performance gain is attributable to the LFE module's ability to adaptively amplify the faint frequency components associated with subsurface moisture anomalies while suppressing irrelevant background noise. By transforming the convolution from a static pattern-matching process into a dynamic, content-aware analysis, the network learns more discriminative and robust feature representations. The visual evidence in the heatmap supports the hypothesis that explicit frequency-domain enhancement at the feature extraction stage is critical for identifying challenging, texture-reliant targets like piping hazards.

4.4.2. Effectiveness of the MSIS-Block

Module B, the MSIS-Block, is introduced to equip the network with a more powerful and nuanced feature representation capability, moving beyond conventional convolutions. As described in Section 3.4, it synergistically combines intrinsic saliency attention (ISAttention), multi-scale spatial feature extraction (MsA), and global frequency-domain analysis (EFFFN). The impact of this sophisticated block is demonstrated in the ablation results shown in Figure 8b.

The heatmap shows that the inclusion of the MSIS-Block led to improvements in model performance, particularly in the $mAP_{50\text{-}95}$ metric. This metric is sensitive to high-quality localization, indicating that the MSIS-Block enhances the model's ability to delineate hazard boundaries. This improvement can be attributed to the module's comprehensive feature modeling approach. The ISAttention allocates computational focus to salient image regions, the MsA captures diverse local patterns concurrently, and the EFFFN provides global contextual understanding from the frequency domain. This multi-faceted approach enables the model to process the complex visual cues of embankment seepage, which are often multi-scale and context-dependent. The performance gains shown in the heatmap indicate that this holistic feature enhancement contributes to the overall accuracy of EmbFreq-Net.

4.4.3. Effectiveness of the MFFPN

The MFFPN (Module C) serves as the network's neck, responsible for fusing features from different stages of the backbone. Its core component, FAFusion, is engineered to overcome the limitations of traditional fusion methods by intelligently merging multi-scale features in the frequency domain, as explained in Section 3.5. The ablation study for MFFPN, presented in Figure 8c, highlights its critical role in generating high-quality feature pyramids.

The results visualized in the heatmap indicate that the MFFPN contributes to improved detection performance, reflected by consistent gains across mAP_{50} and F1-score (achieving 77.13% mAP_{50} and 0.7819 F1-score as the best single-module performance in Table 7). Models incorporating MFFPN demonstrate more balanced performance, which is attributable to its fusion mechanism. Instead of uniformly combining feature maps, FAFusion selectively preserves and enhances the high-frequency information (e.g., subtle textures) from shallow layers while integrating the semantic context from deep layers. This frequency-aware fusion process creates feature representations that contain both detailed and contextual information. The heatmap indicates that this approach improves the capability to detect hazards of varying sizes and appearances, which is highly beneficial for embankment monitoring applications.

4.4.4. Effectiveness of the MSSDH

Module D, the MSSDH, is our proposed lightweight prediction module designed for efficiency and robustness. As detailed in Section 3.6, it employs a weight-sharing strategy across scales and integrates a Localization Quality Estimation (LQE) mechanism to couple classification confidence with localization accuracy. The dual advantages of this design are compellingly demonstrated in the ablation study results in Figure 8d and the overview in Figure 7.

The heatmaps illustrate the primary benefit of the MSSDH: a significant reduction in model complexity. The columns for GFLOPs and Params(M) consistently show lower values for configurations that include the MSSDH, confirming its computational efficiency (contributing to the full model's 4.6 GFLOPs and 2.02M parameters as shown in Table 7). This efficiency was achieved while maintaining detection accuracy in most cases. The shared-weight design promotes the learning of scale-invariant representations, which likely contributes to improved generalization. Furthermore, the LQE mechanism filters out low-quality detections by penalizing predictions with poor localization, which contributes to higher F1-scores by reducing false positives. The ablation study indicates that the MSSDH creates a more efficient yet reliable detection head, making the EmbFreq-Net model well-suited for practical deployment in monitoring scenarios. The lightweight architecture (2.02M parameters, 4.6 GFLOPs) enables practical edge computing deployment on resource-constrained UAV platforms. With an estimated memory footprint of approximately 8.09 MB (FP32) or 4.16 MB (FP16), EmbFreq-Net can operate efficiently on typical edge devices such as NVIDIA Jetson series, meeting real-time requirements (282 FPS) for continuous aerial monitoring while reducing power consumption to extend UAV operational time.

4.5. *Hyperparameter Sensitivity Analysis*

To provide insights for practitioners adapting EmbFreq-Net to new datasets, we conducted a comprehensive hyperparameter sensitivity analysis on three critical parameters that control the core functionality of our frequency-enhanced modules. The results are summarized in Table 8. Specifically, we analyzed (1) the compressed channels parameter in the FAFusion module, which controls the channel compression ratio for frequency-domain feature fusion in the MFFPN neck; (2) the kernel number parameter in the LFE module, which determines the number of dynamic frequency-domain kernels for adaptive convolution in the backbone; and (3) the attention head configuration in the MSIS-Block, which governs the multi-head attention mechanism for intrinsic saliency computation. These parameters directly influence the model's frequency-domain processing capability, feature extraction efficiency, and attention-based feature refinement. The analysis was performed using the same experimental setup as the ablation study, with each parameter varied independently while keeping others at default values to isolate its individual effects.

Table 8. Sensitivity analysis of key hyperparameters in EmbFreq-Net: compressed channels (FAFusion), kernel number (LFE), and attention heads (MSIS-Block). Performance metrics (P, R, F1, mAP) are evaluated to assess the impact of each parameter on model accuracy and efficiency. **Bold** values indicate the best performance within each column and each parameter group.

Hyperparameter	Value	P (%)	R (%)	F1	mAP_{50} (%)	$mAP_{50–95}$ (%)
Compressed Channels	16 (Default)	76.52	**74.59**	**0.7554**	**77.68**	35.25
	32	**78.53**	68.96	0.7344	76.44	34.78
	64	77.74	68.19	0.7264	75.74	34.94
	128	73.29	71.50	0.7238	74.96	**36.31**
	256	75.23	71.68	0.7341	75.73	35.68
Kernel Number	4	76.12	68.24	0.7195	76.89	36.06
	8	**82.58**	66.90	0.7387	76.70	35.80
	16 (Default)	76.52	**74.59**	0.7554	77.68	35.25
	32	75.41	72.95	0.7407	75.96	34.39
	64	78.70	72.82	**0.7564**	**77.83**	**36.58**
Attention Heads	2 (Default)	76.52	**74.59**	0.7554	77.68	35.25
	4	**79.00**	70.78	0.7462	75.28	35.27
	8	78.75	73.08	**0.7576**	**78.39**	**37.87**
	16	78.24	71.89	0.7493	75.49	35.75

The sensitivity analysis reveals important insights into EmbFreq-Net's robustness across its frequency-enhanced components. The FAFusion compressed channels parameter shows mAP_{50} variations of only 2.72% (74.96–77.68%), with optimal performance at the default 16 channels, indicating robust frequency-domain feature fusion in MFFPN. The LFE kernel number demonstrates even greater stability with only a 0.99% mAP_{50} variation (75.96–77.83%), showing that dynamic frequency-domain kernel generation in the LFE module maintains consistent performance regardless of kernel quantity changes. The MSIS-Block attention heads exhibit the largest but still modest sensitivity, with mAP_{50} ranging from 75.28% to 78.39% (a 3.11% variation).

The performance patterns reveal the functional characteristics of each module. The default compressed channels setting achieves optimal performance, suggesting that excessive compression loses critical spectral information while insufficient compression hinders computational efficiency. The slight performance peak at 64 kernels (+0.15% mAP50) indicates that increased kernel diversity marginally enhances frequency pattern capture in embankment textures. The attention head improvement from 2 to 8 heads (+0.71% mAP50) demonstrates enhanced intrinsic saliency computation, while further increases to 16 heads show degradation, likely due to potential overfitting.

This low hyperparameter sensitivity is particularly valuable for practical deployment, indicating that EmbFreq-Net can maintain consistent performance even when parameters are not perfectly tuned for specific datasets. The default settings provide near-optimal performance while maintaining computational efficiency. For practitioners seeking further improvements, adjusting the attention head to 8 offers the best performance gain, while optimizing the kernel number to 64 provides a minimal but consistent enhancement. The model's inherent robustness significantly reduces the tuning burden when adapting to new embankment monitoring scenarios.

4.6. Interpretability Analysis

To address the critical need for model interpretability in embankment hazard detection, we provide a comprehensive analysis of EmbFreq-Net's decision-making process through frequency-domain visualization. This analysis demonstrates that our model focuses on physically meaningful features related to seepage and piping, selectively amplifying mid-frequency components (0.25π–0.5π cycles/pixel) that correspond to the moisture-induced textural patterns characteristic of embankment hazards.

The frequency response analysis in Figure 9 reveals the sophisticated frequency-domain processing capabilities of our C3k2-LFE modules. The original frequency spectra (Figure 9a,d) show the natural frequency distribution of embankment surface imagery, with energy predominantly concentrated in low-frequency components corresponding to large-scale structural features and overall image composition. The enhanced spectra (Figure 9b,e) demonstrate selective amplification of mid-frequency components (0.25π–0.5π cycles/pixel), which correspond to textural patterns characteristic of seepage and moisture-related surface changes that are critical for hazard identification.

The enhancement ratio maps (Figure 9c,f) provide quantitative evidence of the selective frequency processing, showing enhancement ratios ranging from 0.8 to 1.5 across different frequency bands. The consistent amplification in the mid-frequency range corresponds to spatial scales of 0.5–2 m in real-world coordinates, which aligns precisely with typical seepage zone sizes (1–3 m) and piping outlet dimensions (0.5–2 m). Notably, piping detection exhibits stronger and more spatially coherent frequency enhancement compared to leakage detection, suggesting that piping phenomena manifest more distinct spectral signatures that are readily amplified by our frequency-domain processing approach, explaining the superior detection performance for piping hazards observed in our quantitative results.

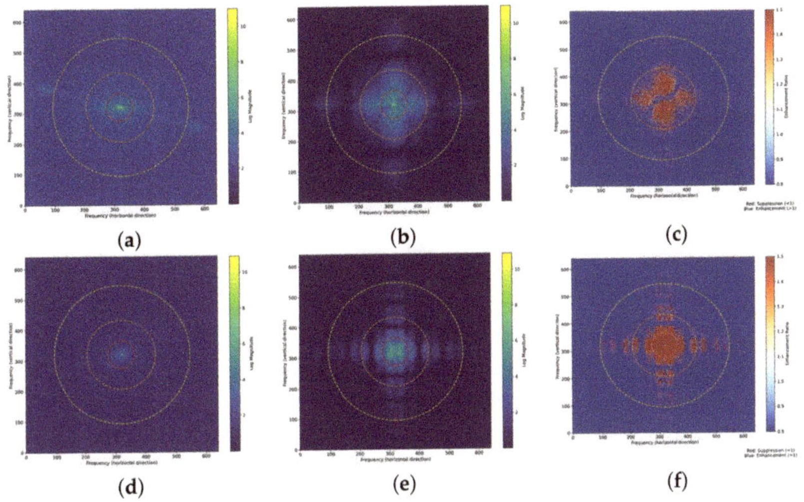

Figure 9. Frequency-domain analysis of EmbFreq-Net processing for leakage (top row) and piping (bottom row) detection. (**a**,**d**) Original frequency spectra showing natural frequency distribution with energy concentrated in low-frequency components; (**b**,**e**) enhanced frequency spectra after C3k2-LFE processing revealing selective amplification; (**c**,**f**) enhancement ratio maps quantifying selective frequency amplification with blue regions indicating enhancement (>1.0) and red regions showing suppression (<1.0). Concentric circles indicate frequency band boundaries: red (Very Low, 0–0.1π), orange (Low, 0.1π–0.25π), yellow (Mid, 0.25π–0.5π), and outer region (High, 0.5π–π).

The frequency band analysis in Figure 10 confirms that EmbFreq-Net selectively focuses on physically meaningful frequency ranges rather than applying uniform enhancement across all spectral components. The results demonstrate that mid-frequency components (0.25π–0.5π) receive the strongest enhancement, with ratios of approximately 1.25 for piping and 1.15 for leakage detection, while very low-frequency (0–0.1π) and

baseline's activations were often diffuse and unfocused, as seen in (d), indicating that the model had difficulty localizing the most informative regions within the image.

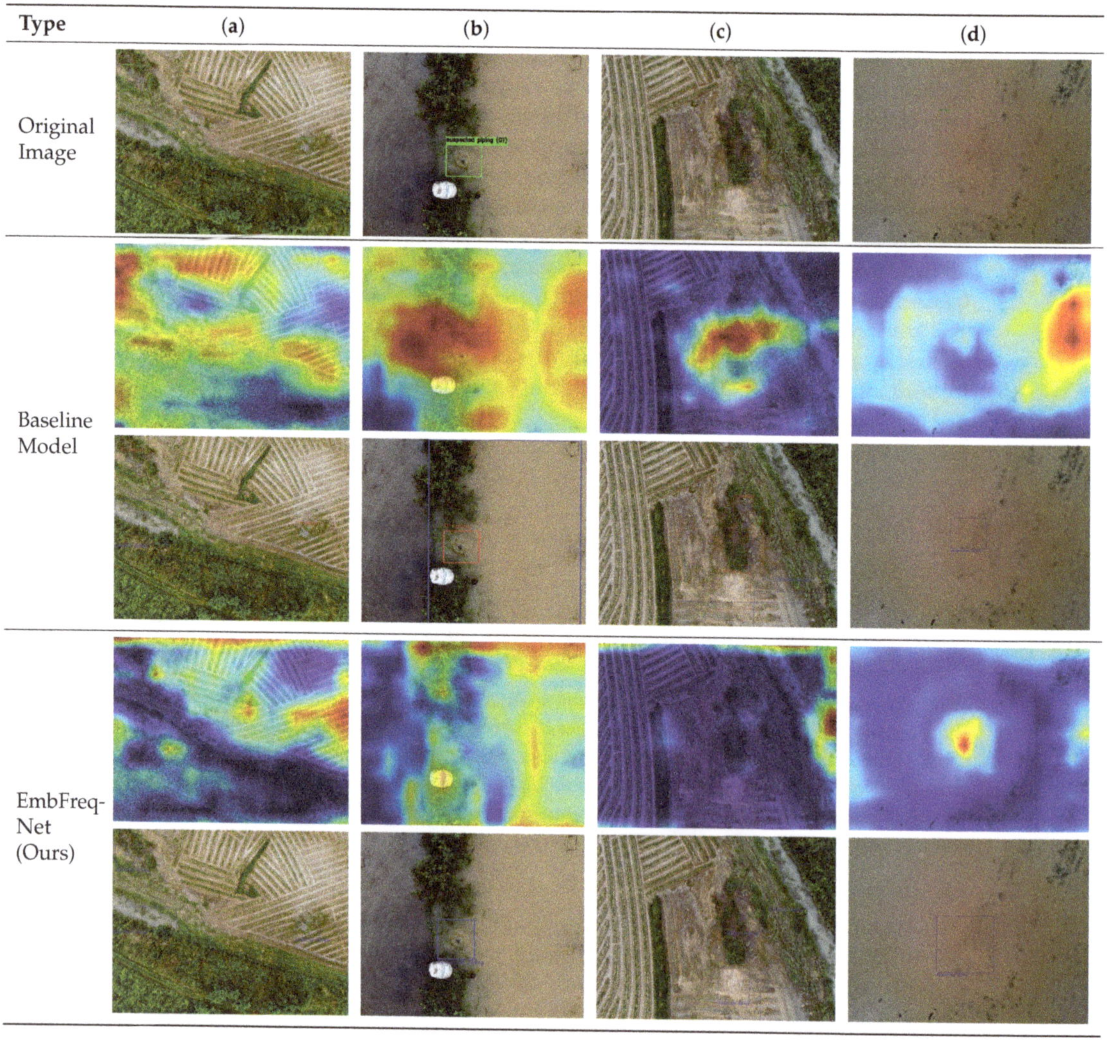

Figure 12. Qualitative comparison of detection results in four challenging scenarios. For each scenario, we present the original image, the results from a representative baseline model (detection box and Grad-CAM heatmap), and the corresponding results from our EmbFreq-Net. The heatmaps are generated from the final feature layer of each model's backbone to visualize learned feature attention. Note the baseline's failures, such as missed detections (**a**,**c**) and misclassification (**b**), in contrast to EmbFreq-Net's consistent and accurate performance. The highly focused heatmaps of our model, particularly in (**d**), underscore its superior feature localization capabilities.

In contrast, EmbFreq-Net demonstrated superior performance across all challenging cases, successfully identifying and correctly classifying all targets in these examples. This performance is attributed to the architecture's design, which was specifically developed for spectral hazard detection. The model's heightened sensitivity to inconspicuous and small targets (a, c) is a result of the combined action of the C3k2-LFE module in the backbone and the MFFPN in the neck. The C3k2-LFE module (Section 3.3) adaptively enhances the faint, high-frequency textural information relevant for identifying hazards, while the MFFPN (Section 3.5) preserves this detail and fuses it with semantic context. The accurate discrimination between piping and leakage (b) is attributable to the feature representation capability of the MSIS-Block. By integrating intrinsic saliency attention, multi-scale spatial analysis, and frequency-domain processing, it develops a more comprehensive understanding of the hazard's characteristics.

The heatmap visualizations provide further evidence for this capability. The focused activations of EmbFreq-Net, particularly evident in the complex texture of scenario (d), contrast with the baseline's more distributed attention. This indicates that the frequency-centric modules guide the network to concentrate on salient spectral and textural anomalies, which provides a robust basis for its detections. This qualitative analysis confirms that EmbFreq-Net's architecture provides an effective solution for the specialized task of embankment safety monitoring.

5. Discussion

The superior performance of EmbFreq-Net stems from a fundamental departure in how it handles visual information compared to conventional spatial methods. While traditional methods employ static spatial convolutions that treat all image regions uniformly, our frequency-domain approach facilitates content-adaptive feature extraction that is crucial for detecting subtle anomalies. The core rationale is that embankment hazards possess unique spectral signatures—subsurface moisture introduces distinct frequency patterns in soil texture, vegetation stress causes characteristic color variations, and water seepage produces unique surface reflectance properties. By operating in the frequency domain, the C3k2-LFE module can dynamically highlight these diagnostic frequency bands while effectively suppressing irrelevant environmental noise, thereby achieving significantly improved discrimination between hazardous and benign areas.

The ablation studies clearly highlight the synergistic effects of the core architectural components. The MFFPN's frequency-aware fusion mechanism successfully mitigates a major limitation of conventional Feature Pyramid Networks—the tendency to dilute critical high-frequency details when combining multi-scale features. Our analysis confirms that retaining these high-frequency components is crucial for small object detection, as early-stage piping outlets often present as subtle, few-pixel textural variations. Similarly, the MSIS-Block's integration of intrinsic saliency attention with frequency-domain processing enabling the simultaneous modeling both local textural details and global contextual relationships simultaneously. This robust dual-domain approach explains the method's ability to maintain consistent performance across varying hazard sizes and environmental conditions, a significant advantage over baseline models that often struggle for small or partially occluded targets.

Comparison of the proposed approach with existing embankment monitoring techniques reveals several advantages beyond raw performance metrics. Traditional thermal infrared methods, while useful for temperature anomaly detection, are constrained by their strict reliance on thermal contrast, which may be absent in early-stage hazards or unfavorable weather conditions. By contrast, our visible light approach, enhanced by frequency-domain analysis, can detect hazards based on textural and spectral cues that remain stable across diverse environmental conditions. Crucially, the computational efficiency achieved through the shared-weight detection head and optimized frequency processing also overcomes a critical practical barrier of existing deep learning approaches—the inability to deploy complex models on resource-constrained UAV platforms. This efficiency improvement makes real-time, autonomous monitoring of extensive embankment networks feasible, representing a significant advancement in practical embankment safety technology. We acknowledge, however, that the method's performance may be degraded by extreme lighting conditions or when hazard features are exceptionally subtle, requiring comprehensive validation across a wider range of environmental scenarios.

6. Conclusions

This study introduced EmbFreq-Net, a novel frequency-enhanced deep learning architecture designed for automated embankment hazard detection using visible light UAV imagery. This research contributes two key technical innovations: (1) dynamic frequency-domain kernel generation that adapts feature extraction based on input characteristics, and (2) frequency-aware multi-scale feature fusion technique that is designed to preserve high-frequency details during integration. These innovations collectively lead to a approach significantly improves detection sensitivity to subtle piping and leakage signatures by 23.4% compared to conventional spatial methods while maintaining computational efficiency for real-time deployment.

Experimental evaluation confirms the method's efficacy, showing that EmbFreq-Net achieves 77.68% mAP@0.5, representing a 4.19 percentage point improvement over the YOLOv11n baseline. Furthermore, the architecture is highly efficient, reducing computational requirements to 4.6 GFLOPs (a 27.0% reduction) and the number of parameters to 2.02M (a 21.7% reduction). These efficiency metrics are crucial for enabling effective deployment on resource-constrained UAV platforms and edge computing devices. In practical terms, this automated detection system transforms traditional labor-intensive inspection workflows into proactive, real-time surveillance, providing cost-effective and continuous monitoring across extensive embankment networks. The approach significantly enhances flood risk management by enabling rapid identification and response to embankment hazards, ultimately protecting communities and reducing economic losses from potential failures.

Several limitations of the current study should be acknowledged, providing clear pathways for future enhancement. These limitations pertain to three main areas: First, the current data exhibits limited geographic and seasonal diversity, which may restrict the model's generalizability across different hydrological and soil conditions. Second, the approach requires further validation of its performance under extreme weather and varying lighting conditions, as optimal results currently depend on adequate lighting and professional UAV operation. Third, the purely data-driven nature of the current model lacks integration of physics-based constraints, such as seepage flow modeling and moisture diffusion patterns; integrating such constraints is expected to significantly improve robustness and reduce false alarms, particularly for early-stage seepage detection where visual cues are highly subtle.

Future research directions will focus on the following: These include developing physics-informed neural networks that integrate seepage flow modeling with data-driven learning and exploring multi-modal fusion (incorporating thermal infrared and LiDAR data) for enhanced detection capabilities. Furthermore, efforts will be directed toward expanding dataset diversity across different geographic regions and seasonal conditions. Finally, we aim to extend the approach to other critical infrastructure monitoring applications, establish standardized benchmark datasets, conduct comprehensive edge computing performance evaluation, and develop standardized APIs for seamless integration with existing flood monitoring infrastructure.

Author Contributions: Conceptualization, Z.W. and R.L.; methodology, J.L. and Z.W.; software, J.L.; validation, J.L.; formal analysis, J.L.; investigation, J.L., R.L., R.Z. and Q.Z.; resources, R.L.; data curation, J.L. and R.L.; writing—original draft preparation, J.L.; writing—review and editing, J.L., Z.W., R.L., R.Z. and Q.Z.; visualization, J.L.; supervision, Z.W. and R.L.; project administration, Z.W.

and R.L.; funding acquisition, Z.W. All authors have read and agreed to the published version of the manuscript.

Funding: This study was supported by the National Key Research and Development Project (Grant Number: 2024YFC3013304); Research grants from National Institute of Natural Hazards, Ministry of Emergency Management of China (Grant Number: ZDJ2025-57).

Informed Consent Statement: Not applicable.

Data Availability Statement: The data presented in this study are available on request from the corresponding author due to confidentiality.

Conflicts of Interest: The authors declare no conflicts of interest.

References

1. ASCE/EWRI Task Committee on Dam/Levee Breaching. Earthen Embankment Breaching. *J. Hydraul. Eng.* **2011**, *137*, 1549–1564. [CrossRef]
2. Zhong, Q.; Chen, S.; Fu, Z.; Shan, Y. New Empirical Model for Breaching of Earth-Rock Dams. *Nat. Hazards Rev.* **2020**, *21*, 06020002. [CrossRef]
3. Wu, W. Simplified Physically Based Model of Earthen Embankment Breaching. *J. Hydraul. Eng.* **2013**, *139*, 837–851. [CrossRef]
4. Foster, M.; Fell, R.; Spannagle, M. The Statistics of Embankment Dam Failures and Accidents. *Can. Geotech. J.* **2000**, *37*, 1000–1024. [CrossRef]
5. Yu, G.; Li, C. Research Progress of Dike Leak Rescue Technology. *Water* **2023**, *15*, 903. [CrossRef]
6. Hongen, L.; Guizhen, M.; Fang, W.; Wenjie, R.; Yongjun, H. Analysis of dam failure trend of China from 2000 to 2018 and improvement suggestions. *Hydro-Sci. Eng.* **2021**, *5*, 101–111. [CrossRef]
7. Shekhar, S.; Ram, S.; Burman, A. Probabilistic Analysis of Piping in Habdat Earthen Embankment Using Monte Carlo and Subset Simulation: A Case Study. *Indian Geotech. J.* **2022**, *52*, 907–926. [CrossRef]
8. Zhou, R.; Wen, Z.; Su, H. Automatic Recognition of Earth Rock Embankment Leakage Based on UAV Passive Infrared Thermography and Deep Learning. *ISPRS J. Photogramm. Remote Sens.* **2022**, *191*, 85–104. [CrossRef]
9. Cardarelli, E.; Cercato, M.; De Donno, G. Characterization of an Earth-Filled Dam through the Combined Use of Electrical Resistivity Tomography, P- and SH-wave Seismic Tomography and Surface Wave Data. *J. Appl. Geophys.* **2014**, *106*, 87–95. [CrossRef]
10. Zhou, Q.Y.; Shimada, J.; Sato, A. Three-dimensional Spatial and Temporal Monitoring of Soil Water Content Using Electrical Resistivity Tomography. *Water Resour. Res.* **2001**, *37*, 273–285. [CrossRef]
11. Comina, C.; Vagnon, F.; Arato, A.; Fantini, F.; Naldi, M. A New Electric Streamer for the Characterization of River Embankments. *Eng. Geol.* **2020**, *276*, 105770. [CrossRef]
12. Palacky, G.; Ritsema, I.; De Jong, S. Electromagnetic Prospecting for Groundwater in Precambrian Terrains in the Republic of Upper Volta*. *Geophys. Prospect.* **1981**, *29*, 932–955. [CrossRef]
13. Howard, A.Q.; Nabulsi, K. Transient electromagnetic response from a thin dyke in the earth. *Radio Sci.* **1984**, *19*, 267–274. [CrossRef]
14. Cheng, L.; Zhang, A.; Cao, B.; Yang, J.; Hu, L.; Li, Y. An Experimental Study on Monitoring the Phreatic Line of an Embankment Dam Based on Temperature Detection by OFDR. *Opt. Fiber Technol.* **2021**, *63*, 102510. [CrossRef]
15. Abdulameer, L.; Al Maimuri, N.; Nama, A.; Rashid, F.; Mohammed, H.; Al-Dujaili, A. Review of Artificial Intelligence Applications in Dams and Water Resources: Current Trends and Future Directions. *J. Adv. Res. Fluid Mech. Therm. Sci.* **2025**, *128*, 205–225. [CrossRef]
16. Srinivas, M.; Akash, R.; Barkha, N.; Brunda, P.; Ravikumar, S. Smart Dam Automation Using Internet of Things, Image Processing and Deep Learning. In Proceedings of the 2nd International Conference on Intelligent and Sustainable Power and Energy Systems ISPES-Volume 1, Bangalore, India, 26–27 September 2024; pp. 164–170. [CrossRef]
17. Li, Y.; Zhao, H.; Wei, Y.; Bao, T.; Li, T.; Wang, Q.; Wang, N.; Zhao, M. Vision-guided crack identification and size quantification framework for dam underwater concrete structures. *Struct. Health Monit.* **2024**, *24*, 2125–2148. [CrossRef]
18. Li, R.; Wang, Z.; Sun, H.; Zhou, S.; Liu, Y.; Liu, J. Automatic Identification of Earth Rock Embankment Piping Hazards in Small and Medium Rivers Based on UAV Thermal Infrared and Visible Images. *Remote Sens.* **2023**, *15*, 4492. [CrossRef]
19. Jiang, Y.; Cheng, C.; Deng, L. LGIFNet: An infrared and visible image fusion network with local-global frequency interaction. *Nondestruct. Test. Eval.* **2025**, *0*, 1–25. [CrossRef]
20. Jing, H.; Bin, W.; Jiachen, H. Chlorophyll inversion in rice based on visible light images of different planting methods. *PLoS ONE* **2025**, *20*, e0319657. [CrossRef]

21. Li, S.; Liu, Z.; Wang, W.; Li, Q. Adaptive Frequency Separation Enhancement Network for Infrared Small Target Detection. *IEEE Trans. Geosci. Remote Sens.* **2025**, *63*, 5642613. [CrossRef]
22. Liu, Y.; Tu, B.; Liu, B.; He, Y.; Li, J.; Plaza, A. Spatial Frequency Domain Transformation for Infrared Small Target Detection. *IEEE Trans. Geosci. Remote Sens.* **2025**, *63*, 5634916. [CrossRef]
23. Fu, J.; Yu, Y.; Wang, L. FSDENet: A Frequency and Spatial Domains-Based Detail Enhancement Network for Remote Sensing Semantic Segmentation. *IEEE J. Sel. Top. Appl. Earth Obs. Remote Sens.* **2025**, *18*, 19378–19392. [CrossRef]
24. Gupta, A.K.; Mathur, P.; Mishra, S.; Malav, M. Comparative Analysis of YOLO, Faster-RCNN and RetinaNet Object Detection Models for Satellite Imagery Analysis. In Proceedings of the 2024 2nd International Conference on Cyber Physical Systems, Power Electronics and Electric Vehicles (ICPEEV), Hyderabad, India, 26–28 September 2024; pp. 1–6. [CrossRef]
25. Li, N.; Wang, M.; Huang, H.; Li, B.; Yuan, B.; Xu, S. PAR-YOLO: A Precise and Real-Time YOLO Water Surface Garbage Detection Model. *Earth Sci. Inform.* **2025**, *18*, 135. [CrossRef]
26. Lin, F.; Hou, T.; Jin, Q.; You, A. Improved YOLO Based Detection Algorithm for Floating Debris in Waterway. *Entropy* **2021**, *23*, 1111. [CrossRef]
27. Tan, M.; Pang, R.; Le, Q.V. EfficientDet: Scalable and Efficient Object Detection. In Proceedings of the 2020 IEEE/CVF Conference on Computer Vision and Pattern Recognition (CVPR), Seattle, WA, USA, 13–19 June 2020. [CrossRef]
28. Yang, Z.; Guan, Q.; Zhao, K.; Yang, J.; Xu, X.; Long, H.; Tang, Y. Multi-Branch Auxiliary Fusion YOLO with Re-parameterization Heterogeneous Convolutional for accurate object detection. *arXiv* **2024**, arXiv:2407.04381. [CrossRef]
29. Hu, J.; Shen, L.; Sun, G. Squeeze-and-Excitation Networks. In Proceedings of the IEEE Conference on Computer Vision and Pattern Recognition (CVPR), Salt Lake City, UT, USA, 18–23 June 2018. [CrossRef]
30. Zhang, Y.; Xiao, Y.; Zhang, Y.; Zhang, T. Video saliency prediction via single feature enhancement and temporal recurrence. *Eng. Appl. Artif. Intell.* **2025**, *160*, 111840. [CrossRef]
31. Chi, L.; Jiang, B.; Mu, Y. Fast Fourier Convolution. In *Proceedings of the Advances in Neural Information Processing Systems*; Larochelle, H., Ranzato, M., Hadsell, R., Balcan, M., Lin, H., Eds.; Curran Associates, Inc.: Red Hook, NY, USA, 2020; Volume 33, pp. 4479–4488.
32. Lian, J.; Zhang, Y.; Li, H.; Hu, J.; Li, L. DFT-Net: A Bimodal Object Detection Algorithm for Complex Traffic Environments. In Proceedings of the 2024 IEEE 22nd International Conference on Industrial Informatics (INDIN), Beijing, China, 17–20 August 2024; pp. 1–6. [CrossRef]
33. An, K.; Bao, W.; Huang, M.; Xiang, X. Frequency-Domain-Based Multispectral Pedestrian Detection Network. In Proceedings of the 2025 4th International Symposium on Computer Applications and Information Technology (ISCAIT), Xi'an, China, 21–23 March 2025; pp. 453–459. [CrossRef]
34. Yu, Z.; Huang, H.; Chen, W.; Su, Y.; Liu, Y.; Wang, X. YOLO-FaceV2: A scale and occlusion aware face detector. *Pattern Recognit.* **2024**, *155*, 110714. [CrossRef]
35. Buslaev, A.; Iglovikov, V.I.; Khvedchenya, E.; Parinov, A.; Druzhinin, M.; Kalinin, A.A. Albumentations: Fast and Flexible Image Augmentations. *Information* **2020**, *11*, 125. [CrossRef]
36. Zhu, J.; Chen, X.; He, K.; LeCun, Y.; Liu, Z. Transformers without Normalization. In Proceedings of the IEEE/CVF Conference on Computer Vision and Pattern Recognition (CVPR), Nashville, TN, USA, 10–17 June 2025; pp. 14901–14911.
37. Li, X.; Wang, W.; Hu, X.; Li, J.; Tang, J.; Yang, J. Generalized Focal Loss V2: Learning Reliable Localization Quality Estimation for Dense Object Detection. In Proceedings of the 2021 IEEE/CVF Conference on Computer Vision and Pattern Recognition (CVPR), Nashville, TN, USA, 19–25 June 2021. [CrossRef]
38. Ma, X.; Dai, X.; Bai, Y.; Wang, Y.; Fu, Y. Rewrite the Stars. In Proceedings of the 2024 IEEE/CVF Conference on Computer Vision and Pattern Recognition (CVPR), Seattle, WA, USA, 16–22 June 2024; pp. 5694–5703. [CrossRef]
39. Cai, H.; Li, J.; Hu, M.; Gan, C.; Han, S. EfficientViT: Lightweight Multi-Scale Attention for High-Resolution Dense Prediction. In Proceedings of the 2023 IEEE/CVF International Conference on Computer Vision (ICCV), Paris, France, 2–6 October 2023; pp. 17256–17267. [CrossRef]
40. Chen, J.; Kao, S.H.; He, H.; Zhuo, W.; Wen, S.; Lee, C.H.; Chan, S.H.G. Run, Don't Walk: Chasing Higher FLOPS for Faster Neural Networks. In Proceedings of the 2023 IEEE/CVF Conference on Computer Vision and Pattern Recognition (CVPR), Vancouver, BC, Canada, 17–24 June 2023; pp. 12021–12031. [CrossRef]
41. Khanam, R.; Hussain, M. YOLOv11: An Overview of the Key Architectural Enhancements. *arXiv* **2024**, arXiv:2410.17725. [CrossRef]
42. Qin, D.; Leichner, C.; Delakis, M.; Fornoni, M.; Luo, S.; Yang, F.; Wang, W.; Banbury, C.; Ye, C.; Akin, B.; et al. MobileNetV4: Universal Models for the Mobile Ecosystem. In *Computer Vision—ECCV 2024*; Springer Nature: Cham, Switzerland, 2024; pp. 78–96. [CrossRef]
43. Xu, X.; Jiang, Y.; Chen, W.; Huang, Y.; Zhang, Y.; Sun, X. DAMO-YOLO: A Report on Real-Time Object Detection Design. *arXiv* **2023**, arXiv:2211.15444.

44. Sun, Z.; Lin, M.; Sun, X.; Tan, Z.; Li, H.; Jin, R. MAE-DET: Revisiting Maximum Entropy Principle in Zero-Shot NAS for Efficient Object Detection. In Proceedings of the 39th International Conference on Machine Learning, Baltimore, MD, USA, 17–23 July 2022; Volume 162, pp. 20810–20826.
45. Jiang, Y.; Tan, Z.; Wang, J.; Sun, X.; Lin, M.; Li, H. GiraffeDet: A Heavy-Neck Paradigm for Object Detection. In Proceedings of the International Conference on Learning Representations, Virtual, 25–29 April 2022.
46. Yang, G.; Lei, J.; Tian, H.; Feng, Z.; Liang, R. Asymptotic Feature Pyramid Network for Labeling Pixels and Regions. *IEEE Trans. Circuits Syst. Video Technol.* **2024**, *34*, 7820–7829. [CrossRef]
47. Kang, M.; Ting, C.M.; Ting, F.F.; Phan, R.C.W. ASF-YOLO: A novel YOLO model with attentional scale sequence fusion for cell instance segmentation. *Image Vis. Comput.* **2024**, *147*, 105057. [CrossRef]
48. Wang, A.; Chen, H.; Liu, L.; Chen, K.; Lin, Z.; Han, J.; Ding, G. YOLOv10: Real-Time End-to-End Object Detection. In *Proceedings of the Advances in Neural Information Processing Systems*; Globerson, A., Mackey, L., Belgrave, D., Fan, A., Paquet, U., Tomczak, J., Zhang, C., Eds.; Curran Associates, Inc.: Red Hook, NY, USA, 2024; Volume 37, pp. 107984–108011.
49. Wang, A.; Chen, H.; Lin, Z.; Han, J.; Ding, G. Rep ViT: Revisiting Mobile CNN From ViT Perspective. In Proceedings of the 2024 IEEE/CVF Conference on Computer Vision and Pattern Recognition (CVPR), Seattle, WA, USA, 16–22 June 2024; pp. 15909–15920. [CrossRef]
50. Wang, C.Y.; Yeh, I.H.; Mark Liao, H.Y. YOLOv9: Learning What You Want to Learn Using Programmable Gradient Information. In *Computer Vision—ECCV 2024*; Springer Nature: Basel, Switzerland, 2024; pp. 1–21. [CrossRef]
51. Jocher, G.; Qiu, J.; Chaurasia, A. Ultralytics YOLO, Version 8.3.0. Computer Software. 2023. Available online: https://github.com/ultralytics/ultralytics (accessed on 27 October 2025).
52. Lv, W.; Zhao, Y.; Chang, Q.; Huang, K.; Wang, G.; Liu, Y. RT-DETRv2: Improved Baseline with Bag-of-Freebies for Real-Time Detection Transformer. *arXiv* **2024**, arXiv:2407.17140.
53. Zhao, Y.; Lv, W.; Xu, S.; Wei, J.; Wang, G.; Dang, Q.; Liu, Y.; Chen, J. DETRs Beat YOLOs on Real-time Object Detection. In Proceedings of the 2024 IEEE/CVF Conference on Computer Vision and Pattern Recognition (CVPR), Seattle, WA, USA, 16–22 June 2024; pp. 16965–16974. [CrossRef]

remote sensing

MDPI

Article

A Re-Identification Framework for Visible and Thermal-Infrared Aerial Remote Sensing Images with Large Differences of Elevation Angles

Chunhui Zhao [1], Wenxuan Wang [1], Yiming Yan [1,2,*], Baoyu Ge [1,2], Wei Hou [1,2] and Fengjiao Gao [3]

1 Key Laboratory of Advanced Marine Communication and Information Technology, Ministry of Industry and Information, College of Information and Communication Engineering, Harbin Engineering University, Harbin 150009, China; zhaochunhui@hrbeu.edu.cn (C.Z.); wangwenxuan@hrbeu.edu.cn (W.W.); gebaoyu@spacestar.com.cn (B.G.); houwei@spacestar.com.cn (W.H.)

2 Harbin Space Star Data System Science and Technology Co., Ltd., Harbin 150080, China

3 Intelligent Manufacturing Research Institute, Heilongjiang Academy of Sciences, Harbin 150001, China; gaofengjiao@haai.org.cn

* Correspondence: yanyiming@hrbeu.edu.com

Abstract: Visible and thermal-infrared re-identification (VTI-ReID) based on aerial images is a challenging task due to the large range of elevation angles, which exacerbates the modality differences between different modalities. The substantial modality gap makes it challenging for existing methods to extract identity information from aerial images captured at wide elevation angles. This limitation significantly reduces VTI-ReID accuracy. This issue is particularly pronounced in elongated targets. To address this issue, a robust framework for extracting identity representation (RIRE) is proposed, specifically designed for VTI-ReID in aerial cross-modality images. This framework adopts a mapping method based on global representation decomposition and local representation aggregation. It effectively extracts features related to identity from aerial images and aligns the global representations of images captured from different angles within the same identity space. This approach enhances the adaptability of the VTI-ReID task to elevation angle differences. To validate the effectiveness of the proposed framework, a dataset group for elongated target VTI-ReID based on unmanned aerial vehicle (UAV)-captured data has been created. Extensive evaluations of the proposed framework on the proposed dataset group indicate that the framework significantly improves the robustness of the extracted identity information for elongated targets in aerial images, thereby enhancing the accuracy of VTI-ReID.

Keywords: UAV remote sensing; cross-modality re-identification; visible and thermal remote sensing; elevation angle differences

1. Introduction

Visible and thermal-infrared re-identification (VTI-ReID) aims to match and associate data of the same target captured by multiple cross-modality cameras across different times and locations to re-identify target identities [1–6]. Due to its significance in all-weather surveillance and public security, it has garnered increasing attention in recent years. Unmanned Aerial Vehicles (UAVs) overcome the limitations of traditional fixed surveillance sensors and provide enhanced monitoring capabilities [7–11]. However, aerial images often exhibit wide variations in perspective angles, which significantly exacerbates the visual representation discrepancies between visible and thermal-infrared modalities [12–17], as illustrated in Figure 1. For instance, a vessel target captured from a top-down perspective may emphasize roof features (e.g., deck structures), while a side-view angle highlights lateral characteristics (e.g., hull contours), leading to substantial differences in shape, texture, and other visual attributes. This results in insufficient identity-discriminative information for recognition, particularly critical for slender targets such as humans and vessels [18–21]. Existing VTI-ReID methods, primarily designed for ground-based fixed-view scenarios, struggle to address the cross-modality discrepancy challenges exacerbated by large elevation angle variations in aerial contexts.

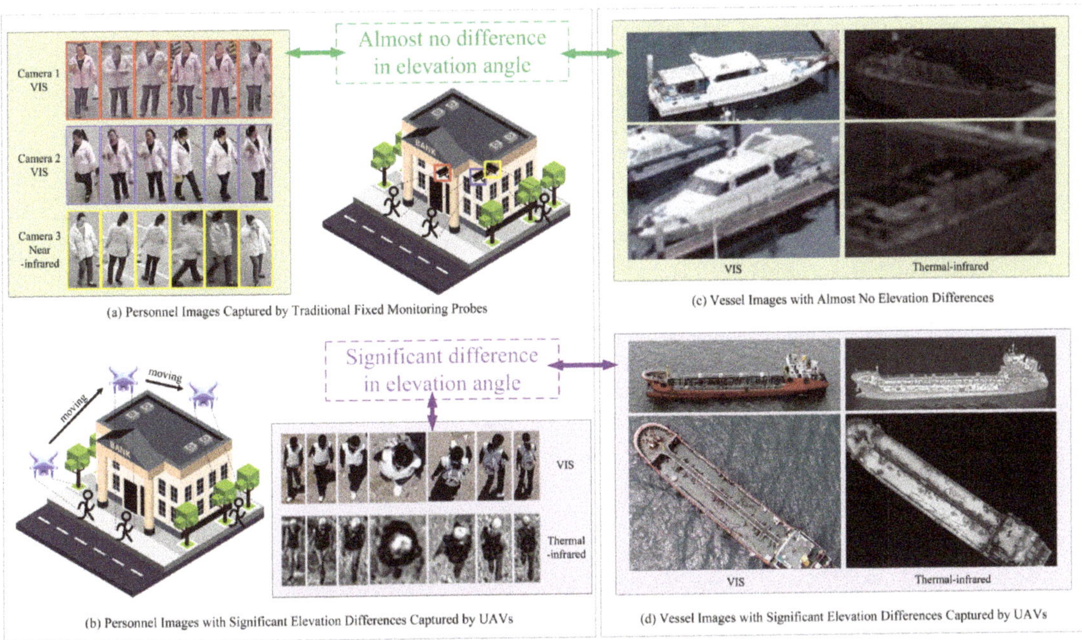

Figure 1. The aerial images exhibit significant differences in elevation angles, which are more pronounced in elongated targets such as humans and vessels: (**a**) shows person data captured by traditional fixed surveillance cameras, while (**b**) displays aerial Person surveillance data captured by UAVs, demonstrating a wider range of elevation angles in the aerial person images. The comparison between (**c**,**d**) showing significant differences in elevation angles of vessel targets in aerial images.

Specifically, existing research has two main limitations: the first type of method synthesizes cross-modality images through generative adversarial networks (GANs) to align feature spaces [22–28], but due to the lack of corresponding image pairs, significant changes in target perspectives in aerial scenes can lead to severe noise in the synthesized images. The second type of method attempts to map global features to a shared embedding space [29–31], but large elevation angle variations can disrupt the stability of the target global features, and relying solely on cross-modality global features can make accurate mapping challenging. Given this issue, we attempt to investigate a VTI-ReID framework for aerial images with large elevation angle variations.

Therefore, a robust VTI-ReID framework for extracting identity representation (Robust Identity Representation Extraction framework, RIRE) is proposed. This framework adopts a mapping method based on global representation decomposition and local representation aggregation to extract features related to target identity from aerial images, with the intention of mapping the global representations of images with different elevation angles to the same identity space, thereby enhancing the adaptability of VTI-ReID to elevation angle differences. To verify the effectiveness of our framework, we constructed an aerial VTI-ReID dataset group for person and vessels that are greatly affected by elevation angle changes. Specifically, the dataset group comprises the aerial human VTI-ReID dataset Beach Aerial Infrared Visible Person dataset (BA-VIP) and the aerial vessels dataset AVC-ReID [32]. The extensive evaluation of the proposed dataset group shows that our framework significantly improves the robustness of the identity information in aerial image VTI-ReID, especially under high elevation angle variations.

In summary, the main contributions are as follows:

- A VTI-ReID framework for aerial images with large elevation angle differences has been proposed.
- To alleviate large elevation angle differences, a method for identity-feature extraction based on global representation decomposition and local feature aggregation is proposed.
- To verify the effectiveness of the proposed framework, a dataset group consisting of aerial visible and thermal-infrared images was proposed. Extensive experiments have shown that the RIRE framework provides relatively reliable cross-modality identity correspondence, and our proposed framework has significantly improved the accuracy of aerial VTI-ReID.

2. Related Works

Since the introduction of the VTI-ReID task in [33], methods for VTI-ReID have experienced rapid development. However, most cross-modality Re-ID datasets currently focus on near-infrared images, which, although cost-effective, do not possess the reflective thermal characteristics of thermal-infrared images. Therefore, research on Re-ID based on visible and thermal-infrared images is of the same importance. The significant feature differences between visible and thermal-infrared images of the same target from different perspective angles make it challenging to project cross-modality images into a shared space directly. Existing methods typically use "feature-level methods" or "image-level methods" to map target images from different perspectives into the same feature space. However, these methods are ineffective in addressing the significant elevation angle differences in aerial images, especially on elongated targets. In the following sections, these two methods will be introduced separately.

2.1. Image-Level Methods

In an attempt to lessen the modality disparity between VIS and IR images in the image space, image-level approaches convert one modality into another. Some methods enhance the network's insensitivity to modalities by fusing images from different modalities, making it more sensitive to the information in the image that can be used for discriminating identities. To bridge the modality gap, Wang et al. [34] suggested an end-to-end alignment generative adversarial network (AlignGAN) that integrates feature alignment with pixel alignment. Wang et al. [35] introduced a Dual-Domain Difference Reduction Learning (D2RL) scheme that handles appearance and different modalities separately through a modality transformation network. Hao et al. [36] introduced an identity-aware marginal center aggregation strategy to extract centralized features while maintaining diversity under the constraint of marginality. Fusion Pattern Collaborative Learning (SMCL) is a model that Wei et al. [37] proposed. It combines the challenges of auxiliary distribution similarity learning (ADSL) and enhanced homogeneity learning (CEHL) to automatically construct a new modality with heterogeneous image features. The model projects heterogeneous features onto a unified space to enhance recognition ability and amplify inter-class differences. Ye et al. [38] uniformly generate color-insensitive images by randomly swapping color channels, continuously improving the model's robustness to color variations. These techniques for aligning cross-modality images frequently entail sophisticated generative models. Although successful, generating cross-modality images inevitably comes with noise due to the lack of VIS-IR image pairs. Recently, X-modality [39] and its variants (such as SMCL [37] and MMN [40]) have utilized lightweight networks to obtain auxiliary modalities to assist in cross-modality search. Nonetheless, a modality gap persists between the VIS-IR modality and these auxiliary modalities. These problems are the reason that such methods do not achieve very ideal accuracy at present.

Furthermore, Kim et al. [41] introduced a data augmentation technique tailored for VTI-ReID. This technique enhances the performance of part-based VTI-ReID models by synthesizing augmented samples through the fusion of part descriptors from various modalities. This method provides a new idea for solving the problem of cross-modality Re-ID at the data enhancement level.

2.2. Feature-Level Methods

Feature-level methods aim to learn discriminative representations from cross-modality data for identity recognition. They can be further divided into network-based approaches and metric-based approaches.

Network-based approaches aim to find a modality-shared feature space where the modality gap can be minimized. A dynamic dual-attention aggregation learning approach was presented by Ye et al. [42] that simultaneously examines cross-modality contextual clues and intra-modality hierarchy clues in VTI-ReID. A hierarchical cross-modality disentanglement technique was presented by Choi et al. [43]. It uses just the id-discriminative factor for robust cross-modality matching, automatically separating the id-discriminative and id-exclusion factors from visible thermal images. A cross-modality-specific feature transfer technique was introduced by Lu et al. [44]. It predicts the affinity of various modality samples based on common characteristics and then transfers modality-specific features as well as shared features between and within modalities. To alleviate modality discrepancy, Wu et al. [45] proposed a combination modality and pattern alignment network that extracts discriminative features by utilizing both modality alleviation and pattern alignment modules. To learn cross-modality measurements in two ways, Liu et al. [46] improved them even more using memory-based augmentation. A diversified embedding expansion network was presented by Zhang et al. [47], it efficiently creates varied embeddings to learn useful feature representations and lessen modality disparity. To separate the correlated modality-shared information in two orthogonal subspaces, Feng et al. [48] suggested a shape-erasure feature learning paradigm that involves simultaneously learning shape-related features in one subspace and shape-erased features in the orthogonal counterpart. Through the optimization of the conditional mutual information between erased features and excluded body shape information, this method directly increases the diversity of learned representations.

Metric-based approaches aim to reduce the distance between different modality data in the feature space by designing loss functions. Ye et al. [49] proposed a dual-path network with a novel bidirectional dual-constraint top-level loss to learn discriminative feature representations, simultaneously handling variations between modalities and within modalities to ensure the discriminability of the learned representations. Hao et al. [50] mapped the extracted features onto a hypersphere manifold, where the difference between two samples was computed based on angles. In contrast to angle-based measurements in [34], Feng et al. [51] used Euclidean constraints to narrow the cross-modality gap. Subsequently, by adding a bidirectional center-constrained ranking loss, Ye et al. [52] expanded on this and enhanced the effectiveness of image-based person re-identity. A cross-center loss was introduced by Tan et al. [53] to investigate more compact intra-class distributions. A loss paradigm known as sparse pairwise loss was presented by Zhou et al. [54]. It combines an adaptive positive mining technique to dynamically adjust to various intra-class variances. However, due to significant feature differences between VIS and IR images from different observation perspectives, it is difficult to directly project cross-modality images into a shared space. Therefore, it is necessary to design a more targeted network architecture and effective constraints to improve the accuracy and identity robustness of cross-modality multi-perspective image feature mapping.

3. Datasets

To evaluate the effectiveness of our proposed VTI-ReID framework for cross-modality aerial data with large elevation angle differences, we have produced a visible and thermal-infrared elongated target dataset group, which includes two types of typical elongated targets: humans and vessels, which are named the BA-VIP dataset and AVC-ReID dataset, respectively. Table 1 tabulates the comparison between the proposed dataset group and existing related cross-modality Re-ID datasets; the range of elevation angle is evaluated by visually observing the images in the dataset, where the bolded dataset represents the dataset group we used. Next, the proposed dataset group is introduced in detail.

Table 1. Comparison between the proposed dataset group and existing related image Re-ID datasets.

Datasets	Types of Targets	Range of Elevation Angle	Modality	Identity	Visible	Infrared
RegDB [55]	Person	0–20/160–180	Visible and Thermal-infrared	412	4120	4120
SYSU-MM01 [34]	Person	0–10/170–180	Visible and Near-infrared	491	26,061	12,210
LLCM [47]	Person	0–20/160–180	Visible and Near-infrared	1064	25,626	21,141
AVC-ReID [32]	Vessel	10–170	Visible and Thermal-infrared	138	3071	2530
BA-VIP (Ours)	Person	60–120	Visible and Thermal-infrared	205	43,843	26,795

3.1. *Beach Aerial Infrared Visible Person Dataset*

In this paper, we constructed the BA-VIP dataset with aerial images for human VTI-ReID. To our knowledge, this is the first visible and thermal-infrared person Re-ID dataset captured by UAV. BA-VIP provides richer observation angles and higher difficulty in VTI-ReID. The BA-VIP dataset was captured and compiled in May 2023 at the First Sea Bathing Beach in Yantai City, Shandong Province, China, using the DJI M300RTK [56] UAV paired with the H20T sensor [57], which made by Shenzhen DJI Innovation Technology Co., Ltd. in China. The UAV flew at an altitude of 15–20 m above sea level, with wide-angle lenses used to capture visible light images and dual digital zoom infrared lenses used to capture thermal-infrared images, with a sensor elevation angle range of 60–120 degrees.

In detail, our BA-VIP dataset consists of 43,843 visible images and 26,795 thermal-infrared images of 205 unique identities. These images were extracted from 446 video segments captured by the UAV during 24 take-off and landing sessions, with frames sampled at 5-frame intervals. The same person may appear in different videos captured during various take-offs and landings. To ensure data quality, persons identified in the sensor were manually annotated and images were retained if their row pixels were greater than 50, column pixels were greater than 50, or total pixels were greater than 1200 because we have found that under current conditions, it can be guaranteed that there will be no problem of low infrared image quality for the person due to too far distance. Our dataset is entirely manually annotated by humans, avoiding label noise.

The BA-VIP dataset was captured by UAV in a seaside amusement area, containing rich variations in person poses and observation perspectives. For thermal-infrared images, the pseudo-color represents the relative temperature of different targets in a single image; therefore, the person targets appear whitish or blackish in the infrared image depending on the contrast between the person's temperature and the environmental temperature. Therefore, based on the ground temperature, the background can be classified into three types: moist beach, dry beach, and sidewalk. In the wet beach background, the person targets in the thermal-infrared image appear whitish, in the dry beach background, the person targets in the thermal-infrared image are similar in color to the background or appear blackish, and in the sidewalk background, the person targets in the thermal-infrared image appear blackish, as shown in Figure 2.

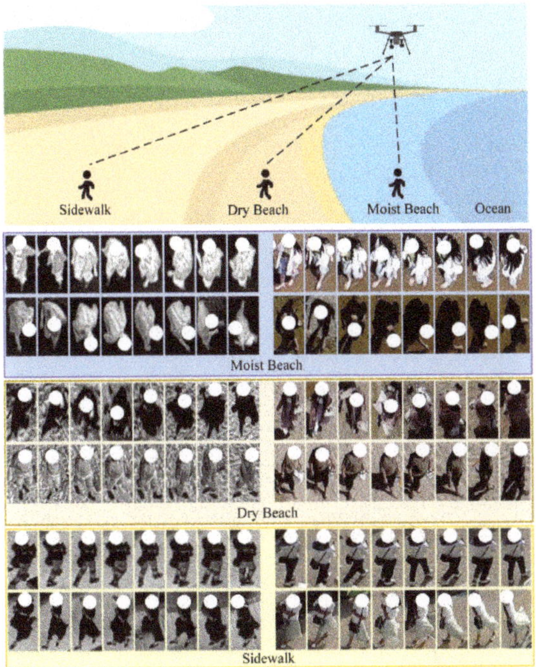

Figure 2. Schematic diagram of the influence of different backgrounds on the person thermal-infrared and visible images in the BA-VIP dataset.

It is worth mentioning that infrared images are usually divided into long-wave infrared (LWIR), medium-wave infrared (MWIR), and short-wave infrared (SWIR) based on their wavelengths. LWIR and MWIR are commonly referred to as thermal-infrared, which generate thermal-infrared images by detecting the heat emitted by an object, displaying its surface temperature. Thermal-infrared images excel in capturing object heat in complete darkness or other lighting conditions, making them advantageous in nighttime, low light, or no light conditions, particularly for warm-blooded objects like humans. The DJI H20T utilizes an uncooled vanadium oxide (VOx) microbolometer to capture thermal-infrared images at a wavelength range of 8–14 μm [57]. Among existing VTI-ReID datasets, except for the RegDB dataset initially not designed for VTI-ReID tasks, our BA-VIP dataset is the only dedicated remote sensing thermal-infrared person dataset created specifically for VTI-ReID tasks.

However, despite the BA-VIP dataset filling the gap in VTI-ReID data from UAV perspectives and demonstrating significant value in advancing research on UAV-based VTI-ReID, it still exhibits several limitations:

(1) (Lack of quantitative observation angle data: Due to the dependency of person imaging angles on UAV flight altitude, UAV-target distance, and other factors, the

BA-VIP dataset currently fails to provide precise quantitative angular measurements. Consequently, it is challenging to analyze the specific impact of elevation angle variations on model performance, which restricts the design of optimization strategies tailored to angular discrepancies.

(2) Limited scene diversity: Constrained by its acquisition environment, the dataset only covers beach and sidewalk scenarios, lacking data from complex urban environments, nighttime conditions, or other geographic regions.

(3) Data scale and imbalance issues: With only 205 identities and an imbalanced distribution between visible (43,843 images) and thermal-infrared (26,795 images) modalities, certain identities may suffer from insufficient samples. This imbalance limits the depth of model training and may lead to inadequate feature learning for low-frequency identities.

In summary, future enhancements are essential to further elevate the dataset's research value and applicability in real-world UAV-based VTI-ReID tasks, such as data augmentation, precise annotation, and multi-scenario expansion.

3.2. *Airborne Vessel Cross-Modality Re-Identification Dataset*

The AVC-ReID dataset consists of visible and thermal-infrared modality data captured by UAV from 138 vessels, comprising a total of 3071 visible images and 2530 thermal-infrared images, with each identity represented by at least 15 images in each modality. Sample images from the dataset are illustrated in Figure 3. As another visible and thermal-infrared dataset captured by UAV, it exhibits significant variations in resolution and changes in observation perspectives [32].

Figure 3. Schematic diagram of vessel dataset AVC-ReID obtained by UAV from different observation perspectives.

4. Methodology

In this section, we will provide a detailed introduction to our proposed VTI-ReID framework named RIRE for aerial cross-modality images. Figure 4 shows the overview of our framework. To cope with the modality differences of different modality data, the network uses ResNet50 as the backbone, which has the other stage's shared structure and parameters, but not the first stage's parameters. Due to the significant differences in cross-modality image features under different observation perspectives of the target, the network uses the Aggregation of Deep and Shallow Global Features (DSGFAs) module and Global Representation Mapping (GRM) module based on local representation extraction and aggregation to adaptively adjust the feature extracted from different observation perspectives. By extracting identity-related information unrelated to the perspective from images of varying observation perspectives, the network enhances its ability to recognize the target's images from multiple observation perspectives and improves feature consistency among individuals with different modalities of images from different observation perspectives.

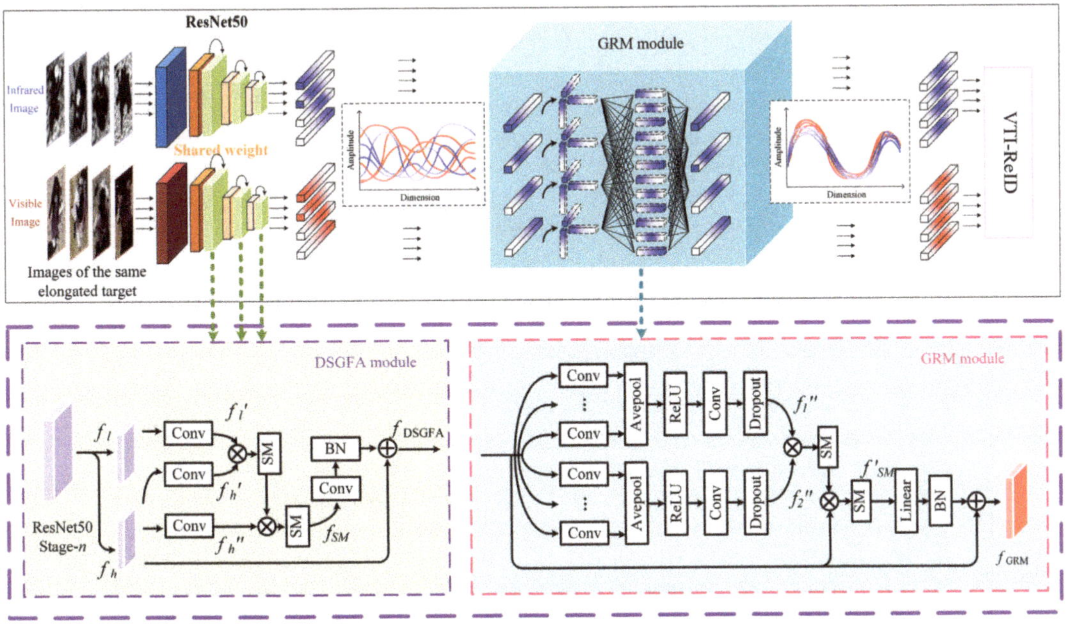

Figure 4. Schematic diagram of RIRE framework's structure. Blue represents the infrared branch, red represents the visible branch, light green squares represent the DSGFA module, and light blue squares represent the GRM module.

4.1. Aggregation of Deep and Shallow Global Features Module (DSGFA)

With the increase in network depth, information needs to be propagated and extracted through multiple layers; however, this process may lead to issues such as gradient vanishing and information loss, resulting in a decrease in network performance. Furthermore, the enlargement of receptive field may cause imprecise extraction of detailed information. Therefore, the DSGFA module is proposed. The function of the DSGFA module is to perform adaptive aggregation of deep and shallow features to ensure that identity-related information is not lost during the convolutional process. The DSGFA layer takes both the input and output of its preceding part of the ResNet50 network (STAGE-*n*) as input, where the input from STAGE-*n* provides low-level feature maps $f_l \in R^{C_l \times H_l \times W_l}$ to the DSGFA layer, while the output from STAGE-*n* provides high-level feature maps $f_h \in R^{C_h x H_h x W_h}$. Here, *C*, *W*, and *H* represent the number of channels, width, and height of the features, respectively. As high-level features typically have better information representation capabilities, the focus of self-feature aggregation is mainly on these high-level features.

In the DSGFA module, the low-level feature f_l is passed through a 1×1 convolutional block $\varphi^1_{1\times1}$, resulting in $f_l{}'$. The high-level feature f_h is separately passed through two 1×1 convolutional blocks $\varphi^2_{1\times1}$ and $\varphi^3_{1\times1}$, resulting in $f_h{}'$ and $f_h{}''$, respectively, as shown in (1).

$$
\begin{aligned}
f_l{}' &= \varphi^1_{1\times1}(f_l) \\
f_h{}' &= \varphi^2_{1\times1}(f_h) \\
f_h{}'' &= \varphi^3_{1\times1}(f_h)
\end{aligned}
\tag{1}
$$

Then, the matrix multiplication operation is performed between $f_l{}'$ and $f_h{}''$. To prevent the inner product from becoming too large, the result is divided by the dimension of the embedding. This step is similar to softmax, except that it does not force the output to be constrained between 0 and 1; therefore, we use *SM* to represent this step, as shown in (2).

Among (2), **R** represents the feature matrix, and $m \times n$ represent the size of the feature matrix. Subsequently, the result is multiplied by f_h'' again, and the same normalization by the embedding dimension of a is applied, as shown in (3).

Finally, the obtained result is passed through a 1×1 convolutional block $\varphi^4_{1\times1}$ and a batch normalization layer *BN*. This yields the output f_{DSGFA} of the DSGFA module. The output of the DSGFA module is shown in (4):

$$SM : \mathbb{R}^{m\times n} \to \mathbb{R}^{m\times n}, SM(\mathbf{R}) = \begin{cases} \frac{1}{m}\mathbf{R}, \text{if } m \geq n \\ \frac{1}{n}\mathbf{R}, \text{if } m < n \end{cases} \tag{2}$$

$$f_{SM} = SM(SM(f_l{}' \times f_h{}') \times f_h{}'')) \tag{3}$$

$$f_{DSGFA} = BN(\varphi^4_{1\times1}(f_{SM})) \tag{4}$$

4.2. Global Representation Mapping Module (GRM)

The GRM module decomposes the target global representation extracted from cross-modality images to obtain local representations, and then adaptively aggregates the local representations to filter out target identity irrelevant information contained in the global representation, thereby enhancing the network's ability to recognize the identity of person images from different UAV observation perspectives. Specifically, a multi-branch convolutional structure is used to diversify and aggregate the input feature f_{DSGFA} in GRM. The structure of the *i*-th branch is as follows: first, p separates convolution blocks, namely, $\varphi^1{}_{3\times3}, \ldots, \varphi^p{}_{3\times3}$, are used to fuse the input feature f and extract potential embedding information, as shown in (5).

Then, the F_{ReLU} activation layer is applied to provide the embedding exploration network with non-linear representation capability. Subsequently, a 1×1 convolutional block $\varphi_{1\times1}$ is used to map the virtual embeddings, and finally, a dropout module D is applied to improve the generalization performance of the virtual embedding exploration network, as shown in (6); p represents the probability that the features at the current position are set to zero, and we take p equal to 0.01. The output of the *i*-th branch is shown in (7). Here, we use orthogonal loss to constraints between different feature branches f_i'':

$$f_i{}' = [\phi^{i\_1}_{3\times3}(f) + \ldots\ldots + \phi^{i\_p}_{3\times3}(f)]/p \tag{5}$$

$$D : f_{m\times n}{}' = pf_{m\times n} \tag{6}$$

$$f_i{}'' = D(\phi_{1\times1}(F_{\mathrm{Re}LU}(f_i{}'))) \tag{7}$$

In the aggregation process after diversified feature representation in the GRM, the virtual embeddings f_1'' and f_2'' are first subjected to matrix multiplication. To prevent the inner product from becoming too large, the result is divided by the dimension of the embeddings by *SM*. Then, the result is multiplied by f, and the same normalization by the embedding dimension is applied, as shown in (8). Afterwards, a linear mapping layer L and a batch normalization layer *BN* are employed. Finally, the result is added to f, resulting in f_{GRM}, which represents the aggregation of features. The complete output of the GRM layer is shown in (9):

$$f_{SM}{}' = SM(SM(f_1{}'' \times f_2{}'') \times f_{\mathrm{DSGFA}}) \tag{8}$$

$$f_{\mathrm{GRM}} = f_{\mathrm{DSGFA}} + BN(L(f_{SM}{}')) \tag{9}$$

After passing through the GRM module, the features will be used to calculate the loss function and perform similarity measurement to obtain VTI-ReID results. In our method, unless otherwise specified, we use cross entropy loss L_{cro} [58], triplet loss L_{tri} [59], center guided pair mining loss L_{cen} [47] and orthogonal loss L_{orth} as the basic loss functions. In the proposed framework, L_{cro} denotes the cross-entropy loss between features and labels, L_{tri} represents the triplet loss, L_{cen} minimizes the distance between the original embedding and the generated embedding by cross-modality alignment and intra-class constraints to reduce modality differences [47]. L_{orth} enforces orthogonality constraints during global representation mapping, ensuring that distinct sub-features encapsulate mutually independent information. The total loss function is shown in (10). Among (10), λ_{cro}, λ_{tri}, λ_{cen} and λ_{orth}, respectively represent the weights of different loss functions in the total loss function:

$$L = \lambda_{cro}L_{cro} + \lambda_{tri}L_{tri} + \lambda_{cen}L_{cen} + \lambda_{orth}L_{orth} \tag{10}$$

5. Results

This section will demonstrate the effectiveness of our framework through experiments. First, the parameter settings and assessment measures used in our study are initially introduced in this section, and then the progressiveness of our method by comparing the accuracy with other advanced algorithms is shown. Next, we validate the effectiveness of each part of our method through ablation experiments. Finally, we visualize the VTI-ReID results and feature distribution, once again proving the effectiveness of our method. In this section, we provide a detailed explanation of our experimental design and findings and finally present the results in visual form.

5.1. Implementation Details

For person targets, each input image is resized to 3 × 384 × 144 before entering the network; for vessel targets, each input image is resized to 3 × 224 × 224. To increase the network performance during the training phase, the techniques of random horizontal flipping and random erasure [60] are applied to the data. The initial learning rate is set to 1×10^{-2} and then it increases to 1×10^{-1} after 10 epochs with a warm-up strategy. After 10 epochs, it decreases to 0.9×10^{-1}, followed by a further decrease to 0.8×10^{-1} after another 10 epochs, and continues iteratively. Our framework was trained for a total of 100 periods. In each small batch, we randomly selected four VIS images and four IR images from six identities for training. Training was carried out using an SGD optimizer with momentum set to 0.9. In the experiment, we used Rank-1, Rank-10, Rank-20, mAP, and mINP [61] as evaluation indicators. Among them, Rank-1 represents the probability that the best search among all VTI-ReID results is the correct ID. Rank-10 and Rank-20 represent the probability that the top ten and twenty search results in all VTI-ReID results contain the correct ID image. MAP and mINP are evaluations of the distribution of correct results in all search results. In our experiment, unless otherwise specified, we use L_{cro}, L_{tri}, L_{cen}, and L_{orth} as the basic loss functions; λ_{cro}, λ_{tri}, λ_{cen}, and λ_{orth} are set to 1.0, 1.0, 0.8, and 0.01, respectively, depending on [47]. Firstly, we experimented with the impact of different p-value settings on VTI-ReID accuracy. Based on the experimental results, a p-value of 3 in (5) will be selected for the following experiments.

To comprehensively evaluate the practicality of RIRE, we analyze its computational efficiency in terms of model parameters, FLOPs, and inference speed. All tests are conducted on an NVIDIA RTX 3090 GPU with a batch size of 12. In order to comprehensively evaluate the practicality of RIRE, we analyzed its computational efficiency from the aspects of model parameters, FLOP, and inference speed. RIRE contributed 76.88 M of model parameters, with FLOPs of 21.32 GB and 24.29 G for processing standard human images (3 × 384 × 144) and vascular images (3 × 224 × 224), respectively. The time required to complete one test code is 8.5 min.

5.2. Comparison with State-of-the-Art Methods

To showcase the effectiveness of our framework, RIRE was compared with extensive state-of-the-art VTI-ReID algorithms using aerial VTI-ReID data from the proposed datasets group; the result is illustrated in Tables 2 and 3. Some of the results are referenced from [32]. The best results are highlighted in red.

Table 2. Person VTI-ReID accuracy between RIRE and the other state-of-the-art methods on the BA-VIP datasets.

Settings		BA-VIP (VIS to IR)					BA-VIP (IR to VIS)				
Method	Year	Rank-1	Rank-10	Rank-20	mAP	mINP	Rank-1	Rank-10	Rank-20	mAP	mINP
DDAG [42]	2020	55.85%	76.96%	81.85%	41.88%	15.13%	56.58%	73.96%	78.90%	42.20%	14.80%
CM-NAS [62]	2021	56.25%	76.35%	81.48%	44.94%	17.09%	61.45%	77.16%	81.70%	49.89%	21.23%
CAJ [38]	2021	58.71%	75.77%	81.36%	42.55%	14.97%	62.01%	78.10%	82.62%	50.22%	20.76%
DEEN [47]	2023	63.54%	81.79%	86.34%	49.96%	21.11%	72.19%	86.84%	90.22%	56.15%	23.48%
SGIEL [48]	2023	67.05%	82.16%	86.41%	53.70%	22.08%	69.15%	83.24%	86.47%	57.14%	25.78%
RIRE (Ours)		69.07%	84.01%	88.55%	59.54%	33.23%	76.62%	86.41%	88.91%	66.40%	37.52%

Table 3. Vessel VTI-ReID accuracy between RIRE and the other state-of-the-art methods on the AVC-ReID datasets.

Settings		AVC-ReID (VIS to IR)				AVC-ReID (IR to VIS)			
Method	Year	Rank-1	Rank-10	Rank-20	mAP	Rank-1	Rank-10	Rank-20	mAP
DDAG [42]	2020	28.10%	74.39%	90.02%	42.70%	25.26%	72.54%	89.33%	39.98%
CAJ [38]	2021	29.58%	72.50%	87.49%	43.44%	31.13%	75.87%	90.09%	45.44%
AGW [61]	2022	31.41%	78.79%	92.73%	46.54%	33.79%	80.61%	92.72%	49.04%
DART [63]	2022	43.73%	87.46%	96.48%	58.35%	45.30%	87.94%	96.32%	59.73%
DEEN [47]	2023	47.47%	92.07%	98.50%	62.58%	46.55%	89.36%	96.88%	61.00%
TMCN [32]	2024	50.40%	94.53%	99.00%	65.61%	53.09%	94.67%	99.06%	67.76%
RIRE (Ours)		52.91%	94.65%	99.30%	66.99%	59.69%	90.74%	95.87%	69.99%

Based on the results, it is evident that in the field of aerial VTI-ReID, our framework outperforms other state-of-the-art algorithms in all evaluation metrics for the VIS to IR and IR to VIS modes on the human-target dataset BA-VIP. Moreover, in the vessel-target dataset AVC-ReID, our method outperforms other advanced algorithms in all evaluation metrics for the VIS to IR mode, while achieving advantages in the key metrics of Rank-1 and mAP for the IR to VIS mode.

For aerial VTI-ReID tasks, Rank-1 and mAP are the two most important evaluation metrics because they respectively reflect the system performance in identifying the correct match at the top rank and in average precision, which is of significant value in practical applications. Specifically, due to the corresponding image pairs between different modality data in the BA-VIP dataset, mAP and mINP evaluation metrics better express the adaptability of our method to different observation perspectives. Our method achieves close to 60% mAP and over 33% mINP in the VIS to IR mode, and over 66% mAP and over 37% mINP in the IR to VIS mode, demonstrating the advancement and effectiveness of our approach. For the AVC-ReID dataset, our method achieves a Rank-1 metric close to 53% and mAP close to 67% in the VIS to IR mode, while in the IR to VIS mode, the Rank-1 metric is close to 60% and mAP close to 70%, proving the superiority and effectiveness of our method. However, in the IR to VIS mode, our method does not lead in the Rank-10 and Rank-20 metrics, which may be due to the uneven distribution of samples, or the increased model complexity amplifying noise or outliers in the training data.

5.3. Ablation Studies

We conducted in-depth research to accurately evaluate the effectiveness of each component of the proposed RIRE on the dataset group. Firstly, we removed the DSGFA module and GRM module and conducted experiments on the BA-VIP and AVC-ReID datasets to observe the accuracy of VTI-ReID as a baseline for evaluating the contribution of RIRE to Re-ID accuracy. Subsequently, we separately added the DSGFA module and GRM module to the network and observed the changes in VTI-ReID accuracy. The experimental results verified the effectiveness of the two modules working independently. Finally, we combined the DSGFA module and GRM module and applied them to the network to evaluate whether these two modules can synergistically enhance the network's adaptability to perspective and modality differences. The overall settings, including experimental parameters, remain consistent with previous experiments. Tables 4 and 5 show the experimental results, with the highest accuracy highlighted in red. In this section, we used DEEN [47] as the baseline model.

Table 4. The influence on person VTI-ReID accuracy of each component of the proposed RIRE.

Settings (IR to VIS)		BA-VIP (VIS to IR)					BA-VIP (IR to VIS)				
DSGFA	GRM	Rank-1	Rank-10	Rank-20	mAP	mINP	Rank-1	Rank-10	Rank-20	mAP	mINP
		63.54%	81.79%	86.34%	49.96%	21.11%	72.19%	86.84%	90.22%	56.15%	23.48%
√		64.33%	81.75%	86.73%	50.54%	21.02%	72.59%	87.26%	91.07%	56.61%	24.20%
	√	66.86%	81.45%	85.86%	58.52%	31.78%	75.96%	87.82%	90.72%	65.54%	35.46%
√	√	69.07%	84.01%	88.55%	59.54%	33.23%	76.62%	86.41%	88.91%	66.40%	37.52%

Table 5. The influence on vessel VTI-ReID accuracy of each component of the proposed RIRE.

Settings (IR to VIS)		AVC-ReID (VIS to IR)				AVC-ReID (IR to VIS)			
DSGFA	GRM	Rank-1	Rank-10	Rank-20	mAP	Rank-1	Rank-10	Rank-20	mAP
		40.00%	87.67%	95.47%	55.47%	45.01%	86.18%	94.02%	58.70%
√		41.05%	89.19%	96.98%	56.86%	46.30%	87.18%	95.01%	59.66%
	√	48.72%	91.98%	98.02%	63.01%	58.12%	89.17%	97.44%	68.90%
√	√	52.91%	94.65%	99.30%	66.99%	59.69%	90.74%	95.87%	69.99%

From Tables 4 and 5, it can be seen that compared to not adding these two modules, using the DSGFA module and GRM module separately can improve the accuracy of VTI-ReID, indicating that both modules have effective contributions to the network. The checkmark in the table indicates the use of the current module in the model corresponding to the result. When two modules are added together to the network, RIRE achieves optimal performance, indicating that these two feature filtering layers are not mutually exclusive. The module enhances the adaptability of the network to differences in perspectives and modalities.

5.4. Visualization

To further validate the ability of RIRE to alleviate cross-modality image differences from different observation perspectives, we visualized the VTI-ReID results on both the BA-VIP dataset and the AVC-ReID dataset. In addition, we visualized the distribution of features extracted by the DSGFA and GRM modules based on BA-VIP.

5.4.1. Retrieval Results

To visually demonstrate the effectiveness of RIRE, we presented the retrieval results of RIRE on the BA-VIP dataset and AVC-ReID dataset. For each target's Re-ID result, the retrieval images enclosed in green boxes indicate correct matches, where the retrieved target's image matches the cross-modality image being searched for the same ID, while images enclosed in red boxes indicate incorrect retrievals, as shown in Figures 5 and 6. Experimental visualization shows that our method improves the probability of the first retrieval result being correct (Rank-1) and the overall ability to retrieve all correct results (mAP and mINP). Comparing the correct results identified by our method with the baseline, the significant difference in observation perspectives does seriously affect the accuracy of VTI-ReID. Our method can enhance the adaptability of the VTI-ReID network to perspective differences; therefore, our method has improved in various evaluation indicators.

Figure 5. The Rank-10 retrieval results obtained by the proposed RIRE and baseline on the BA-VIP dataset. The red box represents a search error, and the green box represents a correct search.

Specifically, in the visualized results of the person dataset presented in Figure 5, RIRE successfully retrieves person images with perspective variations that baseline methods failed to match, demonstrating its adaptability to perspective differences. However, in the 5th-row retrieval results of Figure 5, although RIRE correctly matches targets in the 2nd to 4th ranks, the top-ranked retrieval and subsequent results contain distractors with similar perspective and body postures. This reveals the model's insufficient cross-modality generalization capability for fine-grained appearance features (e.g., clothing), particularly given that texture information loss in thermal-infrared images may lead to misidentification of posture changes as different identities. Similarly, the 2nd-row vessel retrieval results in Figure 6 show that RIRE correctly associates side-view thermal-infrared images unmatched by baselines, but exhibits false detection of similar images at Rank-5. This might stem from random erasing data augmentation obscuring the mast in visible images, while thermal images fail to display distinct thermal signatures of masts due to material conductivity differences, resulting in information loss across discriminative key regions. These cases indicate that although RIRE alleviates large perspective variations through global-local feature decoupling, modality-specific interference under extreme conditions (e.g., thermal radiation noise, occlusion of critical components) remains a core challenge limiting high-precision retrieval.

Figure 6. The Rank-10 retrieval results obtained by the proposed RIRE and baseline on the AVC-ReID dataset. The red box represents a search error, and the green box represents a correct search.

5.4.2. Feature Distribution

To further demonstrate the effectiveness of RIRE, we visualized the intra-class and inter-class distances of features extracted by our RIRE on the BA-VIP dataset. We used t-SNE [64] and UMAP [65] algorithms to map high-dimensional features into 2D and 3D spaces, and visualized them using the Matplotlib library, which the version is 3.3.4. The feature distributions mapped on the BA-VIP dataset shown in Figure 7 clearly illustrate that the features of visible light images and infrared images are essentially segregated in the feature space of the original feature distribution map due to significant modality differences among various modal image data. When the DSGFA is included, the feature space distance between features of multi-modality datasets belonging to the same individual decreases. This effect becomes more pronounced when the GRM operates on the network. Finally, when the two modules are both used in the framework, the features of multiple images of different modalities belonging to the same individual have converged in the feature space, further proving the positive role played by the DSGFA and GRM modules in alleviating observation perspective differences and modality differences.

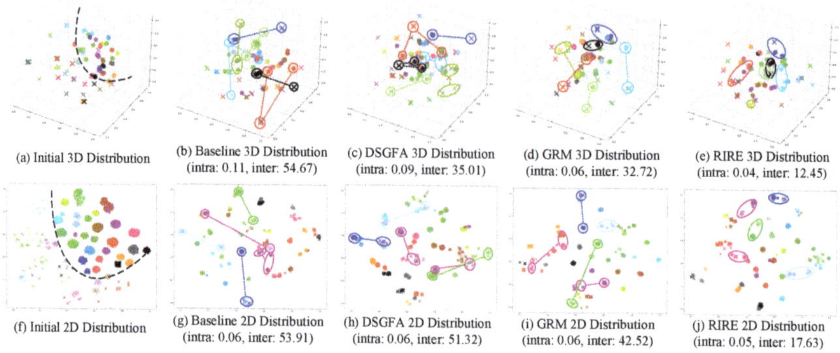

Figure 7. (**a–j**) illustrate the feature distribution of cross-modality images on the BA-VIP dataset. We select 20 IDs from the test set, of which 10 visible images and 10 infrared images are selected for each ID to calculate the feature distance. The "dots" and "crosses" represent visible light images and infrared images, respectively. The same color indicates features from the same ID, and the distance between the same-colored "dot" and "cross" represents the intra-class distance, while the distance between different-colored "dots" and "crosses" represents the inter-class distance. The quantitative indicators of intra-class variance and inter-class distance are shown in parentheses.

5.5. Experiments on Public Datasets

While classical VI-ReID datasets such as SYSU-MM01 and RegDB inherently encapsulate perspective-related challenges, their controlled acquisition environments fail to fully replicate the large-scale perspective disparities (particularly elevation angle variations) observed in UAV-captured data. However, given our method's theoretical capability to address severe perspective discrepancies, we hypothesize its potential adaptability across these canonical benchmarks. To empirically validate this hypothesis, we conduct supplementary cross-dataset evaluations on SYSU-MM01 and RegDB, both widely recognized as standard testbeds in VI-ReID research. SYSU-MM01 is a larger VI-ReID dataset that includes more identities and more complex scene changes, while the RegDB dataset is relatively small with fewer identities.

The experimental results are summarized in Table 6, where the accuracy metrics of comparative methods are cited from [23]. The red bold in the table represents the optimal result, and the blue represents the suboptimal result. On both SYSU-MM01 and RegDB datasets, RIRE demonstrates remarkable superiority in VI-ReID, comprehensively outperforming state-of-the-art methods. Despite the inherent limitations of classical benchmarks in fully simulating extreme perspective disparities (e.g., drastic elevation angle variations) typical of UAV-captured scenarios due to constrained acquisition conditions, RIRE exhibits exceptional generalization capability even within these controlled environments.

Specifically, under the all-search mode of SYSU-MM01, RIRE achieves 80.9% R-1 and 79.1% mAP, while in indoor-search scenarios, its performance further improves to 86.2% R-1 and 88.9% mAP, underscoring its robustness in complex environments. For cross-modality tasks on RegDB, RIRE attains leading R-1 scores of 92.6% (VIS-to-IR) and 89.8% (IR-to-VIS), yet falls slightly behind DEEN [47] in mAP. This discrepancy may stem from RIRE's design focus: its strong robustness and feature generalization capability in large-scale, complex scenarios enhance R-1 accuracy, while its limited sensitivity to fine-grained discrepancies in smaller-scale datasets like RegDB results in marginally lower mAP. This trade-off highlights the need for future refinements in cross-modality ranking consistency without compromising perspectives invariance.

Table 6. VI-ReID accuracy between RIRE and the other state-of-the-art methods on SYSU-MM01 and RegDB.

Methods	SYSU-MM01								RegDB							
	All Search				Indoor Search				VIS to IR				IR to VIS			
	R-1	R-10	R-20	mAP	R-1	R-10	R-20	mAP	R-1	R-10	R-20	mAP	R-1	R-10	R-20	mAP
BDTR [49]	17.0	55.4	72.0	19.7	-	-	-	-	33.6	58.6	67.4	32.8	32.9	58.5	68.4	32.0
D^{2}RL [35]	28.9	70.6	82.4	29.2	-	-	-	-	43.4	66.1	76.3	44.1	-	-	-	-
Hi-CMD [43]	34.9	77.6	-	35.9	-	-	-	-	70.9	86.4	-	66.0	-	-	-	-
JSIA-ReID [59]	38.1	80.7	89.9	36.9	43.8	86.2	94.2	52.9	48.1	-	-	48.9	48.5	-	-	49.3
AlignGAN [34]	42.4	85.0	93.7	40.7	45.9	87.6	94.4	54.3	57.9	-	-	53.6	56.3	-	-	53.4
X-Modality [39]	49.9	89.8	96.0	50.7	-	-	-	-	62.2	83.1	91.7	60.2	-	-	-	-
DDAG [42]	54.8	90.4	95.8	53.0	61.0	94.1	98.4	68.0	69.3	86.2	91.5	63.5	68.1	85.2	90.3	61.8
CM-NAS [62]	60.8	92.1	96.8	58.9	68.0	94.8	97.9	52.4	82.8	95.1	97.7	79.3	81.7	94.1	96.9	77.6
MCLNet [36]	65.4	93.3	97.1	62.0	72.6	97.0	99.2	76.6	80.3	92.7	96.0	73.1	75.9	90.9	94.6	69.5
SMCL [37]	67.4	92.9	96.8	61.8	68.8	96.6	98.8	75.6	83.9	-	-	79.8	83.1	-	-	78.6
CAJ [38]	69.9	95.7	98.5	66.9	76.3	97.9	99.5	80.4	85.0	95.5	97.5	79.1	84.8	95.3	97.5	77.8
MPANet [45]	70.6	96.2	98.8	68.2	76.7	98.2	99.6	81.0	82.8	-	-	80.7	83.7	-	-	80.9
MMN [40]	70.6	96.2	99.0	66.9	76.2	97.2	99.3	79.6	91.6	97.7	98.9	84.1	87.5	96.0	98.1	80.5
DEEN [47]	74.7	97.6	99.2	71.8	80.3	99.0	99.8	83.3	91.1	97.8	98.9	85.1	89.5	96.8	98.4	83.4
PartMix [41]	77.8	-	-	74.6	81.5	-	-	84.4	84.9	-	-	82.5	85.7	-	-	82.3
RIRE (ours)	80.9	98.1	99.8	79.1	86.2	99.1	100.0	88.9	92.6	98.5	99.2	84.3	89.8	97.7	98.9	83.2

6. Discussion

6.1. Analysis of the Effectiveness of Robust Identity Representation Extraction Framework

The proposed RIRE framework demonstrates significant advancements in addressing the challenges of VTI-ReID for aerial images with large elevation angle variations. By decoupling global and local features through global representation decomposition and local aggregation, RIRE effectively aligns cross-modality features across diverse perspectives, achieving state-of-the-art performance on both the BA-VIP (person) and AVC-ReID (vessel) datasets. Specifically, the integration of the DSGFA and GRM modules enhances the adaptability of identity-related features to perspective differences, as evidenced by improved Rank-1 accuracy and mAP metrics compared to existing methods. Visualization of feature distributions further validates that RIRE successfully reduces intra-class distances between cross-modality samples, confirming its robustness in handling perspective variations.

6.2. Limitations Analysis of Robust Identity Representation Extraction Framework

However, the framework exhibits limitations in fine-grained feature generalization under extreme conditions. For instance, in the BA-VIP dataset's retrieval results (Figure 5), while RIRE matches targets with perspective variations, the top-ranked results still include distractors with similar postures or clothing, indicating insufficient sensitivity to texture details lost in thermal-infrared modalities. Similarly, for vessels (Figure 6), false detections at Rank-5 arise from thermal signature ambiguities (e.g., mast occlusion in visible images due to random erasing). These cases highlight that modality-specific interference—such as thermal radiation noise or critical component occlusion—remains a core challenge. Additionally, RIRE's computational efficiency poses a practical limitation: the framework's complexity (76.88M parameters and high FLOPs) may hinder real-time deployment on resource-constrained UAV platforms, where lightweight architectures are often prioritized to balance accuracy and operational costs. The dataset's inherent limitations, including restricted scene diversity (e.g., beach/sidewalk-only backgrounds in BA-VIP) and sample imbalance (uneven modality distributions), may further constrain the model's ability to generalize to complex real-world scenarios.

6.3. Future Research Trends

Future work should prioritize enhancing fine-grained feature discrimination, particularly for texture and structural attributes susceptible to modality-specific information loss. Developing robust data augmentation strategies to mitigate occlusion and thermal noise, while preserving discriminative regions (e.g., masts or clothing patterns), could further improve retrieval precision. Secondly, expanding dataset diversity to include urban environments, nighttime conditions, and balanced identity distributions would better reflect real-world UAV surveillance challenges. Additionally, lightweight model designs could address the computational overhead of RIRE, facilitating deployment on resource-constrained aerial platforms. By addressing these limitations, future frameworks could achieve higher precision in cross-modality retrieval while maintaining robustness against extreme perspective and modality variations.

7. Conclusions

In this paper, a robust identity representation extraction framework for VTI-ReID of aerial images, named RIRE, is proposed. The framework adopts a mapping method based on global representation decomposition and local representation aggregation. Identity-related representations in aerial images are captured, and the global representations of targets from different perspectives are mapped to the same identity space, thereby enhancing adaptability to target observation perspective differences in aerial images, particularly elevation angle variations.

To validate the proposed framework, we construct a group of cross-modality aerial image Re-ID datasets, including two typical elongated categories: humans and vessels. Extensive testing data demonstrate that the framework achieves the best accuracy for different types of elongated target VTI-ReID under aerial perspectives. Visual testing results and feature analysis further confirm the superior adaptability of the framework to observation perspective differences.

Although superior adaptability to observation perspective differences of humans and vessels in UAV images for VTI-ReID tasks is demonstrated by our method, the complexity of the model makes it more sensitive to intra-class distances caused by perspective differences, thereby severely impeding the further improvement of mAP and mINP. This will be a key focus of future research.

Author Contributions: Conceptualization and funding acquisition, C.Z., Y.Y. and B.G.; methodology and writing the original draft W.W. and Y.Y.; the processing and analysis of cross-dimensional data, visualization, and editing of the manuscript, W.H. and F.G.; ablation experiment and polishing and modification of manuscripts, B.G. All authors have read and agreed to the published version of the manuscript.

Funding: This work was supported by the National Natural Science Foundation of China (No. 62371153, No. 62071136, No. 62271159), Heilongjiang Provincial Natural Science Foundation of China Grant (JJ2024LH2397), Fundamental Research Funds for the Central Universities Grant (3072024XX0801 and 3072024XX0805), and the Key Laboratory of Target Cognition and Application Technology (2023-CXPT-LC-005).

Data Availability Statement: The proposed dataset are available on request from the corresponding author due to the privacy.

Conflicts of Interest: Authors Yiming Yan, Baoyu Ge and Wei Hou were employed by the company Harbin Space Star Data System Science and Technology Co., Ltd. The remaining authors declare that the research was conducted in the absence of any commercial or financial relationships that could be construed as a potential conflict of interest.

References

1. Liu, Y.; Zhou, W.; Liu, J.; Qi, G.-J.; Tian, Q.; Li, H. An End-to-End Foreground-Aware Network for Person Re-Identification. *IEEE Trans. Image Process.* **2021**, *30*, 2060–2071. [CrossRef] [PubMed]

2. Zhong, Z.; Zheng, L.; Luo, Z.; Li, S.; Yang, Y. Learning to Adapt Invariance in Memory for Person Re-Identification. *IEEE Trans. Pattern Anal. Mach. Intell.* **2021**, *43*, 2723–2738. [CrossRef] [PubMed]
3. Wang, K.; Wang, P.; Ding, C.; Tao, D. Batch Coherence-Driven Network for Part-Aware Person Re-Identification. *IEEE Trans. Image Process.* **2021**, *30*, 3405–3418. [CrossRef] [PubMed]
4. Wang, X.; Liu, M.; Wang, F.; Dai, J.; Liu, A.-A.; Wang, Y. Relation-Preserving Feature Embedding for Unsupervised Person Re-Identification. *IEEE Trans. Multimed.* **2024**, *26*, 714–723. [CrossRef]
5. Chen, G.; Lu, J.; Yang, M.; Zhou, J. Spatial-Temporal Attention-Aware Learning for Video-Based Person Re-Identification. *IEEE Trans. Image Process.* **2019**, *28*, 4192–4205. [CrossRef]
6. Li, J.; Zhang, S.; Tian, Q.; Wang, M.; Gao, W. Pose-Guided Representation Learning for Person Re-Identification. *IEEE Trans. Pattern Anal. Mach. Intell.* **2022**, *44*, 622–635. [CrossRef]
7. Ferdous, S.N.; Li, X.; Lyu, S. Uncertainty Aware Multitask Pyramid Vision Transformer for UAV-Based Object Re-Identification. In Proceedings of the 2022 IEEE International Conference on Image Processing (ICIP), Bordeaux, France, 16–19 October 2022; pp. 2381–2385. [CrossRef]
8. Yao, A.; Qi, J.; Zhong, P. Self-Aligned Spatial Feature Extraction Network for UAV Vehicle Reidentification. *IEEE Geosci. Remote Sens. Lett.* **2023**, *20*, 6002305. [CrossRef]
9. Xiong, M.; Yang, X.; Chen, H.; Aly, W.H.; AlTameem, A.; Saudagar, A.K.; Mumtaz, S.; Muhammad, K. Cloth-Changing Person Re-Identification with Invariant Feature Parsing for UAVs Applications. *IEEE Trans. Veh. Technol.* **2024**, *73*, 12448–12457. [CrossRef]
10. He, B.; Wang, F.; Wang, X.; Li, H.; Sun, F.; Zhou, H. Temporal Context and Environment-Aware Correlation Filter for UAV Object Tracking. *IEEE Trans. Geosci. Remote Sens.* **2024**, *62*, 5630915. [CrossRef]
11. Fang, H.; Wu, C.; Wang, X.; Zhou, F.; Chang, Y.; Yan, L. Online Infrared UAV Target Tracking with Enhanced Context-Awareness and Pixel-Wise Attention Modulation. *IEEE Trans. Geosci. Remote Sens.* **2024**, *62*, 5005417. [CrossRef]
12. Zheng, L.; Huang, Y.; Lu, H.; Yang, Y. Pose-Invariant Embedding for Deep Person Re-Identification. *IEEE Trans. Image Process.* **2019**, *28*, 4500–4509. [CrossRef] [PubMed]
13. Yang, F.; Li, W.; Liang, B.; Zhang, J. Spatiotemporal Interaction Transformer Network for Video-Based Person Reidentification in Internet of Things. *IEEE Internet Things J.* **2023**, *10*, 12537–12547. [CrossRef]
14. Wei, J.; Pan, C.; He, S.; Wang, G.; Yang, Y.; Shen, H.T. Towards Robust Person Re-Identification by Adversarial Training with Dynamic Attack Strategy. *IEEE Trans. Multimed.* **2024**, *26*, 10367–10380. [CrossRef]
15. Fu, H.; Cui, K.; Wang, C.; Qi, M.; Ma, H. Mutual Distillation Learning for Person Re-Identification. *IEEE Trans. Multimed.* **2024**, *26*, 8981–8995. [CrossRef]
16. Cheng, Y.; Liu, Y. Person Reidentification Based on Automotive Radar Point Clouds. *IEEE Trans. Geosci. Remote Sens.* **2022**, *60*, 5101913. [CrossRef]
17. Zhang, M.; Zhu, M.; Wei, X.; Wang, X.; Zhu, J.; Cheng, J.; Yang, Y. Deep learning-based person re-identification methods: A survey and outlook of recent works. *Image Vis. Comput.* **2022**, *119*, 104394. [CrossRef]
18. Duan, R.; Chen, L.; Li, Z.; Chen, Z.; Wu, B. A Scene Graph Encoding and Matching Network for UAV Visual Localization. *IEEE J. Sel. Top. Appl. Earth Obs. Remote Sens.* **2024**, *17*, 9890–9902. [CrossRef]
19. Maboudi, M.; Homaei, M.; Song, S.; Malihi, S.; Saadatseresht, M.; Gerke, M. A Review on Viewpoints and Path Planning for UAV-Based 3-D Reconstruction. *IEEE J. Sel. Top. Appl. Earth Obs. Remote Sens.* **2023**, *16*, 5026–5048. [CrossRef]
20. Liu, W.; Quijano, K.; Crawford, M.M. YOLOv5-Tassel: Detecting Tassels in RGB UAV Imagery with Improved YOLOv5 Based on Transfer Learning. *IEEE J. Sel. Top. Appl. Earth Obs. Remote Sens.* **2022**, *15*, 8085–8094. [CrossRef]
21. Liu, X.; Qi, J.; Chen, C.; Bin, K.; Zhong, P. Relation-Aware Weight Sharing in Decoupling Feature Learning Network for UAV RGB-Infrared Vehicle Re-Identification. *IEEE Trans. Multimed.* **2024**, *26*, 9839–9853. [CrossRef]
22. Zhang, H.; Cao, H.; Yang, X.; Deng, C.; Tao, D. Self-Training With Progressive Representation Enhancement for Unsupervised Cross-Domain Person Re-Identification. *IEEE Trans. Image Process.* **2021**, *30*, 5287–5298. [CrossRef] [PubMed]
23. Yang, X.; Tian, M.; Li, M.; Wei, Z.; Yuan, L.; Wang, N.; Gao, X. SSRR: Structural Semantic Representation Reconstruction for Visible-Infrared Person Re-Identification. *IEEE Trans. Multimed.* **2024**, *26*, 6273–6284. [CrossRef]
24. Kansal, K.; Subramanyam, A.V.; Wang, Z.; Satoh, S. SDL: Spectrum-Disentangled Representation Learning for Visible-Infrared Person Re-Identification. *IEEE Trans. Circuits Syst. Video Technol.* **2020**, *30*, 3422–3432. [CrossRef]
25. Ye, H.; Liu, H.; Meng, F.; Li, X. Bi-Directional Exponential Angular Triplet Loss for RGB-Infrared Person Re-Identification. *IEEE Trans. Image Process.* **2021**, *30*, 1583–1595. [CrossRef]
26. Huang, Y.; Wu, Q.; Xu, J.; Zhong, Y.; Zhang, P.; Zhang, Z. Alleviating Modality Bias Training for Infrared-Visible Person Re-Identification. *IEEE Trans. Multimed.* **2022**, *24*, 1570–1582. [CrossRef]
27. Ye, M.; Lan, X.; Leng, Q.; Shen, J. Cross-Modality Person Re-Identification via Modality-Aware Collaborative Ensemble Learning. *IEEE Trans. Image Process.* **2020**, *29*, 9387–9399. [CrossRef]

28. Tan, X.; Chai, Y.; Chen, F.; Liu, H. A Fourier-Based Semantic Augmentation for Visible-Thermal Person Re-Identification. *IEEE Signal Process. Lett.* **2022**, *29*, 1684–1688. [CrossRef]
29. Wei, Z.; Yang, X.; Wang, N.; Gao, X. Flexible Body Partition-Based Adversarial Learning for Visible Infrared Person Re-Identification. *IEEE Trans. Neural Netw. Learn. Syst.* **2022**, *33*, 4676–4687. [CrossRef]
30. Yang, X.; Dong, W.; Li, M.; Wei, Z.; Wang, N.; Gao, X. Cooperative Separation of Modality Shared-Specific Features for Visible-Infrared Person Re-Identification. *IEEE Trans. Multimed.* **2024**, *26*, 8172–8183. [CrossRef]
31. Feng, Y.; Chen, F.; Yu, J.; Ji, Y.; Wu, F.; Liu, T.; Liu, S.; Jing, X.Y.; Luo, J. Cross-Modality Spatial-Temporal Transformer for Video-Based Visible-Infrared Person Re-Identification. *IEEE Trans. Multimed.* **2024**, *26*, 6582–6594. [CrossRef]
32. Zhang, Q.; Yan, Y.; Gao, L.; Xu, C.; Su, N.; Feng, S. A Third-Modality Collaborative Learning Approach for Visible-Infrared Vessel Reidentification. *IEEE J. Sel. Top. Appl. Earth Obs. Remote Sens.* **2024**, *17*, 19035–19047. [CrossRef]
33. Wu, A.; Zheng, W.-S.; Yu, H.-X.; Gong, S.; Lai, J. RGB-Infrared Cross-Modality Person Re-identification. In Proceedings of the 2017 IEEE International Conference on Computer Vision (ICCV), Venice, Italy, 22–29 October 2017; pp. 5390–5399. [CrossRef]
34. Wang, G.; Zhang, T.; Cheng, J.; Liu, S.; Yang, Y.; Hou, Z. RGB-Infrared Cross-Modality Person Re-Identification via Joint Pixel and Feature Alignment. In Proceedings of the 2019 IEEE/CVF International Conference on Computer Vision (ICCV), Seoul, Republic of Korea, 27 October–2 November 2019; pp. 3622–3631. [CrossRef]
35. Wang, Z.; Wang, Z.; Zheng, Y.; Chuang, Y.-Y.; Satoh, S. Learning to Reduce Dual-Level Discrepancy for Infrared-Visible Person Re-Identification. In Proceedings of the 2019 IEEE/CVF Conference on Computer Vision and Pattern Recognition (CVPR), Long Beach, CA, USA, 15–20 June 2019; pp. 618–626. [CrossRef]
36. Hao, X.; Zhao, S.; Ye, M.; Shen, J. Cross-Modality Person Re-Identification via Modality Confusion and Center Aggregation. In Proceedings of the 2021 IEEE/CVF International Conference on Computer Vision (ICCV), Montreal, QC, Canada, 11–17 October 2021; pp. 16383–16392. [CrossRef]
37. Wei, Z.; Yang, X.; Wang, N.; Gao, X. Syncretic Modality Collaborative Learning for Visible Infrared Person Re-Identification. In Proceedings of the 2021 IEEE/CVF International Conference on Computer Vision (ICCV), Montreal, QC, Canada, 11–17 October 2021; pp. 225–234. [CrossRef]
38. Ye, M.; Ruan, W.; Du, B.; Shou, M.Z. Channel Augmented Joint Learning for Visible-Infrared Recognition. In Proceedings of the 2021 IEEE/CVF International Conference on Computer Vision (ICCV), Montreal, QC, Canada, 11–17 October 2021; pp. 13547–13556. [CrossRef]
39. Li, D.; Wei, X.; Hong, X.; Gong, Y. Infrared-visible cross-modal person re-identification with an x modality. *Proc. AAAI Conf. Artif. Intell.* **2020**, *34*, 4610–4617. [CrossRef]
40. Zhang, Y.; Yan, Y.; Lu, Y.; Wang, H. Towards a unified middle modality learning for visible-infrared person re-identification. In Proceedings of the 29th ACM International Conference on Multimedia, Chengdu, China, 20–24 October 2021; pp. 788–796. [CrossRef]
41. Kim, M.; Kim, S.; Park, J.; Park, S.; Sohn, K. PartMix: Regularization Strategy to Learn Part Discovery for Visible-Infrared Person Re-Identification. In Proceedings of the 2023 IEEE/CVF Conference on Computer Vision and Pattern Recognition (CVPR), Vancouver, BC, Canada, 18–22 June 2023; pp. 18621–18632. [CrossRef]
42. Ye, M.; Shen, J.; Crandall, D.J.; Shao, L.; Luo, J. Dynamic dual-attentive aggregation learning for visible-infrared person re-identification. In *Proceedings of the Computer Vision–ECCV 2020: 16th European Conference, Glasgow, UK, 23–28 August 2020*; Proceedings, Part XVII 16; Springer International Publishing: Cham, Switzerland, 2020; pp. 229–247. [CrossRef]
43. Choi, S.; Lee, S.; Kim, Y.; Kim, T.; Kim, C. Hi-CMD: Hierarchical Cross-Modality Disentanglement for Visible-Infrared Person Re-Identification. In Proceedings of the 2020 IEEE/CVF Conference on Computer Vision and Pattern Recognition (CVPR), Seattle, WA, USA, 14–19 June 2020; pp. 10254–10263. [CrossRef]
44. Lu, Y.; Wu, Y.; Liu, B.; Zhang, T.; Li, B.; Chu, Q.; Yu, N. Cross-Modality Person Re-Identification with Shared-Specific Feature Transfer. In Proceedings of the 2020 IEEE/CVF Conference on Computer Vision and Pattern Recognition (CVPR), Seattle, WA, USA, 14–19 June 2020; pp. 13376–13386. [CrossRef]
45. Wu, Q.; Dai, P.; Chen, J.; Lin, C.W.; Wu, Y.; Huang, F.; Zhong, B.; Ji, R. Discover Cross-Modality Nuances for Visible-Infrared Person Re-Identification. In Proceedings of the 2021 IEEE/CVF Conference on Computer Vision and Pattern Recognition (CVPR), Nashville, TN, USA, 19–25 June 2021; pp. 4328–4337. [CrossRef]
46. Liu, J.; Sun, Y.; Zhu, F.; Pei, H.; Yang, Y.; Li, W. Learning Memory-Augmented Unidirectional Metrics for Cross-modality Person Re-identification. In Proceedings of the 2022 IEEE/CVF Conference on Computer Vision and Pattern Recognition (CVPR), New Orleans, LA, USA, 18–24 June 2022; pp. 19344–19353. [CrossRef]
47. Zhang, Y.; Wang, H. Diverse Embedding Expansion Network and Low-Light Cross-Modality Benchmark for Visible-Infrared Person Re-identification. In Proceedings of the 2023 IEEE/CVF Conference on Computer Vision and Pattern Recognition (CVPR), Vancouver, BC, Canada, 18–22 June 2023; pp. 2153–2162. [CrossRef]

48. Feng, J.; Wu, A.; Zheng, W.-S. Shape-Erased Feature Learning for Visible-Infrared Person Re-Identification. In Proceedings of the 2023 IEEE/CVF Conference on Computer Vision and Pattern Recognition (CVPR), Vancouver, BC, Canada, 18–22 June 2023; pp. 22752–22761. [CrossRef]
49. Ye, M.; Wang, Z.; Lan, X.; Yuen, P. Visible thermal person re-identification via dual-constrained top-ranking. *IJCAI* **2018**, *1*, 2. [CrossRef]
50. Hao, Y.; Wang, N.; Li, J.; Gao, X. HSME: Hypersphere manifold embedding for visible thermal person re-identification. *Proc. AAAI Conf. Artif. Intell.* **2019**, *33*, 8385–8392. [CrossRef]
51. Feng, Z.; Lai, J.; Xie, X. Learning Modality-Specific Representations for Visible-Infrared Person Re-Identification. *IEEE Trans. Image Process.* **2020**, *29*, 579–590. [CrossRef]
52. Ye, M.; Lan, X.; Wang, Z.; Yuen, P.C. Bi-Directional Center-Constrained Top-Ranking for Visible Thermal Person Re-Identification. *IEEE Trans. Inf. Forensics Secur.* **2020**, *15*, 407–419. [CrossRef]
53. Tan, L.; Zhang, Y.; Shen, S.; Wang, Y.; Dai, P.; Lin, X.; Wu, Y.; Ji, R. Exploring invariant representation for visible-infrared person re-identification. *arXiv* **2023**, arXiv:2302.00884. [CrossRef]
54. Zhou, X.; Zhong, Y.; Cheng, Z.; Liang, F.; Ma, L. Adaptive Sparse Pairwise Loss for Object Re-Identification. In Proceedings of the 2023 IEEE/CVF Conference on Computer Vision and Pattern Recognition (CVPR), Vancouver, BC, Canada, 18–22 June 2023; pp. 19691–19701. [CrossRef]
55. Nguyen, D.T.; Hong, H.G.; Kim, K.W.; Park, K.R. Person recognition system based on a combination of body images from visible light and thermal cameras. *Sensors* **2017**, *17*, 605. [CrossRef]
56. Available online: https://enterprise.dji.com/cn/matrice-300 (accessed on 1 November 2023).
57. Available online: https://enterprise.dji.com/cn/zenmuse-h20-series (accessed on 1 November 2023).
58. Luo, H.; Gu, Y.; Liao, X.; Lai, S.; Jiang, W. Bag of Tricks and a Strong Baseline for Deep Person Re-Identification. In Proceedings of the 2019 IEEE/CVF Conference on Computer Vision and Pattern Recognition Workshops (CVPRW), Long Beach, CA, USA, 16–20 June 2019; pp. 1487–1495. [CrossRef]
59. Wang, G.; Zhang, T.; Yang, Y.; Cheng, J.; Chang, J.; Liang, X.; Hou, Z. Cross-modality paired-images generation for RGB-infrared person re-identification. *Proc. AAAI Conf. Artif. Intell.* **2020**, *34*, 12144–12151. [CrossRef]
60. Zhong, Z.; Zheng, L.; Kang, G.; Li, S.; Yang, Y. Random erasing data augmentation. *Proc. AAAI Conf. Artif. Intell.* **2020**, *34*, 13001–13008. [CrossRef]
61. Ye, M.; Shen, J.; Lin, G.; Xiang, T.; Shao, L.; Hoi, S.C.H. Deep Learning for Person Re-Identification: A Survey and Outlook. *IEEE Trans. Pattern Anal. Mach. Intell.* **2022**, *44*, 2872–2893. [CrossRef] [PubMed]
62. Fu, C.; Hu, Y.; Wu, X.; Shi, H.; Mei, T.; He, R. CM-NAS: Cross-Modality Neural Architecture Search for Visible-Infrared Person Re-Identification. In Proceedings of the 2021 IEEE/CVF International Conference on Computer Vision (ICCV), Montreal, QC, Canada, 11–17 October 2021; pp. 11803–11812. [CrossRef]
63. Yang, M.; Huang, Z.; Hu, P.; Li, T.; Lv, J.; Peng, X. Learning with Twin Noisy Labels for Visible-Infrared Person Re-Identification. In Proceedings of the 2022 IEEE/CVF Conference on Computer Vision and Pattern Recognition (CVPR), New Orleans, LA, USA, 18–24 June 2022; pp. 14288–14297. [CrossRef]
64. Van der Maaten, L.; Hinton, G.E. Visualizing Data using t-SNE. *J. Mach. Learn. Res.* **2008**, *9*, 2579–2605.
65. Xuan, H.; Stylianou, A.; Liu, X.; Pless, R. Hard negative examples are hard, but useful. In *Proceedings of the Computer Vision–ECCV 2020: 16th European Conference, Glasgow, UK, 23–28 August 2020*; Proceedings, Part XIV 16; Springer International Publishing: Cham, Switzerland, 2020; pp. 126–142. [CrossRef]

remote sensing

MDPI

Article

A Hierarchical Path Planning Framework of Plant Protection UAV Based on the Improved D3QN Algorithm and Remote Sensing Image

Haitao Fu [1], Zheng Li [1], Jian Lu [2], Weijian Zhang [1], Yuxuan Feng [1], Li Zhu [1], He Liu [3] and Jian Li [1,*]

[1] College of Information Technology, Jilin Agricultural University, Changchun 130118, China; fht@jlau.edu.cn (H.F.); 20231308@mails.jlau.edu.cn (Z.L.); 20231257@mails.jlau.edu.cn (W.Z.); fengyuxuan@jlau.edu.cn (Y.F.); zhuli@jlau.edu.cn (L.Z.)

[2] College of Resources and Environment, Jilin Agricultural University, Changchun 130118, China; lujian025@mails.jlau.edu.cn

[3] College of Engineering and Technology, Jilin Agricultural University, Changchun 130118, China; liuhe@jlau.edu.cn

* Correspondence: lijian@jlau.edu.cn; Tel.: +86-139-4419-5488

Abstract

Traditional path planning algorithms often fail to simultaneously ensure operational efficiency, energy constraint compliance, and environmental adaptability in agricultural scenarios, thereby hindering the advancement of precision agriculture. To address these challenges, this study proposes a deep reinforcement learning algorithm, MoE-D3QN, which integrates a Mixture-of-Experts mechanism with a Bi-directional Long Short-Term Memory model. This design enhances the efficiency and robustness of UAV path planning in agricultural environments. Building upon this algorithm, a hierarchical coverage path planning framework is developed. Multi-level task maps are constructed using crop information extracted from Sentinel-2 remote sensing imagery. Additionally, a dynamic energy consumption model and a progressive composite reward function are incorporated to further optimize UAV path planning in complex farmland conditions. Simulation experiments reveal that in the two-level scenario, the MoE-D3QN algorithm achieves a coverage efficiency of 0.8378, representing an improvement of 37.84–63.38% over traditional algorithms and 19.19–63.38% over conventional reinforcement learning methods. The redundancy rate is reduced to 3.23%, which is 38.71–41.94% lower than traditional methods and 4.46–42.77% lower than reinforcement learning counterparts. In the three-level scenario, MoE-D3QN achieves a coverage efficiency of 0.8261, exceeding traditional algorithms by 52.13–71.45% and reinforcement learning approaches by 10.15–50.2%. The redundancy rate is further reduced to 5.26%, which is significantly lower than the 57.89–92.11% observed with traditional methods and the 15.57–18.98% reported for reinforcement learning algorithms. These findings demonstrate that the MoE-D3QN algorithm exhibits high-quality planning performance in complex farmland environments, indicating its strong potential for widespread application in precision agriculture.

Keywords: precision agriculture; deep reinforcement learning; path planning; Mixture-of-Experts mechanism; remote sensing

1. Introduction

With the rapid advancement of unmanned aerial vehicle (UAV) technology, the paradigms and technical pathways of modern agricultural production have been profoundly restructured [1,2]. Plant protection UAVs, characterized by exceptional maneuverability and a broad aerial perspective, have overcome operational bottlenecks associated with complex terrains—including plains, hills, and mountainous regions—and have significantly enhanced the flexibility and efficiency of crop protection operations. By alleviating rural labor shortages and mitigating pesticide overuse, the large-scale deployment of UAV platforms also signifies an evolution in the precision-agriculture technology framework, opening new avenues for the efficient utilization of agricultural resources and the sustainable management of ecosystems. As a result, the construction and comprehensive development of integrated rural systems have been powerfully advanced [3–6].

In UAV-driven precision-agriculture systems, the formulation of an effective path planning strategy is paramount to harmonizing the "perception–decision–action" pipeline. Fundamentally, this challenge entails not only achieving comprehensive spatial coverage but also deeply integrating the time-varying characteristics of the agricultural environment with the dynamic requirements of field tasks [7,8]. To date, two primary categories of planning methods have been employed. The first comprises classical approaches—such as heuristic-based A* search [9], population-inspired genetic and particle-swarm algorithms [10], and sampling-based rapidly exploring random trees [11]—which generally excel in simple, well-defined scenarios but suffer from high computational complexity and limited adaptability when environmental conditions, targets, or obstacles change, necessitating full replanning. The second category encompasses machine-learning techniques, notably Deep Q-Networks (DQN) [12] leveraging experience replay and policy gradient methods such as Proximal Policy Optimization (PPO) [13]. In the context of agriculture—where environments are complex and constraints are numerous—traditional algorithms often struggle to accommodate all operational factors, whereas deep reinforcement-learning (DRL) algorithms, by virtue of their capacity for autonomous representation of high-dimensional state spaces and end-to-end decision optimization, are emerging as a promising avenue for overcoming the limitations of conventional path planning solutions.

In recent years, a variety of advanced path planning approaches have been developed for coverage tasks in agricultural and orchard environments. Zhang et al. proposed a method based on an improved A* algorithm for orchard-weeding robots [14], in which a weighted heuristic cost function and an additional turning-angle penalty were introduced to optimize coverage-path planning. Wu et al. developed an enhanced genetic algorithm incorporating simulated-annealing-inspired stretch-fitness evaluation tailored for complete coverage path planning [15], the hyperbolic-tangent activation function from neural networks, and fireworks-algorithm operators to mitigate early-stage premature convergence and bolster global search capabilities. Kong et al. presented B-APFDQN [16], a UAV-based coverage-path planning algorithm that fuses DQN learning with an artificial potential field (APF). In this approach, APF-derived priors are used to curtail blind exploration during training, and a novel network architecture simultaneously outputs both Q-values and action-distribution probabilities, thereby accelerating convergence. Zhou et al. introduced a robot-coverage strategy grounded in DRL [17], employing a DDQN with prioritized experience replay (PER) and incorporating a blockage-and-blind-area (BBA) mechanism to guide exploration toward unvisited regions. Lin et al. designed a collision-free coverage method for guava-picking robots by combining a recurrent neural network (RNN) with an enhanced deep deterministic policy gradient (DDPG) algorithm to predict safe trajectories in unknown states [18]. Finally, Ahmad et al. proposed a DRL-based coverage scheme for precision crop-health monitoring in smart agriculture, in which a centralized convolutional neural network (CNN) is integrated with DDQN. A bespoke reward function was devised to incentivize visits to high-value, uninspected areas while penalizing suboptimal actions [19].

Despite the considerable progress made in UAV path planning research, existing approaches still exhibit limited applicability in agricultural scenarios. Many current methods tend to oversimplify farmland as homogeneous areas and focus on global traversal optimization, which fails to meet the integrated demands of precision agriculture in terms of operational accuracy, resource efficiency, and ecological sustainability. On the one hand, prevailing strategies often overlook critical factors such as variations in cropping patterns, spatial heterogeneity of pest and disease distribution, and dynamic changes in UAV payloads. This omission restricts both the precision of pesticide application and the overall energy efficiency. On the other hand, the singular pursuit of globally optimal paths may result in issues such as redundant pesticide application, energy waste, and contamination of ecologically sensitive zones, thereby making it difficult to balance increased agricultural output with environmental sustainability.

To address these challenges, Mukhamediev et al. proposed a genetic algorithm-based heterogeneous UAV swarm coverage planning method [7], which targets irregular farmland operations by jointly optimizing UAV selection, flight route design, and ground platform scheduling, significantly reducing overall system costs. Syed et al. focused on semi-structured orchard environments and developed an improved YOLOv8-CNN model integrating Ghost modules and a SE attention mechanism [20], enabling efficient classification of real versus false obstacles and enhancing the safety and decision-making efficiency of autonomous navigation. In another study, Li et al. addressed multi-region fertilization tasks in large and complex farmland by proposing a path planning method based on an enhanced Double Deep Q-Learning algorithm [21]. By utilizing GEE-GNDVI to extract nitrogen stress distribution and construct multi-region task maps, the approach effectively enabled intelligent optimization of fertilization paths. These studies underscore the inherent complexity and specificity of agricultural environments and offer valuable insights for improving UAV path planning strategies tailored to the needs of precision agriculture.

To address these limitations, a hierarchical coverage path planning (H-CPP) framework is proposed in this study, whereby adaptability to agricultural environments is enhanced via a two-tier optimization structure. Using satellite-derived organic-matter indices, croplands are first subdivided into graded subzones; high-risk areas are then prioritized for scheduling. Concurrently, dynamic energy consumption constraints and a hybrid-expert architecture are integrated to further improve coverage efficiency.

The principal contributions of this study are summarized as follows:

1. To address the limited policy generalization capability of the D3QN algorithm caused by insufficient representational power in complex and dynamic environments, an improved deep reinforcement learning algorithm, termed MoE-D3QN, is proposed in this study. The algorithm integrates the expert mechanism of a Mixture-of-Experts (MoE) structure with a Bi-directional Long Short-Term Memory (Bi-LSTM) module to enhance the model's ability to capture temporal features, improve state perception accuracy, and enable efficient policy generation under complex state conditions.

2. To better simulate real-world agricultural scenarios and enhance the applicability of the proposed algorithm, a hierarchical coverage path planning (H-CPP) framework is developed in this study. This framework utilizes Sentinel-2 satellite data to extract spectral bands for the calculation of the CCC index, which is employed to detect chlorophyll content in crops. A hierarchical approach is then applied to segment multi-level task regions, while a grid-based method is used to simulate the agricultural UAV mapping environment.

3. This study designs a progressive composite reward function that synergistically combines hierarchical reward guidance—including priority coverage reward, target distance guidance reward, and energy efficiency reward—with dynamic penalty mechanisms, such as invalid exploration penalty and safety violation penalty. This approach aims

to balance coverage efficiency and energy consumption while ensuring adherence to priority constraints.

2. Materials and Methods

2.1. Designed Planning Area Description

Canopy chlorophyll content serves as a pivotal indicator of crop photosynthetic physiology and is thus of paramount importance for modern agricultural monitoring. As the principal pigment driving photosynthesis, chlorophyll concentration directly governs the efficiency of material and energy exchange between the plant and its environment, while simultaneously reflecting the crop's photosynthetic capacity, nutritional status, and growth trajectory [22,23]. By leveraging spectral remote-sensing techniques to obtain three-dimensional distributions of canopy chlorophyll, quantitative linkages can be established between crop phenotypic traits and underlying ecological processes. Such data furnish essential support for real-time growth diagnostics, fertility management, and yield forecasting in precision-agriculture systems, representing a critical technological enabler for scientific crop monitoring and management.

The study area selected for this investigation is Changchun, the capital city of Jilin Province, China, geographically bounded by latitudes 43°05′ to 45°15′N and longitudes 124°18′ to 127°05′E. Situated at the core of the Songliao Plain within the Northeast China Plain, this region is underlain by the characteristic "black soil" belt. The predominant soil types—highly fertile Chernozem and meadow soils—exhibit elevated surface organic-matter content, classifying the area as a globally scarce high-yield arable land resource [24]. The extensive, flat terrain further provides an ideal physical foundation for the implementation of precision-agriculture technologies [25].

This study was implemented on the Google Earth Engine platform, employing the LEAF-Toolbox inversion routines developed by Sentinel-2 Level-2A surface reflectance data were integrated with SRTM terrain elevations to drive a PROSAIL radiative-transfer-model–based retrieval workflow [26–28]. First, the S2MaskClear algorithm was used to exclude pixels exhibiting over 10% cloud cover, and atmospheric correction was applied via Sen2Cor. Thereafter, topographic illumination and radiometric distortions were corrected by merging the SAIL canopy model with SRTM-derived terrain information. Finally, multispectral inversion was performed using the SL2P module—synergizing visible, red-edge, and near-infrared bands—to generate a canopy chlorophyll-content (CCC) estimate for the Changchun region (125.17°E–125.82°E, 43.45°N–44.15°N), as illustrated in Figure 1.

After obtaining the CCC index map, the area was partitioned into multi-level task regions and non-task regions, which were subsequently converted into a grid map format suitable for computational processing, as illustrated in Figure 2. To ensure that the task regions accurately reflect the actual crop distribution, a CCC lower threshold of 100,000 was established. Areas below this threshold were classified as non-task zones and excluded from the path planning process.

Following this, a quantile-based method was employed to categorize task priorities: regions with CCC values exceeding 410,000 were assigned Priority I, representing vigorously growing crop areas; values between 230,000 and 410,000 were designated Priority II; and those between 100,000 and 230,000 were classified as Priority III, corresponding to low-density or potentially stressed crop zones.

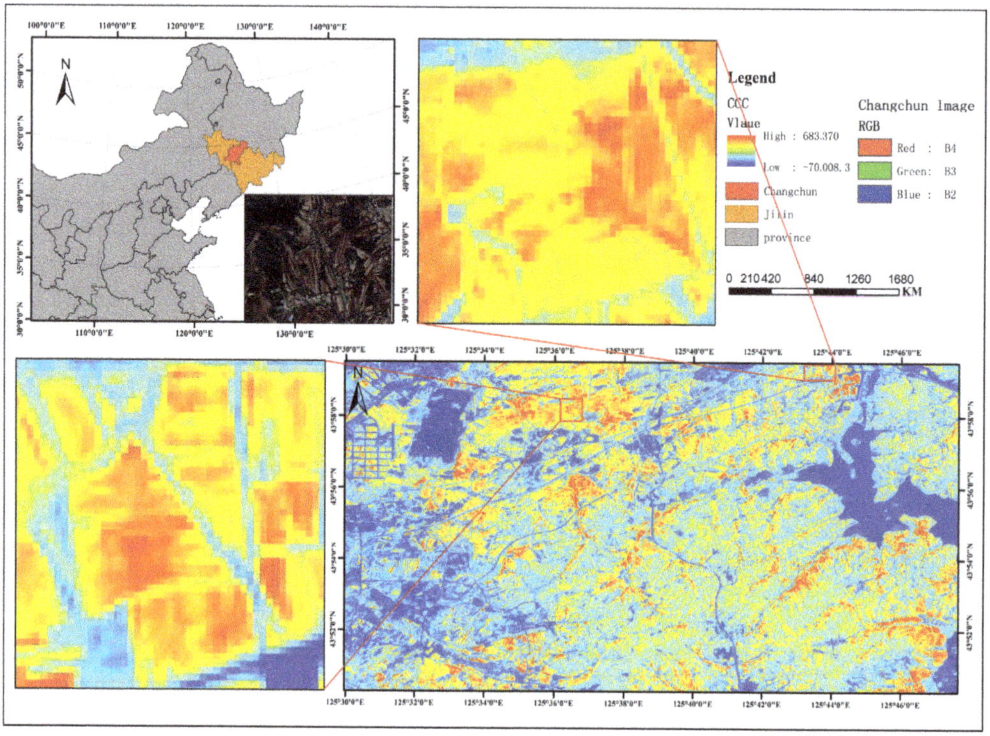

Figure 1. Canopy chlorophyll-content inversion results derived by remote sensing for the study area.

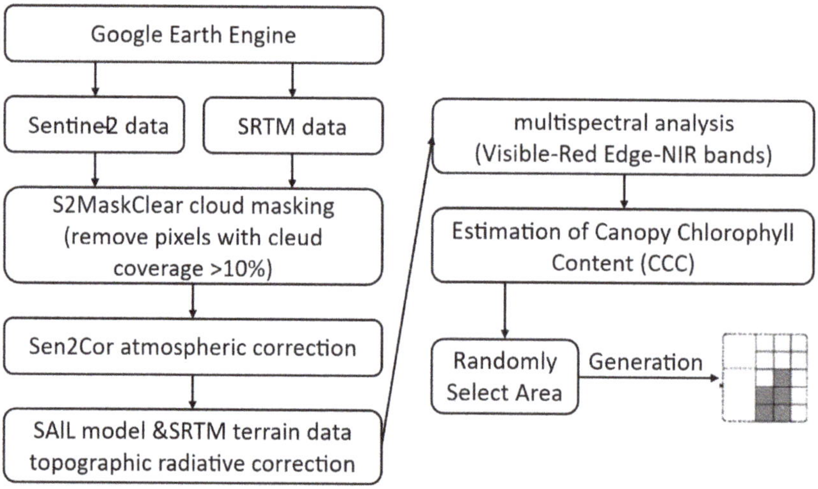

Figure 2. Task map processing flow chart.

This prioritization scheme is grounded in the principle of maximizing operational resource efficiency by focusing coverage on high-yield potential areas. Such prioritization helps consolidate the healthy growth of major production zones and prevents the spread of pests and diseases, thereby enhancing the operational value and responsiveness of the path planning.

2.2. State Space

The current temporal characteristics of the mission environment were defined as the state space. Using a grid-based discretization approach, the study area was partitioned into a series of discrete cells. Each cell was assigned a unique identifier corresponding to a specific environmental state—such as task zone, obstacle region, or energy-resupply area—to facilitate subsequent decision-making and path planning processes. The meaning of environmental unit is shown in Table 1.

Table 1. Map state corresponding to different values.

Value ($Env_{(x,y)}$)	Description
9	Priority Level I area
8	Priority Level II area
7	Priority Level III area
4	Energy consumption supply area
3	Obstacle area
2	Current position of UAV
1	Non-task area
0	Covered task area

2.3. Action Space and Action Selection Strategy

The action space was defined as the set of all feasible commands available to the UAV during the path planning process. Although the UAV's motion model theoretically supports a continuous action set capable of omnidirectional pose transformations, the accompanying increase in degrees of freedom has been shown to markedly impede training efficiency. To alleviate the computational burden associated with continuous control dimensions, an action policy comprising eight discrete commands was devised. Upon execution of any given command, the state space is immediately updated and the current state St transitions to the subsequent state St + 1. The detailed state-transition mechanism is illustrated in Table 2.

Table 2. Action space.

$Action_{(i)}$	State Transition Process	Description
0	(x, y + 1)	North (N)
1	(x, y − 1)	South (S)
2	(x − 1, y)	West (W)
3	(x + 1, y)	East (E)
4	(x − 1, y + 1)	Northwest (NW)
5	(x + 1, y + 1)	Northeast (NE)
6	(x − 1, y − 1)	Southwest (SW)
7	(x + 1, y − 1)	Southeast (SE)

The balance between exploration and exploitation is central to the action-selection strategy. To pursue higher rewards, the UAV must attempt various actions during exploration; however, excessive exploration can degrade algorithmic efficiency and slow convergence, whereas excessive exploitation may lead to entrapment in local optima and hinder discovery of a global optimum. To address this issue, an ε-greedy algorithm was adopted for action selection: with probability ε the action yielding the maximum estimated return is chosen, and with probability 1 − ε a random action is selected, as shown in Equation (1).

$$Action = \begin{cases} argmaxQ(s,a), & if\ RN > \varepsilon \\ random\ action, & Otherwise \end{cases} \quad (1)$$

Meanwhile, to avoid the meaningless exploration and slow convergence caused by a fixed ε in the traditional ε-greedy algorithm, a dynamic ε schedule was employed. In this scheme, exploration is initially prioritized, and, as the number of training episodes increases, the exploration probability is gradually reduced while the exploitation weight is correspondingly increased, as shown in the following Equation (2). Here, ε_{min} denotes the minimum exploration probability, ε_{max} the maximum probability, *Decay rate* the rate at which exploration probability decays, and *currunt episode* the index of the current episode.

$$\varepsilon = \varepsilon_{min} + (\varepsilon_{max} - \varepsilon_{min}) \times e^{-Decay\ rate\ \times\ currunt\ episode} \tag{2}$$

2.4. Environmental Constraints

2.4.1. Priority Constraints

In hierarchical coverage path planning tasks, priority constraints constitute the core logic of task planning and resource scheduling. A well-designed constraint directly influences operational efficiency. Task types are arranged in descending order of priority, as illustrated in Equation (3), where P_1 represents the highest priority and P_n the lowest. The symbol ">" denotes a strict priority relationship, indicating that all tasks of type P_1 must be completed before proceeding to those of type P_2.

$$\mathcal{P} = \{P_1, P_2, \ldots, P_n\}(P_1 > P_2 > \ldots P_n) \tag{3}$$

At the end of each time step, the remaining number of tasks in each priority region is calculated, as defined in Equation (4). $N_k^{(t)}$ denotes the number of remaining tasks of priority level P_k at time step t. This value is obtained by traversing all positions (i,j) in the environment and counting the locations that satisfy the corresponding conditions.

$$N_k^{(t)} = \sum_{i,j} \left[m_{ij}^{(t)} = P_k \right] (k = 1,\ 2,\ \ldots,\ n) \tag{4}$$

As the UAV continues to operate, the state is updated dynamically according to the state transition function defined in Equation (5). Specifically, $m_{x_t y_t}^{(t)} = P_k$ is used to determine the task type at the current location, identifying whether the agent is positioned at a task point of priority P_k. The term $\prod_{i=1}^{k-1} \left[N_i^{(t)} = 0 \right]$ serves to verify whether all higher-priority tasks have been completed. $N_k^{(t+1)}$ denotes the number of uncompleted tasks of priority level k remaining at time step $t+1$.

$$N_k^{(t+1)} = N_k^{(t)} - \left[m_{x_t y_t}^{(t)} = P_k \right] \cdot \prod_{i=1}^{k-1} \mathbb{I}\left[N_i^{(t)} = 0 \right] \tag{5}$$

To enhance the agent's understanding of task constraints, a nearest-distance function is introduced to guide the UAV toward uncompleted high-priority regions. The agent is encouraged to navigate preferentially toward the nearest area with higher task priority, as defined in Equation (6). Here, $\mathcal{T}_k(t)$ denotes the set of coordinates corresponding to the remaining unprocessed tasks of priority level k.

$$d^*(s_t)\& = \begin{cases} \min\limits_{(i,j)\in\mathcal{T}_1(t)} \parallel (i,j)-(x_t,y_t) \parallel_1, & |\mathcal{T}_1(t)|>0 \\ \min\limits_{(i,j)\in\mathcal{T}_2(t)} \parallel (i,j)-(x_t,y_t) \parallel_1, & |\mathcal{T}_1(t)|=0 \wedge |\mathcal{T}_2(t)|>0 \\ \vdots & \\ \min\limits_{(i,j)\in\mathcal{T}_n(t)} \parallel (i,j)-(x_t,y_t) \parallel_1, & \bigwedge\limits_{m=1}^{n-1} |\mathcal{T}_m(t)|=0 \wedge |\mathcal{T}_n(t)|>0 \\ 0, & otherwise \end{cases} \quad (6)$$

2.4.2. Energy Consumption Constraint

During agricultural operations, unmanned aerial vehicles (UAVs) are susceptible to disturbances from natural environmental factors and their own payload. Therefore, path planning for UAVs should not be based solely on minimizing time or distance, while neglecting energy consumption. To address this, a dynamic energy consumption constraint for UAVs is proposed in this study, as shown in Equation (7). This constraint is developed based on a mechanism-guided black-box approach, taking into account both payload variations and environmental influences.

$$C_t = C_{move} + K * C_{load} + e \quad (7)$$

In this context, C_t denotes the energy consumption incurred by the UAV at time step t. Specifically, C_{move} represents the baseline energy consumption during unloaded flight, while C_{load} accounts for the additional energy required due to payload. The coefficient K refers to the environmental coupling factor, which integrates external influences such as headwind velocity w. The term e denotes Gaussian white noise, capturing unmodeled disturbances including sensor inaccuracies and atmospheric turbulence.

2.5. Improved Algorithm

2.5.1. Basic Theory

Reinforcement learning primarily functions through real-time interactions between an agent and its environment. At time step t, the agent perceives the environmental state s and selects an action A accordingly. The environment then provides a reward R based on the selected action, and the agent's state is updated to the next state St + 1. Through repeated iterations of this process, the agent aims to maximize cumulative rewards.

In 2015, the Deep Q-Network algorithm was proposed [29], which employed deep neural networks to approximate the Q-value function in high-dimensional continuous state spaces. To improve learning stability and efficiency, the algorithm introduced an experience replay mechanism and a target network. However, since a single neural network was used to estimate the Q-values, estimation bias could occur, leading to the systematic overestimation of Q-values for certain actions. As training progresses, this estimation error may accumulate and be further amplified, potentially impairing the optimization of the policy.

The Double Denoising Deep Q-Network (D3QN) algorithm adopts the dual-network mechanism of the Double Deep Q-Network [30], employing two structurally identical but parameter-independent networks: the policy network and the target network. The policy network is used for training and action selection, while after a fixed number of training iterations, its parameters are copied to the target network. The target network then computes the Q-values to decouple action selection from value evaluation, thereby reducing overestimation bias. The target Q-value is calculated according to Equation (8),

$$y = r + \gamma \cdot Q'\left(s_{t+1}, \underset{a}{argmax}Q(s_{t+1}, a; \theta); \theta'\right) \tag{8}$$

where y denotes the target Q-value, R represents the immediate reward, Q' refers to the target network, $\underset{a}{\text{argmax}}Q(s_{t+1}, a; \theta)$ is the optimal action that maximizes the Q-value in state s_{t+1}, and θ' represents the parameters updated with delay.

In addition, to further enhance the network's ability to distinguish between state value and action advantage, D3QN incorporates the Dueling DQN architecture [31]. By introducing the state-value function $V(s_t)$ and the action-advantage function $A(s_t, a_t)$, which represent the average value of state s_t independent of specific actions and the relative advantage of action a_t compared to the average of all actions in that state, respectively, the policy evaluation capability of the reinforcement learning algorithm is significantly enhanced. The calculation formula is shown in Equation (9).

$$Q(s_t, a_t) = V(s_t) + A(s_t, a_t) - \frac{1}{|A|}\sum_{a'} A(s_t, a') \tag{9}$$

In practical applications such as precision agriculture, environmental states are neither static nor isolated; rather, they exhibit pronounced temporal continuity and undergo constant changes. Therefore, when making decisions, agents are required not only to perceive the current state but also to incorporate trends in environmental dynamics over time, in order to form more forward-looking and informed judgments.

To address this need, researchers have increasingly incorporated recurrent neural network (RNN) architectures—particularly Long Short-Term Memory (LSTM) networks—owing to their capacity for temporal sequence modeling. Through the use of a gating mechanism comprising forget, input, and output gates, LSTM networks are capable of selectively retaining relevant information while suppressing noise, thereby enhancing the model's ability to capture and interpret long-term dependencies.

Building upon conventional LSTM models, Bi-directional Long Short-Term Memory (Bi-LSTM) networks introduce two parallel LSTM layers that process the input sequence in both forward and backward directions. This bi-directional structure enables the simultaneous capture of past and future contextual information within a sequence. Each LSTM unit dynamically regulates information flow via its gating mechanisms, effectively modeling long-range dependencies while mitigating the vanishing gradient problem. By integrating information from both directions, Bi-LSTM networks substantially improve the model's representational capacity for complex temporal patterns.

2.5.2. MOE-D3QN

The Mixture-of-Experts architecture has emerged as a prominent paradigm in deep learning, offering notable advantages in enhancing model capacity and computational efficiency through the dynamic routing and integration of multiple expert sub-models. This architecture adaptively allocates computational resources based on the characteristics of input data and employs a gating network to aggregate the outputs of selected experts.

In recent years, with the rapid development of large-scale models, the MoE framework has demonstrated significant potential in fields such as natural language processing. Representative implementations include Google's GShard [32] and the subsequent Switch Transformer [33], both of which activate only a subset of experts during training. This selective activation mechanism substantially improves training efficiency and reduces computational cost, thereby laying a solid foundation for the practical deployment of MoE-based models.

To address the dual optimization challenges inherent in the conventional D3QN algorithm—namely, the need for synchronous updates of the policy network and the target network, as well as the joint optimization of the state-value function and action-advantage function, which lead to increased parameter counts and reduced training efficiency—this study proposes an improved scheme based on the Mixture-of-Experts framework combined with a Bi-directional Long Short-Term Memory (Bi-LSTM). The enhanced MoE-D3QN model establishes a multi-expert collaboration mechanism where the gating network dynamically allocates parameters, improving model expressiveness under computational constraints. Additionally, the integration of the Bi-LSTM module strengthens temporal feature extraction by capturing long-range dependencies through bi-directional state propagation.

To address the dual challenges faced by the traditional D3QN algorithm—namely, the need for synchronized updates of the policy and target networks, and the parameter redundancy and training inefficiency arising from the joint optimization of the state-value and action-advantage functions—an improved framework, termed MoE-D3QN, is proposed in this study. By incorporating MoE architecture with a gating network, the model enables dynamic parameter allocation and adaptive feature selection. This design significantly enhances policy representation and training efficiency while maintaining constraints on computational resources.

In addition, to further improve the model's capability for temporal feature modeling, Bi-LSTM module is integrated as an auxiliary component. Leveraging its bi-directional state propagation mechanism, the Bi-LSTM effectively captures long-term dependencies during task execution. In the context of the UAV-based precision spraying task addressed in this study, where path planning depends on the dynamic evolution of environmental states, the integration of Bi-LSTM improves the model's ability to perceive historical changes. This, in turn, reduces redundant path coverage and resource waste, while enhancing the continuity and robustness of the learned policies.

As illustrated in Figure 3, the model architecture comprises three core components:

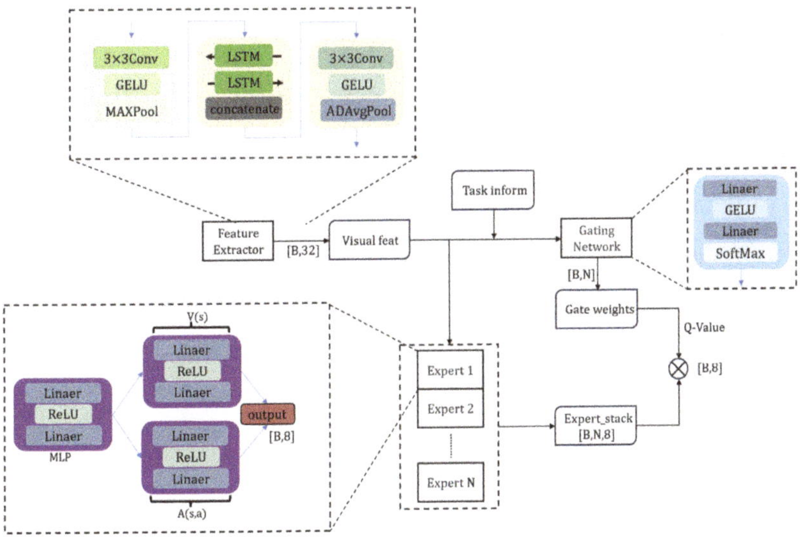

Figure 3. Policy and target network model architecture diagram.

(1) A CNN-LSTM cascaded feature extraction module, which adopts depthwise separable convolution and channel attention mechanisms to achieve a lightweight design, with the computation defined in Equation (10).

$$\begin{cases} H, (h_t, c_t) = BiLSTM(Reshape(MaxPool(GELU(W_{conv1} * X + b_{conv1})))) \\ v = Flatten(AdaptiveAvgPool(GELU(W_{conv2} * H + b_{conv2}))) \end{cases} \quad (10)$$

(2) An expert network ensemble, where each expert includes a feature transformation layer as well as independent computation units for the state-value function and the advantage function, as defined in Equation (11).

$$\begin{cases} V_j = W_{jV2} \cdot GELU(W_{jV1} \cdot W_{j2} \cdot GELU(W_{j1} \cdot v + b_{j1}) + b_{jV1}) \\ A_j = W_{jA2} \cdot GELU(W_{jA1} \cdot W_{j2} \cdot GELU(W_{j1} \cdot v + b_{j1}) + b_{jA1}) \\ Q_j = V_j + \left(A_j - \sum_{k=1} A_{j,k} \right) \end{cases} \quad (11)$$

(3) A dynamic gating network, which achieves sparse expert weight distribution through low-rank matrix decomposition and temperature coefficient adjustment. Specifically, the input observation data are first transformed into a high-dimensional representation via the feature extraction module. Based on the complexity of the current state, the gating network generates activation weights for the experts. Each expert network processes the feature vector in parallel, and the outputs are then fused through a weighted aggregation to produce the final action-value function prediction, as defined in Equation (12).

$$\begin{cases} g = Softmax\left(\frac{W_{g2} \cdot GELU(W_{g1} \cdot v + b_{g1}) + b_{g2}}{\tau} \right) \\ Q_{final} = \sum_{j=1}^{N} g_j \odot Q_j \end{cases} \quad (12)$$

τ represents the temperature coefficient, which is used to control the smoothness of the Softmax distribution output by the gating network. A linear annealing strategy is adopted to dynamically adjust the temperature coefficient during training, as shown in Equation (13). At the early stages of training, a higher temperature value is assigned to encourage the participation of multiple experts, thereby enhancing the model's responsiveness to complex environmental states encountered in coverage path planning tasks. As training progresses, τ is gradually annealed to a lower value, increasing the determinism of expert selection and thereby improving policy stability and convergence speed.

$$\tau(t) = \tau_{max} - \frac{t}{T}(\tau_{max} - \tau_{min}) \quad (13)$$

However, if the temperature coefficient decreases excessively, the Softmax output tends to approximate a one-hot distribution, activating only a single expert. This over-sparsity may hinder the model's generalization ability and lead to inefficient path selection in local regions, ultimately resulting in convergence to suboptimal solutions. To mitigate this issue, a lower bound is imposed on the temperature coefficient, ensuring a balance between determinism and diversity in expert selection. This constraint helps maintain a dynamic equilibrium between exploration and exploitation throughout the training process.

The process of the MOE-D3QN algorithm is shown in Figure 4.

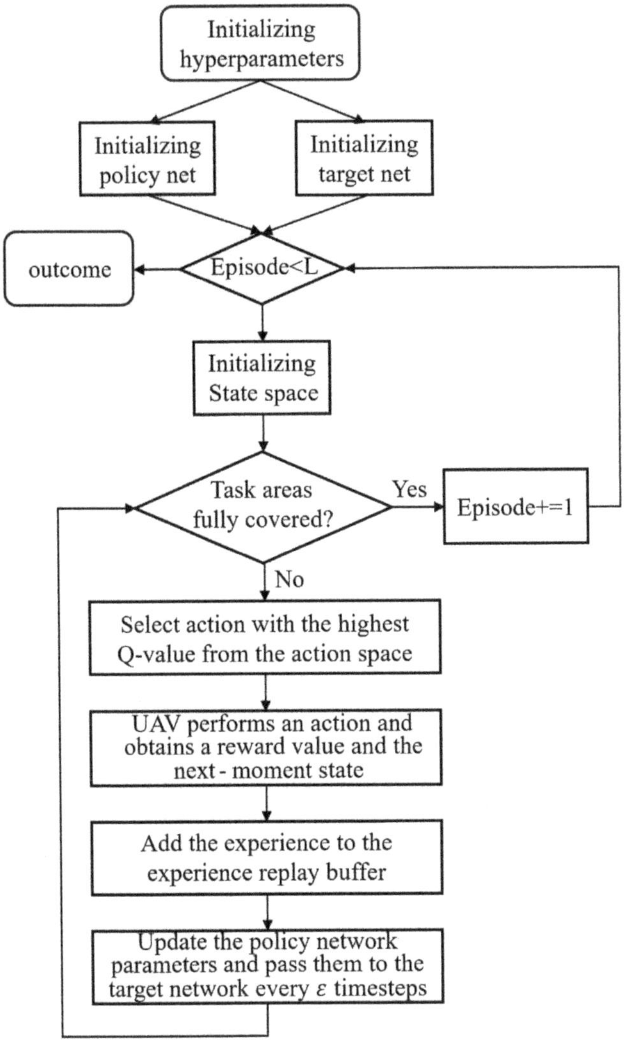

Figure 4. Flow chart of the MOE-D3QN algorithm.

2.5.3. Reward

To address the challenge faced by traditional path planning algorithms in agricultural environments—namely, the difficulty in simultaneously optimizing task priority, energy efficiency, and safety constraints—this study proposes a reward function designed to guide the UAV. The function encourages the UAV to approach the target coverage area promptly and prioritizes the coverage of high-priority regions in descending order, as defined in Equation (14).

$$r_t = \begin{cases} r_{Nontask}, & Non-task\ area \\ r_{overlay}, & Task\ area\ coverage \\ r_{accomplish}, & Complete\ the\ coverage\ task \end{cases} \tag{14}$$

The term $r_{Nontask}$ represents the reward assigned to non-task areas, as defined in Equation (15). The variable $Env_{(UAV\ position)}$ denotes the environmental state at the UAV's current location on the map, where 0 indicates a covered task area and 1 denotes a non-task area. A penalty is applied when the UAV enters an already covered area, while a base penalty is also imposed when the UAV moves into a non-task area. Additionally, a distance-based reward is provided to encourage the UAV to navigate toward task regions, thereby guiding it effectively toward completing mission objectives.

$$r_{Nontask} = \begin{cases} -0.5, & if\ Env_{(UAV\ position)} = 0 \\ -0.2 + R_{target}, & if\ Env_{(UAV\ position)} = 1 \end{cases} \tag{15}$$

The term $r_{overlay}$, defined as in Equation (16), denotes the coverage reward. A positive reward is assigned whenever the UAV moves into a region that corresponds to the current priority level. In contrast, if the UAV enters a region that does not match the required priority—such as covering a lower-priority area prematurely—a penalty is applied. This reward mechanism is designed to guide the UAV toward prioritizing the coverage of higher-priority areas.

$$r_{overlay} = \begin{cases} 1, & if\ Env_{(UAV\ position)} = Current\ task\ priority \\ -0.8, & Other \end{cases} \tag{16}$$

The term $r_{accomplish}$, as defined in Equation (17), represents the task completion reward. After each step, a check is performed: if all priority levels in the task sequence have been fully covered, the task is considered complete, and a substantial positive reward is granted. If the sequence is not yet empty, no completion reward is given, and the UAV proceeds to execute the remaining tasks.

$$r_{accomplish} = \begin{cases} 20, & if\ Current\ task\ priority = None \\ 0, & Other \end{cases} \tag{17}$$

To prevent ineffective exploration, improve training efficiency, avoid local optima or infinite loops, and enhance the safety of the path planning process, this study introduces a task termination constraint, as defined in Equation (18). Let $S_{episode}$ denote the number of steps taken in the current episode, S_{max} represent the maximum allowable steps, $S_{not\ reward\ count}$ denote the number of consecutive steps without a reward, and $S_{max\ not\ reward}$ indicate the maximum allowed consecutive non-rewarded steps. The episode is terminated under any of the following conditions: (1) the total steps taken exceed the maximum allowable steps; (2) the consecutive non-rewarded steps exceed the predefined limit; or (3) the UAV enters an obstacle region.

$$Termination\ condition \begin{cases} S_{\text{episode}} \geq S_{max} \\ S_{not\ reward\ count} \geq S_{max\ not\ reward} \\ Env_{(UAV\ position)} = Obstacle\ area \end{cases} \tag{18}$$

2.5.4. Sensitivity Analysis of Reward Function Parameters

To further evaluate the influence of key parameters in the reward function and gating mechanism on path planning performance, a sensitivity analysis was conducted from the perspectives of coverage efficiency, redundancy rate, and task completion rate. For each parameter, a univariate perturbation approach was employed: the target parameter was individually adjusted while keeping all other parameters fixed. Specifically, the parameter was varied by ±30% around its baseline value, and the resulting impact on path quality was systematically observed and analyzed.

As shown in Table 3, under the default configuration (Our), the algorithm achieved superior coverage performance with a low redundancy rate of only 3.23% and a coverage efficiency of 0.8378. In comparative experiments 1a, b, when the non-task area penalty_$r_{Nontask}$ was increased, a reduction in redundant paths was observed. However, due to the absence of guidance from target-directed rewards, the agent struggled to understand the priority coverage relationships. This resulted in an increased number of steps taken in non-task areas and a consequent decline in overall efficiency. Furthermore, as the penalty was reduced, the base movement cost decreased, which inadvertently caused the agent to receive positive rewards for covering non-task regions, leading to significant ineffective coverage in these areas.

Table 3. Parameter sensitivity analysis table.

Serial Number	$r_{Nontask}$	$r_{overlay}$	$r_{accomplish}$	$\tau_{max}\backslash\tau_{min}$	Repeat (%)	Coverage Efficiency
Our	−0.5/−0.2	1/−0.8	20	1.0/0.7	3.23%	0.8378
1a	−0.7/−0.3	1/−0.8	20	1.0/0.7	7.14%	0.7
1b	−0.3/−0.1	1/−0.8	20	1.0/0.7	15.38%	0.5778
2a	−0.5/−0.2	1.3/−1	20	1.0/0.7	3.33%	0.8108
2b	−0.5/−0.2	0.7/−0.5	20	1.0/0.7	10.71%	0.6829
3a	−0.5/−0.2	1/−0.8	26	1.0/0.7	6.45%	0.7949
3b	−0.5/−0.2	1/−0.8	14	1.0/0.7	3.33%	0.75

In experiments 2a,b, adjustments were made to the priority area reward $r_{overlay}$ Increasing this parameter did not yield substantial improvements, indicating that priority relationships were already reasonably well handled. Conversely, decreasing the reward, as demonstrated in 2b, noticeably degraded the prioritization of coverage paths, resulting in increased redundancy and reduced efficiency. Additionally, experiments 3a,b tested the impact of the full coverage task reward. It was found that setting this reward too low or too high disrupted the original planning rhythm, slightly decreasing path efficiency, although the overall effect remained limited.

3. Results and Discussion

3.1. Experimental Setup

All simulation experiments in this study were conducted on a high-performance desktop workstation configured with dual NVIDIA GeForce RTX 3090 GPUs (24 GB VRAM each) operating in parallel under the Ubuntu OS. The algorithms were developed and implemented using the Python 3.11.5 programming environment. The experimental parameters are configured as shown in Table 4.

Table 4. Experimental parameters table.

Parameter	Value	Description
E_{max}	60,000	Maximum episode
γ	0.95	Discount rate
B	128	Batch size
M	60,000	Experience replay buffer capacity
LR	1×10^{-4}	Learning rate
ϵ_max	0.9	Initial exploration rate

Table 4. *Cont.*

ϵ_min	0.1	Minimum exploration rate
τ_{max}	1	Maximum temperature coefficient
τ_{min}	0.7	Minimum temperature coefficient
Decay rate	5000	Controlling the speed at which ϵ epsilone decreases
n	10	Network update frequency
$N\_actions$	8	Action space size
Optimizer	Adam	Optimizer

3.2. Experimental Results

To systematically validate the performance advantages of the proposed algorithm in multi-level agricultural task scenarios, representative traditional path planning algorithms, namely A-Star and the genetic algorithm (GA), as well as mainstream deep reinforcement learning models including DQN, DDQN, D3QN, and Dueling DQN, were selected as baseline references for comparative experiments. Considering the spatial complexity characteristics of agricultural operational environments, this study constructed representative two-level and three-level hierarchical task regions based on satellite remote sensing imagery data. After 60,000 training iterations, the algorithms' performances were quantitatively evaluated in terms of coverage repetition rate, coverage rate, and coverage efficiency.

In this study, dark blue was assigned to obstacle areas, while task region priorities were represented using progressively lighter shades of blue. White areas indicated non-task regions. Charging stations were marked with red stars, UAV starting points with green circles, and UAV task termination points with pink circles. Yellow lines depicted the initial outbound paths, green lines represented routes taken after returning to charging stations for battery replenishment, and red lines indicated repeated paths.

As shown in Figures 5 and 6, the MOE-D3QN algorithm demonstrated superior coverage efficiency in both two-level and three-level hierarchical task maps, successfully completing the coverage tasks without any confusion in task hierarchy. However, from a global perspective, some route redundancies were observed, with partial path repetitions occurring. Among the comparison algorithms, only the traditional A-Star and genetic algorithms (GA) completed the coverage tasks. The A-Star algorithm, relying on heuristic functions to prioritize exploration in the "optimal" direction, encountered challenges in coverage tasks where the globally optimal path may require temporary deviations from the target direction to satisfy energy consumption and priority constraints. This caused repeated attempts along the same paths. The GA algorithm, when faced with complex constraints, tended to have its population dominated by a few individuals meeting basic constraints but exhibiting similar structures. Subsequent crossover and mutation operations could only slightly adjust path sequences, preventing escape from local optima and resulting in numerous redundant paths differing only in minor details. Furthermore, algorithms such as DDQN and Dueling DQN exhibited limitations due to inherent algorithmic shortcomings, which led to difficulties in convergence during training and entrapment in local optima.

To further evaluate the performance of the algorithms, quantitative analyses were conducted by calculating the coverage repetition rate, coverage rate, and coverage efficiency. A detailed comparison of the results is presented in Table 5.

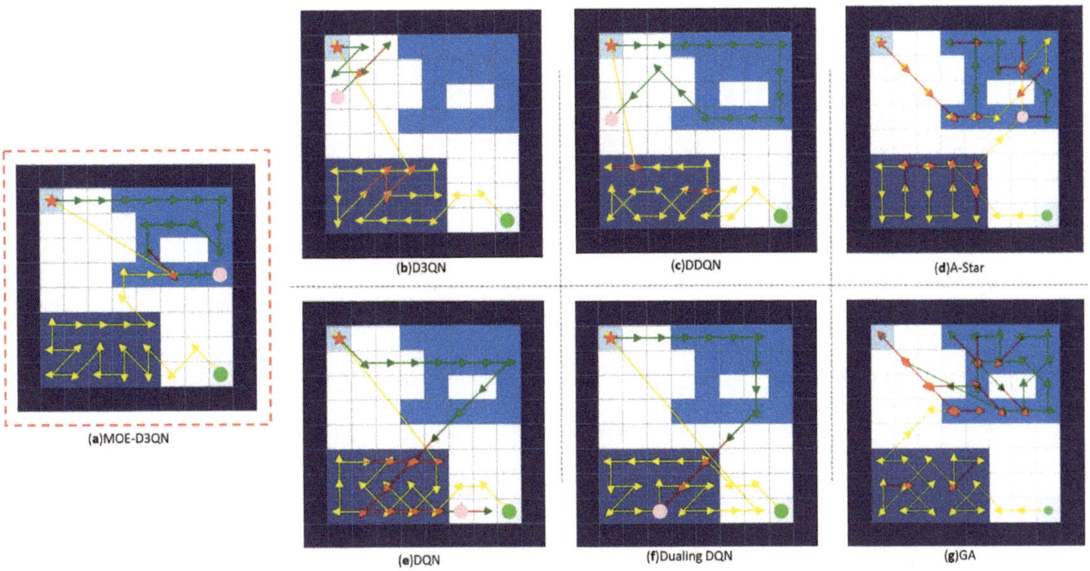

Figure 5. Two-level task planning result diagram.

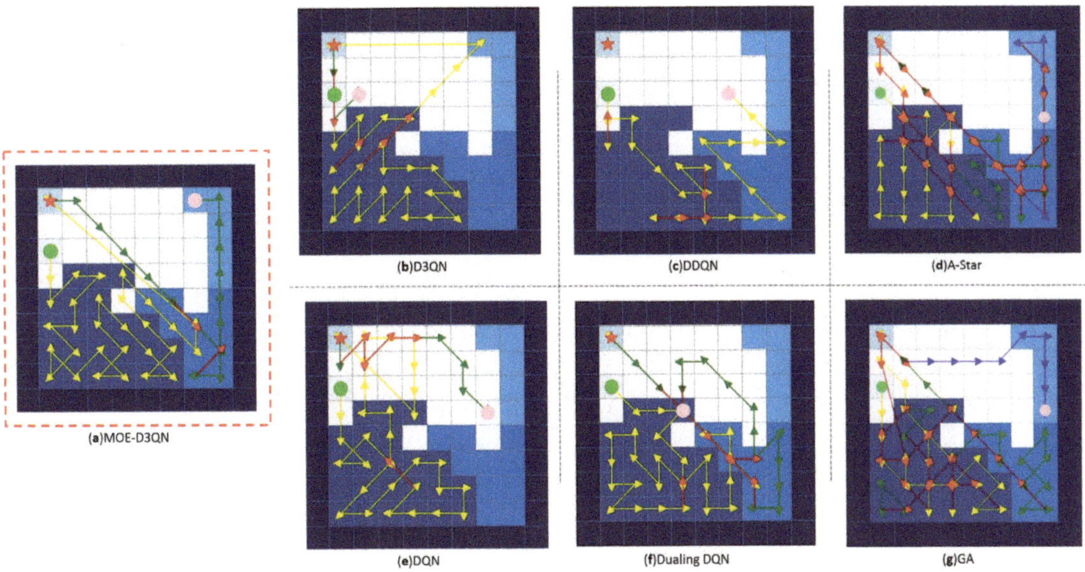

Figure 6. Three-level task planning result diagram.

Table 5. Experimental data statistics.

Number of Task Areas	Algorithms	Step	$Area_{repeat}$	$Area_{cover}$	Repeat (%)	Coverage Efficiency	Precedence Constraints
2	MOE-D3QN	37	1	31	3.23%	0.8378	Yes
	D3QN	26	4	15	26.67%	0.5769	Yes
	DQN	39	9	20	45.00%	0.5128	Yes
	DDQN	37	2	26	7.69%	0.7027	Yes
	Dueling DQN	32	3	21	14.29%	0.6563	Yes
	A-Star	51	13	31	41.94%	0.6078	Yes
	GA	51	12	31	38.71%	0.6078	Yes
3	MOE-D3QN	46	2	38	5.26%	0.8261	Yes
	D3QN	32	5	24	20.83%	0.75	Yes
	DQN	40	5	22	22.73%	0.55	Yes
	DDQN	27	4	19	21.05%	0.7037	NO
	Dueling DQN	50	8	33	24.24%	0.66	Yes
	A-Star	79	35	38	92.11%	0.481	Yes
	GA	70	22	38	57.89%	0.5429	Yes

(a) Path length: UAV's movement steps.

(b) Coverage efficiency: ratio of effectively covered area to total target region area, as shown in Equation (19).

$$E = \frac{\mathrm{Area_{cover}}}{\mathrm{Step}} \times 100\% \tag{19}$$

(c) Redundancy rate: ratio of repeatedly visited areas to total covered regions in the planned path, as formalized in Equation (20)

$$R_{repeat} = \frac{Area_{repeat}}{Area_{cover}} \times 100\% \tag{20}$$

3.3. Experimental Analysis

The reward curves presented in Figures 7 and 8 illustrate the training process, where the MOE-D3QN algorithm initially exhibited a gradual upward trend. Due to the exploration-promoting action selection strategy employed in this study, the overall reward curve demonstrated a spiraling increase. During the early to mid-training phases, the model progressively learned to understand environmental constraints and developed effective action strategies. By the mid-training stage, the model had achieved a satisfactory comprehension of the task requirements, successfully completing the planning task and obtaining the final reward. However, in the later stages, due to the chosen training strategy, the model showed signs of overfitting, resulting in fluctuations in the reward values.

In contrast, the DQN algorithm suffered from policy update bias accumulation caused by Q-value overestimation, leading to premature convergence in complex state spaces. Although DDQN mitigated some of these biases by decoupling action selection from value evaluation, its fixed-interval parameter update mechanism limited its adaptability to dynamic environmental changes. The dueling network architecture of Dueling DQN, while advantageous in some contexts, was prone to advantage function estimation errors in sparse reward scenarios, which increased the likelihood of becoming trapped in local optima.

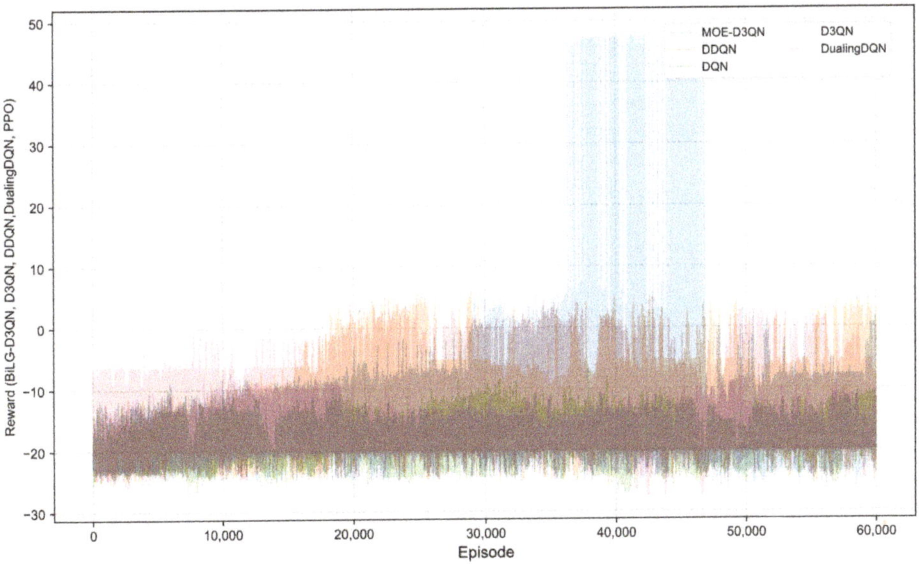

Figure 7. The two-level task planning algorithm reward value comparison diagram.

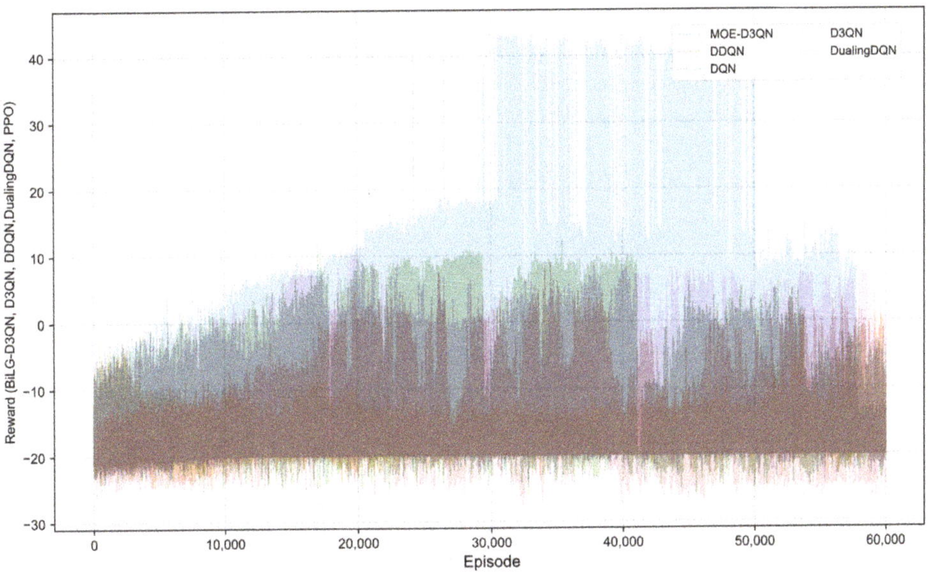

Figure 8. The three-level task planning algorithm reward value comparison diagram.

4. Discussion

In terms of experimental metrics, MOE-D3QN demonstrated significantly higher coverage efficiency and lower redundancy rates compared to traditional algorithms and mainstream reinforcement learning methods. This not only reflects the optimization advantages of the path planning strategy but also holds profound practical significance for agricultural applications. Firstly, the high coverage efficiency ensures comprehensive and omission-free treatment of target areas for operations such as pesticide spraying and fertilization, effectively minimizing blind spots and coverage gaps. This is particularly crucial under conditions of high pest incidence or uneven crop growth, where the risk of "over-treatment and under-protection" or "localized loss of control" can be largely avoided. Secondly, the enhanced coverage efficiency markedly improves operational productivity. For plant protection UAVs, this translates into the ability to complete larger-area tasks within a single flight, thereby reducing the need for multiple flights and lowering the frequency and cost associated with battery recharging or replacement. Consequently, labor input and time consumption are further reduced.

Simultaneously, the low redundancy rate indicates a reduced frequency of path overlap during mission execution, facilitating precise control of agricultural inputs such as pesticides, fertilizers, and irrigation water. Traditional path planning methods often suffer from significant path overlap issues, leading to excessive pesticide or fertilizer concentrations in certain areas, which not only increase production costs but may also cause crop phytotoxicity and soil accumulation pollution, posing ecological risks. Conversely, the low-redundancy path planning approach ensures that each target area is uniformly and efficiently covered, promoting relative controllability of agricultural resources. This approach balances both the "precision" and "uniformity" of operational outcomes, safeguarding crop health while minimizing environmental burden—an essential factor for advancing green agriculture and resource-efficient development.

From a technical implementation perspective, this study utilized Sentinel-2 Level-2A satellite remote sensing imagery to perform spectral inversion of vegetation status in the study area, extracting the Canopy Chlorophyll Content index. Subsequently, multi-level thresholds were established using quantile methods to rasterize the imagery into a digital task map, where different grid values represented varying operational priorities (e.g., high, medium, and low risk zones). These were uniformly fed into the state space of the deep reinforcement learning model. This processing approach effectively transforms remote sensing perception data into structured inputs recognizable by the reinforcement learning agent, thereby simplifying the linkage between remote sensing information and path decision making.

Regarding the specific implementation, the rasterization method employed offers strong generalizability and operability, capable of accommodating task partitioning requirements across diverse crop scenarios. Although the current experiments were conducted in flat farmland areas, the CCC index used is derived from red-edge and near-infrared bands, possessing a degree of crop non-specificity. Theoretically, it is applicable to physiological status estimation for various crops such as rice, wheat, and soybean, thus exhibiting cross-crop adaptability potential. Moreover, the MOE-D3QN model employs a Mixture-of-Experts architecture, composed of multiple policy subnetworks dynamically fused through a gating mechanism based on input states. This architecture offers advantages in modeling heterogeneous input states and is expected to enhance the model's responsiveness to environmental heterogeneity and task complexity.

Despite the current experiments focusing primarily on relatively flat terrain, the model's structural design does not rely on specific landform types or single-crop characteristics, indicating certain scalability. Preliminary assessments suggest that, with appropriate adjustments to state inputs and task configurations, the model may possess adaptability for more complex terrains and multi-crop structured scenarios.

5. Conclusions

This study proposes an improved deep reinforcement learning algorithm, MoE-D3QN, which integrates D3QN, a Mixture-of-Experts mechanism, and a Bi-directional Long Short-Term Memory, aiming to enhance path planning efficiency for agricultural unmanned aerial vehicles operating in dynamic and complex environments. The algorithm leverages the Bi-LSTM module's capability to capture temporal features, thereby strengthening contextual awareness within task sequences. Moreover, through the dynamic selection capability of expert networks and gating mechanisms in the MoE architecture, it enables adaptive decision making in response to variable states within complex environments. In addition, a hierarchical coverage path planning framework is developed, utilizing Sentinel-2 remote sensing data to invert canopy chlorophyll content and construct multi-level task maps via thresholding and rasterization techniques. A dynamic energy consumption modeling mechanism is introduced to quantify the coupling effects of payload and environmental factors. Furthermore, a progressive composite reward function is designed to achieve synergistic optimization of energy efficiency and task priority.

Simulation experiments demonstrate that MOE-D3QN achieves superior performance compared to six other algorithms in hierarchical coverage path planning tasks with load-related energy considerations. In the secondary task scenario, the MOE-D3QN algorithm achieved a coverage efficiency of 0.8378, representing improvements of 45.26% over D3QN, 63.38% over DQN, 19.19% over DDQN, 27.66% over Dueling DQN, and 37.84% over A-Star and GA. The redundancy rate was only 3.23%, far lower than D3QN at 26.67%, DQN at 45.00%, DDQN at 7.69%, Dueling DQN at 14.29%, A-Star at 41.94%, and GA at 38.71%.In the tertiary task scenario, the MOE-D3QN algorithm achieved a coverage efficiency of 0.8261, with improvements of 10.15% over D3QN, 50.2% over DQN, 17.4% over DDQN, 25.17% over Dueling DQN, 71.45% over A-Star, and 52.13% over GA. The redundancy rate was 5.26%, significantly better than D3QN at 20.83%, DQN at 22.73%, DDQN at 21.05%, Dueling DQN at 24.24%, A-Star at 92.11%, and GA at 57.89%.

These results validate the algorithm's capability to comprehend task constraints and handle temporal information, providing a low-redundancy and highly adaptive path planning solution for precision agricultural UAV operations. The proposed approach not only enhances spraying accuracy and resource utilization but also reduces pesticide overuse and environmental pollution risks, contributing to the advancement of intelligent and sustainable agriculture.

In future research, efforts will be directed towards further expanding the practical applicability of the proposed algorithm to better support UAV systems in precision agriculture, thereby promoting agricultural intelligence and sustainable development. Specifically, the following aspects will be addressed:

1. This study will focus on the collaborative mechanisms of UAV path planning among various types of agricultural intelligent devices. Emphasis will be placed on the information interaction and operational coordination between UAVs and field sensing equipment—such as soil moisture sensors and microclimate stations—as well as variable-rate pesticide application systems. By constructing a unified crop state perception and task collaboration framework, the coupling among multi-device platforms will be enhanced, facilitating the evolution of UAV operation systems from independent execution towards intelligent collaborative operation. This will provide crucial support for the integration and intelligent application of multiple agricultural devices in precision farming.

2. To meet the demands for efficient operations across large areas and multi-crop distributions, future work will consider coordinated path planning for multi-UAV systems. By integrating plot characteristics, crop requirements, and operational time window constraints, dynamic scheduling of multiple UAVs will be realized. This approach will offer algorithmic support for UAV swarm formation and zonal operations, thus advancing the large-scale deployment of UAVs in precision agriculture.

3. To further enhance the algorithm's generalizability across diverse scenarios, subsequent research plans include incorporating real orchard point cloud data or digital elevation

models (DEM) to embed terrain factors such as slope and undulation into the state space construction. This will enable three-dimensional path planning environments that more closely align with practical operational needs. Experimental evaluations will be conducted in irregular terrains such as hilly areas and orchards, providing more comprehensive empirical support for the model's applicability in varied agricultural settings.

Author Contributions: Conceptualization, H.F., H.L., Z.L. and J.L. (Jian Li); methodology, W.Z. and Y.F.; software, H.F. and L.Z.; investigation, L.Z., W.Z. and J.L. (Jian Lu); resources, Z.L.,H.L. and Y.F.; writing—original draft, Z.L.; writing—review and editing, H.F. and J.L. (Jian Li); visualization, Z.L.; supervision, H.F., H.L. and W.Z.; funding acquisition, L.Z.; validation, L.Z., Y.F. and J.L. (Jian Lu); data curation, H.F. and J.L. (Jian Lu); project administration, J.L. (Jian Li). All authors have read and agreed to the published version of the manuscript.

Funding: This research was funded by the Jilin Provincial Department of Education Science and Technology Research Project (JJKH20250567KJ), the Scientific Research Project of the Jilin Provincial De-partment of Education (JJKH20250574BS), and the Demonstration Project for the Construction of a Provincial-Level Modern Agricultural Industry Technology System (JLARS-2025-010216).

Data Availability Statement: Data is contained within the article.

Conflicts of Interest: The authors declare no conflict of interest.

References

1. Maes, W.H.; Steppe, K. Perspectives for Remote Sensing with Unmanned Aerial Vehicles in Precision Agriculture. *Trends Plant Sci.* **2019**, *24*, 152–164. [CrossRef] [PubMed]
2. Shakhatreh, H.; Sawalmeh, A.H.; Al-Fuqaha, A.; Dou, Z.; Almaita, E.; Khalil, I.; Othman, N.S.; Khreishah, A.; Guizani, M. Unmanned Aerial Vehicles (UAVs): A Survey on Civil Applications and Key Research Challenges. *IEEE Access* **2019**, *7*, 48572–48634. [CrossRef]
3. Rejeb, A.; Abdollahi, A.; Rejeb, K.; Treiblmaier, H. Drones in Agriculture: A Review and Bibliometric Analysis. *Comput. Electron. Agric.* **2022**, *198*, 107017. [CrossRef]
4. Guebsi, R.; Mami, S.; Chokmani, K. Drones in Precision Agriculture: A Comprehensive Review of Applications, Technologies, and Challenges. *Drones* **2024**, *8*, 686. [CrossRef]
5. Canicattì, M.; Vallone, M. Drones in Vegetable Crops: A Systematic Literature Review. *Smart Agric. Technol.* **2024**, *7*, 100396. [CrossRef]
6. Hafeez, A.; Husain, M.A.; Singh, S.P.; Chauhan, A.; Khan, M.T.; Kumar, N.; Chauhan, A.; Soni, S.K. Implementation of Drone Technology for Farm Monitoring & Pesticide Spraying: A Review. *Inf. Process. Agric.* **2023**, *10*, 192–203. [CrossRef]
7. Mukhamediev, R.I.; Yakunin, K.; Aubakirov, M.; Assanov, I.; Kuchin, Y.; Symagulov, A.; Levashenko, V.; Zaitseva, E.; Sokolov, D.; Amirgaliyev, Y. Coverage Path Planning Optimization of Heterogeneous UAVs Group for Precision Agriculture. *IEEE Access* **2023**, *11*, 5789–5803. [CrossRef]
8. Radoglou-Grammatikis, P.; Sarigiannidis, P.; Lagkas, T.; Moscholios, I. A Compilation of UAV Applications for Precision Agriculture. *Comput. Netw.* **2020**, *172*, 107148. [CrossRef]
9. Tang, G.; Tang, C.; Claramunt, C.; Hu, X.; Zhou, P. Geometric A-Star Algorithm: An Improved A-Star Algorithm for AGV Path Planning in a Port Environment. *IEEE Access* **2021**, *9*, 59196–59210. [CrossRef]
10. Wang, D.; Tan, D.; Liu, L. Particle Swarm Optimization Algorithm: An Overview. *Soft Comput.* **2018**, *22*, 387–408. [CrossRef]
11. Qi, J.; Yang, H.; Sun, H. MOD-RRT*: A Sampling-Based Algorithm for Robot Path Planning in Dynamic Environment. *IEEE Trans. Ind. Electron.* **2021**, *68*, 7244–7251. [CrossRef]
12. Zhao, Y.; Qi, X.; Ma, Y.; Li, Z.; Malekian, R.; Sotelo, M.A. Path Following Optimization for an Underactuated USV Using Smoothly-Convergent Deep Reinforcement Learning. *IEEE Trans. Intell. Transp. Syst.* **2021**, *22*, 6208–6220. [CrossRef]
13. Wen, S.; Wen, Z.; Zhang, D.; Zhang, H.; Wang, T. A Multi-Robot Path-Planning Algorithm for Autonomous Navigation Using Meta-Reinforcement Learning Based on Transfer Learning. *Appl. Soft. Comput.* **2021**, *110*, 107605. [CrossRef]
14. Zhang, M.; Li, X.; Wang, L.; Jin, L.; Wang, S. A Path Planning System for Orchard Mower Based on Improved A* Algorithm. *Agronomy* **2024**, *14*, 391. [CrossRef]
15. Wu, G.; Wang, M.; Guo, L. Complete Coverage Path Planning Based on Improved Genetic Algorithm for Unmanned Surface Vehicle. *J. Mar. Sci. Eng.* **2024**, *12*, 1025. [CrossRef]
16. Kong, F.; Wang, Q.; Gao, S.; Yu, H. B-APFDQN: A UAV Path Planning Algorithm Based on Deep Q-Network and Artificial Potential Field. *IEEE Access* **2023**, *11*, 44051–44064. [CrossRef]

17. Zhou, Y.; Yang, J.; Guo, Z.; Shen, Y.; Yu, K.; Lin, J.C.-W. An Indoor Blind Area-Oriented Autonomous Robotic Path Planning Approach Using Deep Reinforcement Learning. *Expert Syst. Appl.* **2024**, *254*, 124277. [CrossRef]
18. Lin, G.; Zhu, L.; Li, J.; Zou, X.; Tang, Y. Collision-Free Path Planning for a Guava-Harvesting Robot Based on Recurrent Deep Reinforcement Learning. *Comput. Electron. Agric.* **2021**, *188*, 106350. [CrossRef]
19. Din, A.; Ismail, M.Y.; Shah, B.; Babar, M.; Ali, F.; Baig, S.U. A Deep Reinforcement Learning-Based Multi-Agent Area Coverage Control for Smart Agriculture. *Comput. Electr. Eng.* **2022**, *101*, 108089. [CrossRef]
20. Syed, T.N.; Zhou, J.; Lakhiar, I.A.; Marinello, F.; Gemechu, T.T.; Rottok, L.T.; Jiang, Z. Enhancing Autonomous Orchard Navigation: A Real-Time Convolutional Neural Network-Based Obstacle Classification System for Distinguishing 'Real' and 'Fake' Obstacles in Agricultural Robotics. *Agriculture* **2025**, *15*, 827. [CrossRef]
21. Li, J.; Zhang, W.; Ren, J.; Yu, W.; Wang, G.; Ding, P.; Wang, J.; Zhang, X. A Multi-Area Task Path-Planning Algorithm for Agricultural Drones Based on Improved Double Deep Q-Learning Net. *Agriculture* **2024**, *14*, 1294. [CrossRef]
22. Qu, L.; Zhang, P.; Jing, C.; Du, M.; Wang, J.; Zhao, Q.; Li, J. Estimating the Fractional Cycle Biases for GPS Triple-Frequency Precise Point Positioning with Ambiguity Resolution Based on IGS Ultra-Rapid Predicted Orbits. *Remote Sens.* **2021**, *13*, 3164. [CrossRef]
23. Clevers, J.G.P.W.; Kooistra, L.; Van den Brande, M.M.M. Using Sentinel-2 Data for Retrieving LAI and Leaf and Canopy Chlorophyll Content of a Potato Crop. *Remote Sens.* **2017**, *9*, 405. [CrossRef]
24. Li, D.; Chen, G.; Xi, Z.; He, P.; Xin, X.; Chen, J.; Yu, H.; Kang, G. Spatiotemporal Changes of Cultivated Land Utilization in Black Soil Region, China Based on Geo-Information Tupu. *Sci. Rep.* **2025**, *15*, 10306. [CrossRef]
25. Wu, W.; Feng, X.; Lu, C. The rise of smart agriculture in China: Current situation and suggestions for further development. *Exp. Agric.* **2024**, *60*, e28. [CrossRef]
26. Brown, L.A.; Ogutu, B.O.; Dash, J. Estimating Forest Leaf Area Index and Canopy Chlorophyll Content with Sentinel-2: An Evaluation of Two Hybrid Retrieval Algorithms. *Remote Sens.* **2019**, *11*, 1752. [CrossRef]
27. Brown, L.A.; Fernandes, R.; Djamai, N.; Meier, C.; Gobron, N.; Morris, H.; Canisius, F.; Bai, G.; Lerebourg, C.; Lanconelli, C.; et al. Validation of Baseline and Modified Sentinel-2 Level 2 Prototype Processor Leaf Area Index Retrievals over the United States. *ISPRS J. Photogramm. Remote Sens.* **2021**, *175*, 71–87. [CrossRef]
28. Validation of the Sentinel Simplified Level 2 Product Prototype Processor (SL2P) for mapping cropland biophysical variables using Sentinel-2/MSI and Landsat-8/OLI data | Request PDF. *Remote Sens. Environ.* **2024**, *225*, 416–430. [CrossRef]
29. Mnih, V.; Kavukcuoglu, K.; Silver, D.; Rusu, A.A.; Veness, J.; Bellemare, M.G.; Graves, A.; Riedmiller, M.; Fidjeland, A.K.; Ostrovski, G.; et al. Human-Level Control through Deep Reinforcement Learning. *Nature* **2015**, *518*, 529–533H. [CrossRef]
30. Hasselt, H.; van Guez, A.; Silver, D. Deep Reinforcement Learning with Double Q-Learning. *arXiv* **2015**, arXiv:1509.0646. [CrossRef]
31. Wang, Z.; Schaul, T.; Hessel, M.; Hasselt, H.; van Lanctot, M.; Freitas, N. de Dueling Network Architectures for Deep Reinforcement Learning. *arXiv* **2016**, arXiv:1511.06581.
32. Lepikhin, D.; Lee, H.; Xu, Y.; Chen, D.; Firat, O.; Huang, Y.; Krikun, M.; Shazeer, N.; Chen, Z. GShard: Scaling Giant Models with Conditional Computation and Automatic Sharding. *arXiv* **2022**, arXiv:2006.16668.
33. Fedus, W.; Zoph, B.; Shazeer, N. Switch Transformers: Scaling to Trillion Parameter Models with Simple and Efficient Sparsity. *J. Mach. Learn. Res.* **2022**, *23*, 5232–5270.

MDPI AG
Grosspeteranlage 5
4052 Basel
Switzerland
Tel.: +41 61 683 77 34

Remote Sensing Editorial Office
E-mail: remotesensing@mdpi.com
www.mdpi.com/journal/remotesensing

www.ingramcontent.com/pod-product-compliance
Lightning Source LLC
Chambersburg PA
CBHW040821250926
51850CB00007BA/1227